Test File

to accompany

Life

The Science of Biology

Fifth Edition

Test File

to accompany

Life

The Science of Biology

Fifth Edition

Purves, Orians, Heller, and Sadava

Written and compiled by

Charles Herr
Eastern Washington University

Kay Greeson
Eastern Washington University

SINAUER ASSOCIATES, INC.

W. H. FREEMAN AND COMPANY

THE COVER

Grizzly bears (*Ursus arctos*) take many years to mature reproductively, and cubs remain with their mothers for several years. If their populations are to persist, adult bears must have access to rich food resources. Grizzly bears that live in coastal Alaska depend on salmon that swim up the rivers to spawn. Given that abundant, high-quality food, they grow to become the world's largest carnivorous mammal.

Photograph by Michio Hoshino/Minden Pictures.

ACKNOWLEDGMENTS

Erica Berquist, Holyoke Community College, and Nels H. Granholm, South Dakota State University, wrote and compiled extensive questions for the test files that accompanied the fourth edition of LIFE: The Science of Biology. Many of their questions have been used or modified in this edition.

Test File to accompany
LIFE: THE SCIENCE OF BIOLOGY, Fifth Edition

Address editorial correspondence to:
    Sinauer Associates, Inc.
    23 Plumtree Road
    Sunderland, MA 01375 U.S.A.
    www.sinauer.com

Address orders to:
    W. H. Freeman and Company
    Distribution Center, 4419 West 1980 South
    Salt Lake City, UT 84104 U.S.A.
    www.whfreeman.com

ISBN: 0-7167-3219-X

Printed in U.S.A.

Contents

1 An Evolutionary Framework for Biology

Fill in the Blank

1. One can discover the absolute ages of rocks by studying their **radioactivity (or radioactivity rates)**.

2. All scientific study begins with observations and the formation of testable **hypotheses.**

3. An important part of the cell theory is the concept that "all cells come from **preexisting cells.**"

4. The properties exhibited by a group of organisms that are not characteristic of the individual organisms are called **emergent properties.**

5. In contrast to eukaryotic cells, bacteria lack **organelles.**

6. To test the effect of vitamin D on growth, two groups of rats were raised under identical conditions and fed the same diet, but one of the groups also received daily injections of vitamin D. The variable vitamin D is being **controlled.**

7. **Darwin (Wallace)** was the first person to suggest that organisms change gradually through the natural selection of variable characteristics.

8. Single-celled organisms that lack a nucleus belong to the two kingdoms called **Archaebacteria** and **Eubacteria.**

9. Single-celled organisms that have a nucleus belong to the kingdom called **Protista.**

10. Multicellular organisms that are nonphotosynthetic and that digest their food within their body's cells belong to the kingdom called **Fungi.**

11. Multicellular organisms that are photosynthetic belong to the kingdom called **Plantae.**

12. Multicellular organisms that are nonphotosynthetic and that digest their food outside their body's cells and absorb the products belong to the kingdom called **Animalia.**

13. According to Irish Archbishop Ussher's estimate, the age of the earth today would be approximately **6,000 years.**

14. Currently, scientists agree with the estimate that life first appeared approximately **4 billion** years ago.

15. Both cellular and multicellular forms of life attempt to maintain a consistent internal environment in spite of changing external conditions. The term for this process is **homeostasis.**

16. There are three domains used to categorize life forms that have evolved separately for about a billion years: **Archaea, Bacteria,** and **Eukarya.**

17. Life can be categorized into six kingdoms: plants, animals, fungi, **eubacteria, archaebacteria,** and **protists.**

18. The **null** hypothesis states that no difference exists due to the variable under investigation.

Multiple Choice

1. When compared to laboratory experiments, one advantage of field experiments is that they
   a. **better apply to actual events of nature.**
   b. more easily control most factors that may affect the outcome of the experiment.
   c. require less money and use of technology.
   d. achieve results more quickly.
   e. demonstrate organisms' reactions to ideal environments.

2. Before Darwin, one scientist who wrote extensively about evolution was
   a. Hooke.
   b. Leeuwenhoek.
   c. **Lamarck.**
   d. Pasteur.
   e. Virchow.

3. Animals are believed to have arisen most directly from ancestral
   a. archaebacteria.
   b. eubacteria.
   c. fungi.
   d. plants.
   e. **protists.**

4. Darwin noted that all populations have the potential to grow _____, but that in nature most populations _____ over time.
   a. linearly; are stable
   b. exponentially; grow more slowly
   c. linearly; fluctuate unpredictably
   d. **exponentially; are stable**
   e. linearly; decrease slowly

5. Plants are
   a. **eukaryotic, multicellular photosynthesizers.**
   b. eukaryotic, unicellular autotrophs.
   c. eukaryotic, multicellular heterotrophs.
   d. prokaryotic, multicellular autotrophs.
   e. prokaryotic, unicellular heterotrophs.

6. An example of a biome is all the
   a. mice in one meadow.
   b. **tropical rainforest communities on Earth.**
   c. inhabitants of a pond.
   d. species of deer on Earth.
   e. bacteria of the soil.

7. What distinguishes living organisms from nonliving matter?
   a. **Living organisms are characterized by the processes of metabolism and reproduction.**
   b. Only living organisms change in response to their environment.
   c. Only living organisms are composed of molecules.
   d. Living organisms do not obey physical and chemical laws.
   e. Only living organisms increase in size.

8. Metabolism is
   a. the consumption of energy.
   b. the release of energy.
   c. **all conversions of matter and energy taking place in an organism.**
   d. the production of heat by chemical reactions.
   e. the exchange of nutrients and waste products with the environment.

9. All living organisms obtain their energy from
   a. food.
   b. **external sources.**
   c. sunlight.
   d. heterotrophs.
   e. autotrophs.

10. Heterotrophs *cannot* obtain their energy from
    a. foods.
    b. autotrophs.
    c. other heterotrophs.
    d. complex chemical substances.
    e. **sunlight.**

11. The proper order of objects, from simple to complex, is
    a. **molecule, cell, tissue, organ, organism, population, community.**
    b. cell, molecule, tissue, organ, organism, population, community.
    c. molecule, cell, organ, tissue, organism, population, community.
    d. molecule, cell, tissue, organ, population, organism, community.
    e. molecule, tissue, organ, cell, population, organism, community.

12. A cell is composed of compounds that include proteins, nucleic acids, lipids, and carbohydrates. A cell is capable of reproduction, but when the compounds that make up a cell are isolated, none of them can reproduce. Thus cell reproduction is an example of
    a. growth.
    b. a molecule.
    c. **an emergent property.**
    d. adaptation.
    e. metabolism.

13. The fundamental unit of life is the
    a. aggregate.
    b. organelle.
    c. organism.
    d. membrane.
    e. **cell.**

14. A species is
    a. all the organisms that live together in a particular area.

b. a group of similar organisms that cannot interbreed.
   c. **a group of similar organisms capable of interbreeding.**
   d. an adult organism and all of its offspring.
   e. a group of similar organisms that live in the same area.

15. The proper order of objects, from simple to complex, is
    a. species, ecosystem, community, biosphere.
    b. community, species, ecosystem, biosphere.
    c. biosphere, species, community, ecosystem.
    d. **species, community, ecosystem, biosphere.**
    e. biosphere, ecosystem, community, species.

16. In multicellular organisms, cells of specific types are organized to form
    a. molecules.
    b. populations.
    c. organelles.
    d. **tissues.**
    e. communities.

17. The highest level of organization of biological systems is the sum of all of Earth's ecosystems, known as the
    a. community.
    b. **biosphere.**
    c. emergent system.
    d. species.
    e. organism.

18. Which of the following is *not* a major stage of the hypothetico-deductive method?
    a. **Controlling an environment**
    b. Making an observation
    c. Forming a hypothesis
    d. Making a prediction
    e. Testing a prediction

19. Which of the following questions *cannot* be answered using the hypothetico-deductive method?
    a. Are bees more attracted to red roses than to yellow roses?
    b. **Are red roses more beautiful than yellow roses?**
    c. Why are red roses red?
    d. Do red roses bloom sooner than yellow roses?
    e. Are red roses more susceptible to mildew than yellow roses?

20. After observing that fish live in clean water but not in polluted water, you make the statement, "Polluted water kills fish." Your statement is an example of
    a. a controlled experiment.
    b. a field experiment.
    c. a laboratory experiment.
    d. **a hypothesis.**
    e. comparative analysis.

21. Which of the following is *not* a feature of scientific hypotheses?
    a. **They are true.**
    b. They make predictions.
    c. They are based on observations.
    d. They can be tested by experimentation.
    e. They can be tested by observational analysis.

22. Our ability to study events that occur over many generations requires techniques for measuring very long time spans. Which of the following is one way to measure the age of very old materials?
    a. **Measuring the decay of radioactive materials in the Earth's crust**
    b. Measuring the rate of chemical reactions in cells
    c. Measuring the time for an organism to reach sexual maturity
    d. Measuring the time from one generation to the next
    e. Measuring the rate of physiological changes in organisms

23. Which of the following is *not* a major organizing concept in biology?
    a. Properties of living organisms are the result of physical and chemical interactions.
    b. **Living organisms are characterized by a vital force that is beyond our comprehension.**
    c. Living organisms can take energy from the environment and convert it to biologically useful forms.
    d. Evolution by natural selection results in adaptation.
    e. All living organisms are composed of cells.

24. Based on the large numbers of offspring produced by many organisms, Darwin proposed that mortality was high and only a few individuals survived to reproduce. He called the differential reproductive success of individuals with particular variations
    a. evolution.
    b. artificial selection.
    c. the cell theory.
    d. **natural selection.**
    e. inheritance of acquired characteristics.

25. Animal breeders have developed dairy cattle that have increased milk production. In this process, only the cows that produce the most milk are bred to give rise to the next generation of cattle. This is an example of
    a. **artificial selection.**
    b. natural selection.
    c. a paradigm.
    d. a hypothesis.
    e. adaptation.

26. Biologists classify living organisms into six large categories called
    a. biomes.
    b. species.
    c. organelles.
    d. **kingdoms.**
    e. divisions.

27. Members of the kingdom Animalia obtain their energy from
    a. decomposing organic matter.
    b. photosynthesis.
    c. **other organisms.**
    d. sunlight.
    e. inorganic molecules.

28. Organisms are said to be adapted to their environments when they
    a. are at thermal and chemical equilibrium with their environment.
    b. **possess characteristics that enhance their survival and reproductive success within their environment.**
    c. do not change, pollute, or destroy their environment.
    d. are able to move around quickly within their environments.
    e. are able to avoid becoming food for other organisms in their environments.

29. A typical cell
    a. can be composed of many types of tissues.
    b. creates an exact copy of itself when it reproduces.
    c. is the smallest entity studied by biologists.
    d. **is the simplest biological structure capable of independent existence and reproduction.**
    e. is found only in plants and animals.

30. Which of the following is *not* part of the hypothetico-deductive method for doing science?
    a. Making observations
    b. Forming hypothetical explanations
    c. Making predictions based upon hypothetical explanations
    d. **Proving hypotheses to be true**
    e. Testing predictions

31. Differing abilities of organisms of a species to survive and reproduce because of variations among these organisms is an important idea involved in
    a. **natural selection.**
    b. Lamarck's explanation of evolution.
    c. defining the differing kingdoms of biology.
    d. explaining why organisms of the same species look so much alike.
    e. explaining why there are both autotrophs and heterotrophs.

32. One is most likely to identify emergent properties as one studies how
    a. **big things are made up of smaller things working together.**
    b. organelles are composed of cells working together.
    c. important experiments are performed by scientists working together.
    d. paradigms become accepted by scientists working together.
    e. organelles are composed of tissues working together.

33. The microscope was first used to look at cells in the year
    a. 1558.
    b. **1665.**
    c. 1791.
    d. 1861.
    e. 1903.

34. Another more common name for prokaryotic cells is
    a. plants.
    b. animals.
    c. **bacteria.**
    d. viruses.
    e. fungi.

35. Which kingdom contains eukaryotic, unicellular organisms?
    a. Plantae
    b. Archaebacteria
    c. Animalia
    d. Protista
    e. Fungi

36. The different kingdoms of organisms are based upon
    a. the structure of the cells of the organisms and how the organisms obtain nutrition.
    b. the typical habitats of the organisms.
    c. whether or not the organisms are able to move from one place to another.
    d. how the organisms reproduce.
    e. how long the organisms have been present on Earth.

37. An example of an organ is
    a. milk.
    b. a protein.
    c. the heart.
    d. an infant.
    e. wood.

38. The key feature of experimentation is to
    a. obtain very accurate quantitative measurements.
    b. prove unambiguously that some hypothesis is correct.
    c. avoid comparative analysis.
    d. answer as many key questions in one experiment as possible.
    e. control factors that might affect a result so the influence of factors that do vary can be seen more clearly.

39. A key point in Darwin's explanation of evolution is
    a. biological structures most likely inherited are those that have become better suited to the environment by their constant use.
    b. mutations that occur are those that will help future generations fit into their environments.
    c. slight variations among individuals significantly affect the chance that a given individual will survive in its environment and be able to reproduce.
    d. genes change in order to help organisms cope with problems encountered within their environments.
    e. extinction is nature's way to weed out undeserving organisms.

40. The smallest entities studied by biologists are
    a. cells.
    b. tissues.
    c. organelles.
    d. molecules.
    e. membranes.

41. It is thought that some prokaryotes invaded other prokaryotes
    a. about 4,000 years before the present.
    b. about 10,000 years before the present.
    c. more than one million years before the present.
    d. more than one billion years before the present.
    e. more than one trillion years before the present.

42. The kingdom _____ is characterized by multicellular organisms that convert sunlight energy to chemical energy.
    a. Eubacteria
    b. Protista
    c. Fungi
    d. Archaebacteria
    e. Plantae

43. It is thought that some prokaryotes invaded other prokaryotes
    a. about 4,000 years before the present.
    b. about 10,000 years before the present.
    c. more than one million years before the present.
    d. more than one billion years before the present.
    e. more than one trillion years before the present.

44. _____ is the sum total of all the chemistry that occurs within a cell or organism.
    a. Adaptation
    b. Evolution
    c. Growth
    d. Metabolism
    e. Temperature

45. All _____ must obtain their energy from complex chemical substances in their environment.
    a. plants
    b. autotrophs
    c. organisms
    d. heterotrophs
    e. bacteria

46. Heterotrophs obtain their energy from
    a. fungi.
    b. water.
    c. other organisms.
    d. vitamins.
    e. heat.

47. It is thought by scientists that the first energy used by life on Earth was from
    a. the sun.
    b. other organisms.
    c. water.
    d. methane and hydrogen sulfide.
    e. ethanol.

48. The initial accumulation of oxygen in the atmosphere was the result of photosynthesis from an organism most like modern
    a. cyanobacteria.
    b. algae.
    c. mosses.
    d. kelp.
    e. eukaryotes.

49. A prerequisite for life to survive on land was the accumulation of
    a. oxygen.
    b. carbon dioxide.
    c. water vapor.
    d. ozone.
    e. bacteria in the soil.

50. Ozone is an important factor to life on Earth because it
    a. is toxic to all forms of life.
    b. can be used in place of oxygen.
    c. **blocks much ultraviolet radiation.**
    d. provides energy to some basic forms of life.
    e. disinfects.

51. Which of the following is *not* a characteristic of most multicellular organisms.
    a. Cells change structure and function during development.
    b. Cells stick together after division.
    c. Cells specialize.
    d. Certain cells specialize for purposes of sexual reproduction.
    e. **Cells can grow without regulation.**

2 Small Molecules: Structure and Behavior

Fill in the Blank

1. Every atom except **hydrogen** has one or more neutrons in its nucleus.

2. The sum of the atomic weights in any given molecule is called its **molecular weight.**

3. **Radioactive decay** occurs when one atom, such as ^{14}C, is transformed into another atom, such as ^{14}N, with an accompanying emission of energy.

4. The nutritionist's Calorie, which biologists call a kilocalorie, is the equivalent of **1,000** heat-energy calories.

5. The water strider skates along the surface of water due to a property of liquids called **surface tension.**

6. The chemical properties of an element are determined by the number of **electrons** its atoms contain.

7. The attraction between a slight positive charge on a hydrogen atom and the slight negative charge of a nearby atom is a **hydrogen bond.**

8. A chemical reaction that can proceed in either direction is called a **reversible reaction.**

9. A **molecule** is two or more atoms linked by chemical bonds.

10. A **calorie** is the amount of heat needed to raise the temperature of 1g of pure water from 14.5°C to 15.5°C.

Multiple Choice

1. A 1.0 *M* solution of NaOH has a pH of
   a. 1.0.
   b. 7.0.
   c. 14.0.
   d. 11.2.
   e. None of the above

2. The part of the atom of greatest biological interest is the
   a. proton.
   b. electron.
   c. neutron.
   d. innermost shell.
   e. None of the above

3. Hydrogen, deuterium, and tritium all possess the same
   a. atomic weight.
   b. atomic number.
   c. mass.
   d. density.
   e. All of the above

4. Which of the following describes an unusual property of water?
   a. Three different physical phases
   b. A solid state that is less dense than its liquid state
   c. Heat loss when changing from a liquid to a gas
   d. Limited ionization
   e. All of the above

5. In addition to covalent bonds and ionic bonds, which of the following interactions is of interest to biologists?
   a. Van der Waals interactions
   b. Hydrogen bonds
   c. Hydrophobic interactions
   d. *a* and *b*
   e. *a*, *b*, and *c*

6. What is the difference between an atom and an element?
   a. An atom is made of protons, electrons, and sometimes neutrons, and an element is a substance composed of only one kind of atom.
   b. An element is made of protons, electrons, and sometimes neutrons, and an atom is a substance composed of only one kind of element.
   c. An atom does not contain electrons while an element does.
   d. An atom contains protons and electrons while an element contains protons, electrons, and neutrons.
   e. None of the above

7. The number of protons in an atom equals the number of
   a. neutrons.
   b. electrons.
   c. electrons + neutrons.
   d. neutrons – electrons.
   e. quarks.

8. The atomic number of an element is the same as the number of
   a. neutrons in each atom.
   b. protons + electrons in each atom.
   c. protons in each atom.
   d. neutrons + protons in each atom.
   e. *a* and *c*

9. The mass number of an element is the same as the number of
   a. electrons in each atom.
   b. protons in each atom.
   c. neutrons in each atom.
   d. protons + neutrons in each atom.
   e. electrons + neutrons in each atom.

10. The mass number of an element is measured in
    a. **daltons.**
    b. units.
    c. electrical charge.
    d. grams.
    e. kilograms.

11. Which pair has similar chemical properties?
    a. ^1H and ^{22}Na
    b. ^{12}C and ^{28}Si
    c. ^{16}O and ^{32}S
    d. **^{12}C and ^{14}C**
    e. ^1H and ^2He

12. $^{31}_{15}P$ and $^{32}_{15}P$ have virtually identical chemical and biological properties because
    a. they have the same half-life.
    b. they have the same number of neutrons.
    c. they have the same atomic weight.
    d. they have the same mass number.
    e. **they have the same number of electrons.**

13. Why is the atomic weight of hydrogen 1.008 and not exactly its mass number, 1.000?
    a. Atomic weight does not take into account the weight of the rare isotopes of an element.
    b. **Atomic weight is the average of the mass numbers of a representative sample of the element that includes all its isotopes.**
    c. The atomic weight includes the weight of the electrons.
    d. The atomic weight does not include the weight of the protons.
    e. The mass number of an element is always lower than its atomic weight.

14. The half-life of ^{14}C is 5,700 years. How much ^{14}C is left in a sample after 11,400 years?
    a. 3%
    b. 6%
    c. **25%**
    d. 50%
    e. 100%

15. The half-life of ^{32}P is 14.3 days. How much radioactive ^{32}P is left in a sample after 71.5 days?
    a. 1.56%
    b. 3.12%
    c. **6.24%**
    d. 20%
    e. 100%

16. Which of the following elements is the most chemically reactive?
    a. **Hydrogen**
    b. Helium
    c. Neon
    d. Argon
    e. They all have the same chemical reactivity.

17. A single covalent chemical bond represents the sharing of how many electrons?
    a. One
    b. **Two**
    c. Three
    d. Four
    e. Six

18. The molecular weight of water is 18.0154. One mole of water weighs exactly
    a. 9 grams.
    b. 18 grams.
    c. **18.0154 grams.**
    d. 36.031 grams.
    e. 6.023×10^{23} grams.

19. What is the difference between covalent and ionic bonds?
    a. Covalent bonds are the sharing of neutrons, and ionic bonds are the sharing of electrons.
    b. Ionic bonds are the sharing of electrons between atoms, and covalent bonds are the electric attraction between two atoms.
    c. Covalent bonds are the sharing of protons between atoms, and ionic bonds are the electric attraction between two atoms.
    d. Covalent bonds are the sharing of protons between atoms, and ionic bonds are the sharing of electrons between two atoms.
    e. **Covalent bonds are the sharing of electrons between atoms, and ionic bonds are the electric attraction between two atoms.**

20. Which contains more molecules, a mole of hydrogen or a mole of carbon?
    a. A mole of carbon
    b. A mole of hydrogen
    c. **Both contain the same number of molecules.**
    d. Unknown

21. Sweating is a useful cooling device for humans because
    a. **water takes up a great deal of heat in changing from its liquid state to its gaseous state.**
    b. water takes up a great deal of heat in changing from its solid state to its liquid state.
    c. water can exist in three states at temperatures common on Earth.
    d. water is an outstanding solvent.
    e. water ionizes readily.

22. What two characteristics make water different from most other compounds?
    a. **Its solid state is less dense than its liquid state, and it takes up large amounts of heat to change to its gaseous state.**
    b. Its solid state is more dense than its liquid state, and it takes up only small amounts of heat to change to its gaseous state.
    c. Its solid state is less dense than its liquid state, and it takes up only small amounts of heat to change to its gaseous state.
    d. Its solid state is more dense than its liquid state, and it takes up large amounts of heat to change to its gaseous state.
    e. Its solid state is just as dense than its liquid state, and it takes no heat to change to its gaseous state.

23. Which characteristic of water contributes to the relatively constant temperatures of the oceans?
    a. Water ionizes only slightly.
    b. It takes a small amount of heat energy to raise the temperature of water.
    c. Water can contain large amounts of salt.
    d. Water has the ability to ionize readily.
    e. **It takes a large amount of heat energy to raise the temperature of water.**

24. If you place a paper towel in a dish of water, the water will move up the towel by capillary action. What property of water gives rise to capillary action?
    a. Water molecules ionize.
    b. Water is a good solvent.
    c. Water molecules have hydrophobic interactions.
    d. Water can form hydrogen bonds.
    e. Water takes up large amounts of heat when it vaporizes.

25. An alkaline solution contains
    a. more OH⁻ ions than H⁺ ions.
    b. more H⁺ ions than OH⁻ ions.
    c. the same number of OH⁻ ions as H⁺ ions.
    d. no OH⁻ ions.

26. What is the difference between an acid and a base?
    a. An acid undergoes a reversible reaction, while a base does not.
    b. An acid releases OH⁻ ions in solution, while a base accepts OH⁻ ions.
    c. An acid releases H⁺ ions in solution, while a base releases OH⁻ ions.
    d. An acid releases OH⁻ ions in solution, while a base releases H⁺ ions.
    e. An acid releases H⁺ ions in solution, while a base accepts H⁺ ions.

27. Polar molecules
    a. have an overall negative electric charge.
    b. have an equal distribution of electric charge.
    c. have an overall positive electric charge.
    d. have an unequal distribution of electric charge.
    e. are ions.

28. The electron cloud of a water molecule
    a. is equally dense throughout the molecule.
    b. is most dense near the hydrogen atoms.
    c. is most dense near the oxygen atom.
    d. covers only the positive portion of the molecule.
    e. is separated by an angle of 104.5°.

29. Which is *not* a consequence of hydrogen bonding?
    a. The attraction between water molecules
    b. The shape of proteins and DNA
    c. The high solubility of a sugar in water
    d. Capillary action
    e. All of the above are consequences of hydrogen bonding.

30. What is a van der Waals interaction?
    a. The attraction between the electrons of one molecule and the nucleus of a nearby molecule
    b. The attraction between the electrons of one molecule and the nucleus of the same molecule
    c. The attraction between the electrons of one molecule and the electrons of a nearby molecule
    d. The attraction between nonpolar molecules due to the exclusion of water
    e. The attraction between nonpolar molecules because they are surrounded by water molecules

31. Oil remains as a droplet in water because of the
    a. van der Waals interactions of the nonpolar oil molecules.
    b. hydrophobic interactions of the nonpolar oil molecules.

    c. hydrogen bonds formed between the nonpolar molecules of the oil and the water molecules.
    d. covalent bonds formed between the nonpolar molecules of the oil.
    e. covalent bonds formed between the nonpolar molecules of the oil and the water molecules.

32. A drop of oil in water disperses when detergent is added because
    a. the nonpolar parts of the detergent molecules associate with the oil, and the polar parts of the detergent molecules associate with the water.
    b. the polar parts of the detergent molecules associate with the oil, and the nonpolar parts of the detergent molecules associate with the water.
    c. the nonpolar parts of the detergent molecules associate with the oil and with the water.
    d. the polar parts of the detergent molecules associate with the oil and with the water.
    e. the detergent lowers the surface tension of the water.

33. When potassium hydroxide (KOH) is added to water, it ionizes, releasing hydroxide ions. This solution is
    a. acidic.
    b. basic.
    c. neutral.
    d. molar.
    e. a buffer.

34. H_2SO_4 can ionize to yield two H⁺ ions and one SO_4^{2-} ion. H_2SO_4 is
    a. molar.
    b. a base.
    c. a buffer.
    d. a solution.
    e. an acid.

35. Acid rain is a serious environmental problem. A sample of rainwater collected in the Adirondack Mountains had an H⁺ concentration of 10^{-4} mol/L. The pH of this sample was
    a. .0001.
    b. −4.
    c. 4.
    d. 0.
    e. 10,000.

36. Carbonic acid and sodium bicarbonate act as buffers in the blood. When a small amount of acid is added to this buffer, the H⁺ ions are used up as they combine with the bicarbonate ions. When this happens, the pH of the blood
    a. becomes basic.
    b. becomes acidic.
    c. doesn't change.
    d. is reversible.
    e. ionizes.

37. Butane (CH_3—CH_2—CH_2—CH_3) is an example of
    a. an unsaturated hydrocarbon.
    b. a saturated hydrocarbon.
    c. an acid.
    d. an amine.
    e. a base.

38. Cholesterol is composed primarily of carbon and hydrogen atoms. What property would you expect cholesterol to have?
    a. **It is insoluble in water.**
    b. It is soluble in water.
    c. It is a base.
    d. It is an acid.
    e. It is a buffer.

39. The compound inositol has six hydroxyl groups attached to a six-carbon backbone. Thus inositol can be classified as a(n)
    a. amine.
    b. acid.
    c. ketone.
    d. **alcohol.**
    e. buffer.

40. Butane and isobutane have the same chemical formula but different arrangements of atoms. These two compounds are called
    a. ionic.
    b. alcohols.
    c. functional groups.
    d. amines.
    e. **isomers.**

41. One dalton is the same as the mass of one
    a. electron.
    b. carbon atom.
    c. **proton.**
    d. gram.
    e. charge unit.

42. Oxygen and carbon are defined as different elements because they have atoms with a different
    a. number of electrons.
    b. **number of protons.**
    c. number of neutrons.
    d. mass.
    e. charge.

43. The mass of an atom is primarily determined by the
    a. number of electrons it contains.
    b. number of protons it contains.
    c. sum of the number of protons and electrons it contains.
    d. **sum of the number of protons and neutrons it contains.**
    e. number of charges it contains.

44. Two atoms are held together in four covalent bonds because of forces between
    a. electrons and protons.
    b. **electrons and electrons.**
    c. protons and neutrons.
    d. protons and protons.
    e. neutrons and neutrons.

45. Two carbon atoms held together in a double covalent bond share
    a. one electron.
    b. two electrons.
    c. **four electrons.**
    d. six electrons.
    e. eight electrons.

46. Particles having a net negative charge are called
    a. electronegative.
    b. cations.
    c. **anions.**
    d. acids.
    e. bases.

47. Avogadro's number is the number of
    a. grams in a mole.
    b. **molecules in a mole.**
    c. moles in gram molecular weight.
    d. molecules in a gram.
    e. moles in a gram.

48. To determine the number of molecules in a teaspoon of sugar you need
    a. the density of the sugar.
    b. the weight of the sugar.
    c. the molecular weight of the sugar.
    d. Avogadro's number.
    e. **the weight and molecule weight of the sugar, and Avogadro's number.**

49. How would you make 100 ml of an aqueous solution with a 0.25 M concentration of a compound that has a molecule weight of 200 daltons?
    a. Add 0.25 grams of the compound to 100 ml of water.
    b. Add 250 grams of the compound to 100 ml of water.
    c. Take 250 grams of the compound and add water until the volume equals 100 ml.
    d. Take 50 grams of the compound and add water until the volume equals 100 ml.
    e. **Take 5 grams of the compound and add water until the volume equals 100 ml.**

50. Of the following amounts of compounds containing ^1H, ^{12}C, and ^{16}O, the one with the greatest number of molecules is
    a. **2 grams of CO.**
    b. 2 grams of CO_2.
    c. 2 grams of HCOOH.
    d. 2 grams of C_2H_5OH.
    e. 2 grams of $C_6H_{12}O_6$.

51. Which of the following molecules is held together primarily by ionic bonds?
    a. H_2O
    b. $C_6H_{12}O_6$
    c. **NaCl**
    d. H_2
    e. NH_3

52. The hydrogen bond between two water molecules arises because water is
    a. **polar.**
    b. nonpolar.
    c. a liquid.
    d. a small molecule.
    e. hydrophobic.

53. The functional group written as —COOH is called the
    a. hydroxyl group.
    b. carbonyl group.
    c. amino group.
    d. ketone group.
    e. **carboxyl group.**

54. The more acidic of two solutions has
    a. more hydroxyl ions.
    b. more hydrogen acceptors.
    c. more H^+ ions per liter.
    d. a higher pH.

55. Among biologically important molecules, an acid always has
    a. amino groups.
    b. hydrogen atoms.
    c. water.
    d. a pH less than 7.
    e. a pH greater than 7.

56. Which of the following atoms usually has the greatest number of covalent bonds with other atoms?
    a. Carbon
    b. Oxygen
    c. Sulfur
    d. Hydrogen
    e. Nitrogen

57. Of the following atomic configurations, the one that has an atomic mass of 14 is the atom with
    a. 14 neutrons.
    b. 14 electrons.
    c. 7 neutrons, 7 electrons.
    d. 7 protons and 7 electrons.
    e. 6 protons and 8 neutrons.

58. An atom that is neutrally charged contains
    a. only neutrons.
    b. the same number of neutrons as electrons.
    c. the same number of neutrons as protons.
    d. the same number of positive particles as negative particles.
    e. no charged particles.

59. The four elements most common in organisms are
    a. calcium, iron, hydrogen, and oxygen.
    b. water, carbon, hydrogen, and oxygen.
    c. carbon, oxygen, hydrogen, and nitrogen.
    d. nitrogen, carbon, iron, and hydrogen.
    e. phosphorus, water, carbon, and oxygen.

60. The notation [H^+] means the
    a. number of H^+ ions present in a solution.
    b. number of protons in an H^+ ion.
    c. charge of an H^+ ion.
    d. concentration of H^+ ions in moles per liter.
    e. chemical reactivity of H^+ ions.

61. Of the following types of chemical bonds, the strongest is the
    a. hydrogen bond.
    b. van der Waals bond.
    c. ionic bond.
    d. acidic bond.
    e. covalent bond.

62. Select the statement that describes a difference between ionic bonds and covalent bonds.
    a. An ionic bond is stronger.
    b. Electron sharing is more equal in the covalant bond.
    c. An ionic bond occurs more often in aqueous solutions.

d. An ionic bond only occurs in acids.
e. A covalent bond occurs in nonpolar molecules.

63. Which of the following statements about the number of bonds formed by atoms is true?
    a. Oxygen forms one, carbon forms four, and hydrogen forms one.
    b. Oxygen forms four, carbon forms four, and hydrogen forms four.
    c. Oxygen forms two, carbon forms four, and hydrogen forms none.
    d. Oxygen forms two, carbon forms four, and hydrogen forms one.
    e. Oxygen forms two, carbon forms two, and hydrogen forms two.

64. Of the following types of molecules, the one always containing nitrogen is
    a. thiol.
    b. sugar.
    c. hydrocarbon.
    d. alcohol.
    e. amino acid.

65. A typical distance between two atoms held together by a chemical bond is
    a. one millimeter.
    b. two micrometers.
    c. 10^8 meters.
    d. 0.2 nanometer.
    e. 10^{-12} meter.

66. Surface tension and capillary action occur in water because it
    a. is wet.
    b. is dense.
    c. has hydrogen bonds.
    d. is nonpolar.
    e. has ionic bonds.

67. The kingdom _____ is characterized by multicellular organisms that convert sunlight energy to chemical energy.
    a. Eubacteria
    b. Archaebacteria
    c. Fungi
    d. Protista
    e. Plantae

68. _____ is the sum total of all the chemistry that occurs within a cell or organism.
    a. Adaptation
    b. Evolution
    c. Growth
    d. Metabolism
    e. Temperature

69. A cell's rate of exhange of nutrients and water products with its environment is a function of its
    a. ability to move.
    b. surface area.
    c. temperature.
    d. density.
    e. age.

3 Macromolecules: Their Chemistry and Biology

Fill in the Blank

1. Fluidity and melting point of fatty acids are determined in part by **their number of unsaturated bonds**.

2. Many monosaccharides like fructose, mannose, and galactose have the same chemical formula as glucose ($C_6H_{12}O_6$), but the atoms are combined differently to yield different structural arrangements. These varying forms of the same chemical formula are called **isomers**.

3. The highly branched polysaccharide that stores glucose in the muscle and the liver of animals is **glycogen**.

4. In proteins, amino acids are linked together by **peptide** bonds.

5. The linear arrangement of amino acids in the polypeptide chain is referred to as the **primary** structure of the protein.

6. Starch is a polymer of glucose subunits. The subunits of any polymer are called **monomers**.

7. Fatty acids with more than one carbon–carbon double bond are called **polyunsaturated**.

8. Cholesterol, vitamin D, and testosterone all have a multiple-ring structure and are members of a family of lipids known as **steroids**.

9. Carbohydrates made up of two simple sugars are called **disaccharides**.

10. Carbohydrates that contain another element in addition to C, H, and O are called **derivative carbohydrates**.

11. The covalent bond formed between the sulfur atoms of two cysteine side chains is called a **disulfide bridge**.

12. All amino acids have a hydrogen, a carboxyl group, and an amino group attached to a carbon atom. The variability in the 20 different amino acids lies in the structure of their R groups or **side chains**.

13. An **ester** linkage connects the fatty acid molecule to glycerol.

14. The glucose molecules found in humans are **the same as** those found in tomato plants.

15. The bonds that link sugar monomers together in a starch molecule are **glycosidic** bonds.

16. Cholesterol is classified as a(n) **lipid**.

17. The reaction A—H + B—OH → A—B + H_2O represents a **condensation reaction**.

Multiple Choice

1. The major classes of biologically significant large molecules include which of the following?
   a. Proteins
   b. Nucleic acids
   c. Carbohydrates
   d. Lipids
   e. All of the above

2. Lipids are
   a. insoluble in water.
   b. readily soluble in organic solvents.
   c. characterized by their solubility.
   d. important constituents of biological membranes.
   e. All of the above

3. Sequence information in DNA specifies which of the following conformational components of proteins?
   a. Primary
   b. Secondary
   c. Tertiary
   d. a, b, and c
   e. None of the above

4. Which of the following is a characteristic of proteins?
   a. Some function as enzymes
   b. Provide structural units in the form of keratin
   c. Possess glycosidic linkages between amino acids
   d. a and b
   e. a, b, and c

5. Which of the following represents a combination of major classes of cellular molecules?
   a. Glycolipids
   b. Glycoproteins
   c. Nucleoproteins
   d. Lipoproteins
   e. All of the above

6. Molecules with molecular weights greater than 1,000 daltons are usually called
   a. proteins.
   b. polymers.
   c. nucleic acids.
   d. macromolecules.
   e. monomers.

7. Polymerization reactions in which proteins are synthesized from amino acids
   a. require energy.
   b. result in the formation of water.
   c. are condensation reactions.
   d. are dehydration reactions.
   e. All of the above

8. In condensation reactions, the atoms that make up a water molecule are derived from
   a. oxygen.
   b. only one of the reactants.
   c. **both of the reactants.**
   d. carbohydrates.
   e. enzymes.

9. Which of the following is *not* a macromolecule?
   a. RNA
   b. DNA
   c. An enzyme
   d. A protein
   e. **Salt**

10. Which of the following is *not* a characteristic of lipids?
    a. **They are readily soluble in water.**
    b. They are soluble in organic solvents.
    c. They release large amounts of energy when broken down.
    d. They form two layers when mixed with water.
    e. They act as an energy storehouse.

11. You have isolated an unidentified liquid from a sample of beans. You add the liquid to a beaker of water and shake vigorously. After a few minutes, the water and the other liquid separate into two layers. To which class of large biological molecules does the unknown liquid most likely belong?
    a. Carbohydrates
    b. **Lipids**
    c. Proteins
    d. Enzymes
    e. Nucleic acids

12. Lipids form the barriers surrounding various compartments within an organism. Which property of lipids makes them a good barrier?
    a. **Many biologically important molecules are not soluble in lipids.**
    b. Lipids are polymers.
    c. Lipids store energy.
    d. Triglycerides are lipids.
    e. Lipids release large amounts of energy when broken down.

13. You look at the label on a container of shortening and see "hydrogenated vegetable oil." This means that during processing, the number of carbon–carbon double bonds in the oil was decreased. What is the result of decreasing the number of double bonds?
    a. The oil is now a liquid at room temperature.
    b. **The oil is now a solid at room temperature.**
    c. There are more "kinks" in the fatty acid chains.
    d. The oil is now a derivative carbohydrate.
    e. The fatty acid is now a triglyceride.

14. The portion of a phospholipid that contains the phosphorous group has one or more electric charges. That makes this region of the molecule
    a. hydrophobic.
    b. **hydrophilic.**
    c. nonpolar.
    d. unsaturated.
    e. saturated.

15. Cholesterol is soluble in ether, an organic solvent, but is *not* soluble in water. Based on this information, what class of biological macromolecules does cholesterol belong to?
    a. Nucleic acids
    b. Carbohydrates
    c. Proteins
    d. Enzymes
    e. **Lipids**

16. Which of the following is *not* a function in which lipids play an important role?
    a. Vision
    b. Storing energy
    c. Membrane structure
    d. **Storing genetic information**
    e. Chemical signaling

17. In a biological membrane, the phospholipids are arranged with the fatty acid chains facing the interior of the membrane. As a result, the interior of the membrane is
    a. **hydrophobic.**
    b. hydrophilic.
    c. charged.
    d. polar.
    e. filled with water.

18. The monomers that make up polymeric carbohydrates like starch are called
    a. nucleotides.
    b. trisaccharides.
    c. **monosaccharides.**
    d. nucleosides.
    e. fatty acids.

19. The atoms that make up carbohydrates are
    a. C, H, and N.
    b. C and H.
    c. C, H, and P.
    d. **C, H, and O.**
    e. C, H, O, and N.

20. Glucose and fructose both have the formula $C_6H_{12}O_6$, but the atoms in these two compounds are arranged differently. Glucose and fructose are known as
    a. **isomers.**
    b. polysaccharides.
    c. oligosaccharides.
    d. pentoses.
    e. steroids.

21. A nucleotide contains a pentose, a phosphate, and
    a. a lipid.
    b. an acid.
    c. **a nitrogen-containing base.**
    d. an amino acid.
    e. a glycerol.

22. A simple sugar with the formula $C_5H_{10}O_5$ can be classified as a
    a. hexose.
    b. polysaccharide.
    c. disaccharide.
    d. **pentose.**
    e. lipid.

23. Lactose, or milk sugar, is composed of one glucose unit and one galactose unit. It can be classified as a
    a. **disaccharide.**
    b. hexose.
    c. pentose.
    d. polysaccharide.
    e. simple sugar.

24. Two important polysaccharides made up of glucose monomers are
    a. guanine and cytosine.
    b. RNA and DNA.
    c. sucrose and lactose.
    d. **cellulose and starch.**
    e. testosterone and cortisone.

25. Cellulose is the most abundant organic compound on Earth. Its main function is
    a. to store genetic information.
    b. as a storage compound for energy in plant cells.
    c. as a storage compound for energy in animal cells.
    d. as a component of biological membranes.
    e. **to provide mechanical strength to plant cell walls.**

26. In animals, glucose is stored in the compound
    a. cellulose.
    b. amylose.
    c. **glycogen.**
    d. fructose.
    e. cellobiose.

27. Which of the following monomer/polymer pairs is *not* correct?
    a. Monosaccharide/polysaccharide
    b. Amino acid/protein
    c. **Triglyceride/lipid**
    d. Nucleotide/DNA
    e. Nucleotide/RNA

28. Amino acids can be classified by
    a. the number of monosaccharides they contain.
    b. the number of carbon–carbon double bonds in their fatty acids.
    c. the number of peptide bonds they can form.
    d. the number of disulfide bridges they can form.
    e. **the characteristics of their side chains.**

29. During the formation of a peptide linkage, which of the following occurs?
    a. **A molecule of water is formed.**
    b. A disulfide bridge is formed.
    c. A hydrophobic bond is formed.
    d. A hydrophilic bond is formed.
    e. An ionic bond is formed.

30. The side chain of leucine is a hydrocarbon. In a folded protein, where would you expect to find leucine?
    a. In the interior of a cytoplasmic enzyme
    b. On the exterior of a protein embedded in a membrane
    c. On the exterior of a cytoplasmic enzyme
    d. **a and b**
    e. a and c

31. What is the theoretical number of different proteins that you could make from 50 amino acids?
    a. 50^{20}
    b. 20 x 50
    c. **20^{50}**
    d. 10^{50}
    e. 250

32. The shape of a folded protein is often determined by
    a. its tertiary structure.
    b. **the sequence of its amino acids.**
    c. whether the peptide bonds have α or β linkages.
    d. the number of peptide bonds.
    e. the base-pairing rules.

33. The amino acids of the protein keratin are arranged in an α helix. This secondary structure is stabilized by
    a. covalent bonds.
    b. peptide bonds.
    c. glycosidic linkages.
    d. polar bonds.
    e. **hydrogen bonds.**

34. What is the nucleotide sequence of the complementary strand of this DNA molecule: A A T G C G A?
    a. **T T A C G C T**
    b. A A T G C G A
    c. G G C A T A G
    d. C C G T T A T
    e. A G C G T A A

35. Which of the following is *not* a difference between DNA and RNA?
    a. DNA has thymine and RNA has uracil.
    b. DNA usually has two polynucleotide strands and RNA usually has one strand.
    c. DNA has deoxyribose sugar and RNA has ribose sugar.
    d. **DNA is a polymer and RNA is a monomer.**
    e. In DNA, A pairs with T and in RNA, A pairs with U.

36. DNA molecules that carry different genetic information can be distinguished by looking at
    a. the number of strands in the helix.
    b. how much uracil is present.
    c. **the sequence of nucleotide bases.**
    d. differences in the base-pairing rules.
    e. the shape of the helix.

37. The "backbone" of nucleic acid molecules is made of
    a. nitrogenous bases.
    b. **alternating sugars and phosphate groups.**
    c. purines.
    d. pyrimidines.
    e. nucleosides.

38. According to the base-pairing rules for nucleic acids, purines always pair with
    a. deoxyribose sugars.
    b. uracil.
    c. **pyrimidines.**
    d. adenine.
    e. guanine.

39. What type of amino acid side chains would you expect to find on the surface of a protein embedded in a cell membrane?
    a. Cysteine
    b. **Hydrophobic**
    c. Hydrophilic
    d. Charged
    e. Polar, but not charged

40. A molecule with the formula $C_{16}H_{32}O_2$ is a
    a. hydrocarbon.
    b. carbohydrate.
    c. lipid.
    d. protein.
    e. nucleic acid.

41. A molecule with the formula $C_{16}H_{30}O_{15}$ is a
    a. hydrocarbon.
    b. carbohydrate.
    c. lipid.
    d. protein.
    e. nucleic acid.

42. Fatty acids are molecules that
    a. contain fats.
    b. are carboxylic acids.
    c. are carbohydrates.
    d. contain glycerol.
    e. are always saturated.

43. Sucrose is a
    a. hexose.
    b. lipid.
    c. disaccharide.
    d. glucose.
    e. simple sugar.

44. DNA and RNA contain
    a. pentoses.
    b. hexoses.
    c. fructoses.
    d. maltoses.
    e. amyloses.

45. The twenty different common amino acids have different
    a. amino groups.
    b. R groups.
    c. acid groups.
    d. peptide linkages.
    e. primary structures.

46. The primary structure of a protein is determined by its
    a. disulfide bridges.
    b. α helix structure.
    c. sequence of amino acids.
    d. branching.
    e. three-dimensional structure.

47. A β pleated sheet organization in a polypeptide chain is an example of
    a. primary structure.
    b. secondary structure.
    c. tertiary structure.
    d. quaternary structure.
    e. coiled structure.

48. A protein can best be defined as
    a. a polymer of amino acids.
    b. containing one or more polypeptide chains.
    c. containing 20 amino acids.
    d. containing 20 peptide linkages.
    e. containing double helices.

49. The four nitrogenous bases of RNA are abbreviated as
    a. A, G, C, and T.
    b. A, G, T, and N.
    c. G, C, U, and N.
    d. A, G, U, and T.
    e. A, G, C, and U.

50. Polysaccharides, polypeptides, and polynucleotides have in common that they all
    a. contain simple sugars.
    b. are formed in condensation reactions.
    c. are found in cell membranes.
    d. contain nitrogen.
    e. have molecular weights less than 30,000 daltons.

51. DNA carries genetic information in its
    a. helical form.
    b. sequence of bases.
    c. tertiary sequence.
    d. sequence of amino acids.
    e. phosphate groups.

52. A molecule often spoken of as having a head and tail is
    a. a phospholipid.
    b. an oligosaccharide.
    c. an RNA.
    d. a steroid.
    e. a triglyceride.

53. A molecule that has an important role in limiting what gets into and out of cells is
    a. glucose.
    b. maltose.
    c. phospholipid.
    d. fat.
    e. phosphohexose.

54. A molecule that has an important role in long-term storage of energy is a(n)
    a. steroid.
    b. RNA.
    c. glycogen.
    d. amino acid.
    e. hexose.

55. A peptide linkage (peptide bond) holds together two
    a. protein molecules.
    b. amino acid molecules.
    c. sugar molecules.
    d. fatty acid molecules.
    e. phospholipid molecules.

56. In DNA molecules,
    a. purines pair with pyrimidines.
    b. A pairs with C.
    c. G pairs with A.
    d. purines pair with purines.
    e. C pairs with T.

57. A type of molecule very often drawn with a single six-sided ring structure is a(n)
    a. sucrose.
    b. amino acid.
    c. glucose.
    d. fatty acid.
    e. steroid.

58. Maltose and lactose are similar in that they both are
    a. simple sugars.
    b. amino acids.
    c. insoluble in water.
    d. disaccharides.
    e. hexoses.

59. Starch and glycogen are different in that only one of them
    a. is a polymer of glucose.
    b. contains ribose.
    c. **is made in plants.**
    d. is an energy storage molecule.
    e. can be digested by human beings.

60. Enzymes are
    a. DNA.
    b. lipids.
    c. carbohydrates.
    d. **protein.**
    e. amino acids.

61. The type of bond that holds two amino acids together in a polypeptide chain is a(n)
    a. ionic bond.
    b. disulfide bridge.
    c. hydrogen bond.
    d. **peptide linkage.**
    e. dehydration bond.

62. The _____ structure of a protein relates to how separate polypeptides assemble together.
    a. primary
    b. secondary
    c. tertiary
    d. **quaternary**
    e. helical

63. In DNA, A binds with T and G with C; these are examples of a specific type of reaction called
    a. **complementary base pairing.**
    b. a dehydration reaction.
    c. a reduction reaction.
    d. a hydrophobic reaction.
    e. a purine–purine reaction.

64. A fat contains fatty acids and _____.
    a. **a glycerol**
    b. a base
    c. an amino acid
    d. a phosphate

65. There are _____ different types of tripeptides (molecules with three amino acids linked together) that can exist using the 20 common amino acids.
    a. 3
    b. 20
    c. 60
    d. 900
    e. **8,000**

66. Chitin is a polymer of
    a. galactosamine
    b. glucose.
    c. **glucosamine.**
    d. glycine.
    e. All of the above

67. Two common amino sugars are
    a. glucose and fructose.
    b. **glucosamine and galactosamine.**
    c. glycine and glutamine.
    d. trehelose and sucrose.
    e. grapeamine and citricamine.

68. Prosthetic groups are molecules that
    a. **associate with proteins but are not proteins.**
    b. are proteins that associate with lipids.
    c. are lipids that associate with fatty acids.
    d. associate with the surface of cells.
    e. are found free, and rarely associate with other molecules.

69. A type of protein that functions by helping the correct folding of other proteins is called
    a. foldzyme.
    b. renaturing protein.
    c. **chaperone protein.**
    d. hemoglobin.
    e. denaturing protein.

4 The Organization of Cells

Fill in the Blank

1. In biology, we call the basic unit of life the **cell**.

2. Photosynthetic membrane systems and mesosomes are internal membrane components of certain organisms termed **prokaryotes**.

3. The light microscope has glass lenses for focusing light (photons) for imaging, whereas the electron microscope has **magnets** for focusing electrons for imaging.

4. Membranous cellular subsystems are termed **organelles**.

5. **Photosynthesis** is the process whereby light energy is converted into chemical bonds.

6. The **Golgi apparatus** is an organelle that serves as a sort of "postal depot" where some of the proteins synthesized on ribosomes and rough ER are processed.

7. In order to isolate organelles for study, scientists employ methods to separate organelles from cells and from one another. This process is called **cell fractionation**.

8. RNA carries information for protein synthesis from the DNA in the nucleus to the ribosomes in the cytoplasm. To get from the nucleoplasm to the cytoplasm, RNA must pass through **nuclear pores**.

9. All organisms are composed of cells; all cells come from preexisting cells. These statements are called **the cell theory**.

10. When you cut an orange in half, you **increase** the surface area-to-volume ratio.

11. The DNA in a prokaryotic cell can be found in the **nucleoid** region.

12. The **capsules** of some bacteria help them avoid being detected by the human immune system.

13. The meshwork of intermediate filaments found on the interior surface of the nuclear membrane is called **the nuclear lamina**.

14. Steroids, fatty acids, phospholipids, and carbohydrates are synthesized in the **smooth ER**.

15. The side of the Golgi facing the ER is the *cis* face.

16. The substances that enter the Golgi come from the **ER**.

17. Toxic peroxides that are unavoidably formed as side products of important cellular reactions are found and neutralized in **peroxisomes**.

Multiple Choice

1. What must cells do in order to survive?
   a. Obtain and process energy
   b. Convert genetic information into proteins
   c. Keep certain biochemical reactions separate from each other
   d. *a* and *b*
   e. ***a, b,* and *c***

2. Examples of cellular "appendages" include
   a. the Golgi apparatus.
   b. cilia.
   c. flagella.
   d. pili.
   e. ***b, c,* and *d***

3. Roles of biological membranes in eukaryotic cells include which of the functions listed below?
   a. Molecular trafficking of molecules
   b. Serving as staging areas for cellular interaction
   c. Mediating adhesion-recognition reactions between cells
   d. Participating in energy transformations
   e. **All of the above**

4. The utilization of "food" in the mitochondria, with the associated formation of ATP, is termed
   a. **cellular respiration.**
   b. metabolic rate.
   c. diffusion.
   d. metabolic processing of fuels.
   e. catabolism.

5. Components of chloroplasts include
   a. grana.
   b. thylakoids.
   c. cristae.
   d. ***a* and *b***
   e. *a, b,* and *c*

6. It is generally agreed that
   a. eukaryotes and prokaryotes evolved simultaneously.
   b. eukaryotes evolved before prokaryotes.
   c. **eukaryotes evolved from prokaryotes.**
   d. eukaryotes and prokaryotes are not related evolutionarily.
   e. eukaryotes and prokaryotes are examples of parallel evolution.

7. The cell is the basic unit of function and reproduction because
   a. **subcellular components cannot regenerate whole cells.**
   b. cells are totipotent.

c. single cells can sometimes produce an entire organism.

d. cells can only come from preexisting cells.

e. a cell can arise by the fusion of two cells.

8. Which of the following is *not* something a cell must do to survive?
   a. Carry out chemical reactions
   b. Obtain energy from the environment
   c. Allow certain materials to enter through its membrane
   d. Use the information contained in the DNA
   e. **Reproduce**

9. What is the major distinction between prokaryotic and eukaryotic cells?
   a. **A prokaryotic cell does not have a nucleus and a eukaryotic cell does.**
   b. A prokaryotic cell does not have DNA and a eukaryotic cell does.
   c. Prokaryotic cells are smaller than eukaryotic cells.
   d. The prokaryotic cells have not prospered, while eukaryotic cells are evolutionary "successes."
   e. Prokaryotic cells cannot obtain energy from their environment.

10. Which of the following is *not* a characteristic of a prokaryotic cell?
    a. A plasma membrane
    b. **A nuclear membrane**
    c. A nucleoid
    d. Ribosomes
    e. Enzymes

11. All members of the kingdom Eubacteria
    a. have nuclei.
    b. have chloroplasts.
    c. are multicellular.
    d. **are prokaryotes.**
    e. have flagella.

12. Ribosomes are made up of
    a. DNA and RNA.
    b. DNA and proteins.
    c. **RNA and proteins.**
    d. proteins.
    e. DNA.

13. Which of the following is found in prokaryotic cells?
    a. Mitochondria
    b. Chloroplasts
    c. Nuclei
    d. **Enzymes**
    e. Endomembrane system

14. The infoldings of the plasma membrane of a prokaryotic cell can form which of the following structures?
    a. **Photosynthetic system**
    b. Cell wall
    c. Nuclear membrane
    d. Capsule
    e. Ribosome

15. The DNA of prokaryotic cells is found in
    a. the plasma membrane.
    b. the nucleus.
    c. the ribosome.
    d. **the nucleoid.**
    e. mitochondria.

16. Which structure supports the plant cell and determines its shape?
    a. Capsule
    b. Flagellum
    c. **Cell wall**
    d. Cytosol
    e. Cytoplasm

17. Some bacteria are able to propel themselves through liquid by means of a structure called the
    a. **flagellum.**
    b. pili.
    c. cytoplasm.
    d. cell wall.
    e. peptidoglycan molecule.

18. If you removed the pili from a bacterial cell, which of the following would you expect to happen?
    a. The bacterium could no longer swim.
    b. **The bacterium could no longer adhere to other cells.**
    c. The bacterium could no longer regulate the movement of molecules into and out of the cell.
    d. The bacterium would dry out.
    e. The shape of the bacterium would change.

19. Of the objects listed, which is the smallest that you can see with the unaided eye?
    a. DNA molecule
    b. **Human egg**
    c. Virus
    d. Ribosome
    e. Human skin cell

20. Ribosomes are not visible under a light microscope, but can be seen with an electron microscope. This is because
    a. electron beams have more energy than light beams.
    b. electron microscopes focus light with magnets.
    c. **electron microscopes have more resolving power than light microscopes.**
    d. electrons have such high energy that they pass through biological samples.
    e. living cells can be observed under the electron microscope.

21. Using a light microscope it is possible to view cytoplasm streaming around the central vacuole in cells of the green alga *Nitella*. Why would you use a light microscope instead of an electron microscope to study this process?
    a. Electron microscopes have less resolving power than light microscopes.
    b. Structures inside the cell cannot be seen with the electron microscope.
    c. Whole cells cannot be viewed with the electron microscope.
    d. **The electron microscope cannot be used to observe living cells.**
    e. The central vacuole is too small to be seen with a scanning electron microscope.

22. Which of the following is a general function of all cellular membranes?
    a. **They regulate which materials can cross the membrane.**
    b. They support the cell and determine its shape.
    c. They produce energy for the cell.
    d. They produce proteins for the cell.
    e. They move the cell.

23. Which statement about the nuclear envelope is true?
    a. It contains pores for the passage of large molecules.
    b. It is composed of two membranes.
    c. It contains ribosomes on the inner surface.
    d. a and b
    e. All of the above

24. What is the purpose of the folds of the inner mitochondrial membrane?
    a. They increase the volume of the mitochondrial matrix.
    b. They create new membrane-bounded compartments within the mitochondrion.
    c. They increase the surface area for the exchange of substances across the membrane.
    d. They anchor more of the mitochondrial DNA.
    e. The folds have no known purpose.

25. Which type of organelle is found in plants but *not* in animals?
    a. Ribosomes
    b. Mitochondria
    c. Nuclei
    d. Plastids
    e. None of the above

26. Where in the cell do you *not* find DNA?
    a. The mitochondrial matrix
    b. The chloroplast stroma
    c. The cell cytosol
    d. The cell nucleus
    e. You find DNA in all of the above.

27. Which of the following statements is true?
    a. Animal cells do not produce chloroplasts.
    b. Animal cells do not have mitochondria.
    c. All plant cells contain chloroplasts.
    d. Plant cells do not have plastids.
    e. None of the above

28. Which of the following is *not* an argument for the endosymbiotic theory?
    a. Mitochondria and chloroplasts have double membranes.
    b. Mitochondria and chloroplasts cannot be grown in culture free of a host cell.
    c. Mitochondria and chloroplasts have DNA and ribosomes.
    d. Mitochondrial ribosomes are similar to bacterial ribosomes.
    e. All of the above

29. What is the difference between "free" and "attached" ribosomes?
    a. Free ribosomes are in the cytoplasm while attached ribosomes are anchored to the endoplasmic reticulum.
    b. Free ribosomes produce proteins that remain in the cytosol while attached ribosomes produce proteins that are exported from the cell.
    c. Free ribosomes produce proteins that are exported from the cell while attached ribosomes make proteins for mitochondria and chloroplasts.
    d. *a* and *c*
    e. *a* and *b*

30. The carotenoid pigments that give ripe tomatoes their red colors are contained in organelles called
    a. chloroplasts.
    b. proplastids.
    c. protoplasts.
    d. leucoplasts.
    e. chromoplasts.

31. If all the lysosomes within a cell suddenly ruptured, what would be the most likely result?
    a. The macromolecules in the cell cytosol would begin to degrade.
    b. The number of proteins in the cytosol would begin to increase.
    c. The DNA within the mitochondria would begin to degrade.
    d. The mitochondria and chloroplasts would begin to divide.
    e. There would be no change in the normal function of the cell.

32. Which is a function of a plant cell vacuole?
    a. Storage of wastes
    b. Support for the cell
    c. Excretion of wastes
    d. *a* and *b*
    e. *b* and *c*

33. Microtubules are
    a. made of actin and function in locomotion.
    b. made of tubulin and function in locomotion.
    c. made of tubulin and function to change cell shape.
    d. made of actin and function to change cell shape.
    e. made of polysaccharides and function in locomotion.

34. Mitochondria and chloroplasts cannot be separated by differential centrifugation, but they can be separated by equilibrium centrifugation. These observations suggest that
    a. mitochondria and chloroplasts differ in density.
    b. mitochondria and chloroplasts differ in radius.
    c. mitochondria and chloroplasts differ in both density and radius.
    d. mitochondria and chloroplasts have identical densities.
    e. chloroplasts are the site of photosynthesis.

35. How would you estimate the purity of an organelle preparation?
    a. Examine the preparation under the light or electron microscope.
    b. Test for the function of the organelle of interest in the preparation.
    c. *a* and *b*
    d. None of the above
    e. There is no need to check for purity since the methods of isolation are flawless.

36. The width of a typical animal cell is
    a. one millimeter.
    b. 20 micrometers.
    c. 1 micrometer.
    d. 10 nanometers.
    e. 10^{-10} meters.

37. The smallest structure that can be clearly seen in a light microscope is
    a. 1 millimeter.
    b. 0.1 millimeter.
    c. 10 micrometers.
    d. 2 micrometers.
    e. 0.2 micrometers.

38. The two major types of cells are
    a. human and nonhuman.
    b. prokaryotic and eukaryotic.
    c. blood and muscle.
    d. plant and animal.
    e. warm-blooded and cold-blooded.

39. The one type of cell always lacking a cell wall is the
    a. bacterial cell.
    b. plant cell.
    c. animal cell.
    d. fungal cell.
    e. prokaryotic cell.

40. A structure found only in plant cells is the
    a. cilium.
    b. nucleus.
    c. mitochondrion.
    d. glyoxysome.
    e. cell membrane.

41. An organelle found in all eukaryotic cells during some portion of their lives is the
    a. chloroplast.
    b. nucleus.
    c. flagellum.
    d. vacuole.
    e. centriole.

42. Ribosomes are important because they are the structures where
    a. chemical energy is stored in making ATP.
    b. cell division is controlled.
    c. genetic information is used to make proteins.
    d. sunlight energy is captured into chemical energy.
    e. new organelles are made.

43. Chloroplasts are important because they are the structures where
    a. chemical energy is stored by making ATP.
    b. cell division is controlled.
    c. genetic information is used to make proteins.
    d. sunlight energy is converted to chemical energy.
    e. new organelles are made.

44. An organelle consisting of a series of flattened sacks stacked somewhat like pancakes is the
    a. mitochondrion.
    b. choroplast.
    c. Golgi apparatus.
    d. rough endoplasmic reticulum.
    e. flagellum.

45. An organelle with an internal cross section showing a characteristic "9 + 2" morphology is the
    a. mitochondrion.
    b. vacuole.
    c. Golgi apparatus.
    d. flagellum.
    e. cytoskeleton.

46. An organelle bounded by two distinct membranes is the
    a. nucleus.
    b. Golgi apparatus.
    c. endoplasmic reticulum.
    d. flagellum.
    e. lysosome.

47. Chromatin is a series of entangled threads composed of
    a. microtubules.
    b. DNA and protein.
    c. fibrous proteins.
    d. cytoskeleton.
    e. membranes.

48. One place where ribosomes are *not* found is in a
    a. mitochondrion.
    b. choroplast.
    c. rough endoplasmic reticulum.
    d. prokaryotic cell.
    e. Golgi apparatus.

49. The overall shape of a cell is determined by its
    a. cell membrane.
    b. cytoskeleton.
    c. nucleus.
    d. cytosol.
    e. endoplasmic reticulum.

50. Of the following structures of an animal cell, the one with the largest volume is the
    a. cilium.
    b. mitochondrion.
    c. lysosome.
    d. nucleus.
    e. ribosomes.

51. Of the following structures of a plant cell, the one that most often has the greatest volume is the
    a. glyoxysome.
    b. lysosome.
    c. chromosome.
    d. ribosome.
    e. vacuole.

52. Of the following structures, the one that an animal cell will usually have the greatest number of is the
    a. vacuole.
    b. nucleus.
    c. ribosome.
    d. flagellum.
    e. plastid.

53. Of the following structures, the one that contains both a matrix and cristae is the
    a. plastid.
    b. lysosome.
    c. Golgi apparatus.
    d. mitochondrion.
    e. chromatin.

54. Light energy for conversion to chemical energy is trapped in the
    a. mitochondrion.
    b. chromoplast.
    c. thylakoid.
    d. endoplasmic reticulum.
    e. Golgi apparatus.

55. Proteins that will function outside of the cytosol are made by
    a. the Golgi apparatus.
    b. ribosomes within the mitochondrion.
    c. the smooth endoplasmic reticulum.
    d. ribosomes on the rough endoplasmic reticulum.
    e. ribosomes within the nucleus.

56. Cilia contain
    a. microtubules.
    b. microfilaments.
    c. intermediate filaments.
    d. ribosomes.
    e. plasmodesmata.

57. RNA would *not* be expected to be found in which of the following structures?
    a. Nucleus
    b. Mitochondrion
    c. Vacuole
    d. Ribosome
    e. Prokaryotic cell

58. The process of centrifugation is useful in _____ cells.
    a. multiplying
    b. magnifying
    c. resolving
    d. suspending
    e. fractionating

59. The structure causing the movement of organelles within a cell is the _____.
    a. Golgi apparatus
    b. endoplasmic reticulum
    c. mitochondrion
    d. microfilaments
    e. intermediate filaments

60. Chloroplasts develop from _____.
    a. leucoplasts
    b. endoplasmic reticulum
    c. chromoplasts
    d. the Golgi apparatus
    e. proplastids

61. Which of the following is *not* a component of the endomembrane system?
    a. Rough endoplasmic reticulum
    b. Smooth endoplasmic reticulum
    c. Golgi apparatus
    d. lysosomes
    e. Plastids

62. A prokaryotic cell does *not* have a
    a. nucleus or organelles.
    b. nucleus or DNA.
    c. nucleus or ribosomes.
    d. nucleus or membranes.
    e. cell wall or membranes.

63. The pores found in the nuclear membrane are composed of
    a. one large protein.
    b. eight large protein granules.
    c. keratin.
    d. intermediate filaments.
    e. lipids.

64. Starch molecules are stored inside
    a. chromoplasts.
    b. granularplasts.
    c. chloroplasts.
    d. potatoplasts.
    e. leucoplasts.

65. Some organelles in eukaryotic cells are thought to have
    a. originated from extracellular symbiotic relationships.
    b. their own endoplasmic reticulum.
    c. their own mitochondria.
    d. originated from endosymbiotic relationships.
    e. the ability to live free from the host cell.

66. The membranes of the endoplasmic reticulum are continuous with the membranes of the
    a. nucleus.
    b. Golgi apparatus.
    c. nucleolus.
    d. plasma membrane.
    e. mitochondria.

67. The rough ER is the portion of the ER that
    a. has a bearded appearance.
    b. is the older part, and was once the smooth ER.
    c. has ribosomes attached.
    d. is connected to the Golgi apparatus.
    e. is the site of steroid synthesis.

68. The ribosome attaches to the surface of the ER, and the protein is transported into the ER lumen because
    a. all proteins must first be processed in the ER.
    b. the mRNA diffuses to the ribosomes on the surface of the ER.
    c. the ribosome has an affinity for the ER surface.
    d. of a signal sequence on the nascent peptide chain.
    e. it is diffusing down its concentration gradient.

69. The difference in the structure of the Golgi of plants, protists, and fungi when compared to that of vertebrates is that the vertebrates' Golgi
    a. form a large apparatus from a few sacks.
    b. form small widely distributed sacks.
    c. form a single large sack.
    d. connect directly to the ER.
    e. lack integral membrane proteins.

70. Materials that enter and leave the Golgi
    a. are transported by proteins.
    b. are packaged on or in vesicles.
    c. are destined for export from the cell.
    d. require a docking protein.
    e. originated in the nucleus.

71. Proteins from the Golgi get transported to the correct location due to
    a. signals found on the packaged proteins.
    b. the direction all vesicles travel within the cell.
    c. the control provided by the nucleus.
    d. motor proteins.
    e. microtubules.

72. A secondary lysosome is a lysosome that
    a. provides a backup to the primary lysosomes.
    b. is smaller than a primary lysosome.

c. will become a primary lysosome after it fuses with a phagosome.
d. is a primary lysosome that has fused with a phagosome.
e. has exocytosed.

73. Lysosomes are important to eukaryotic cells because they contain
a. photosynthetic pigments.
b. starch molecules for energy storage.
c. their own DNA molecules.
d. the cell's waste materials.
e. digestive enzymes.

74. Which of the following cellular components are most important for stabilizing the shape of an animal cell?
a. The Golgi apparatus
b. The nuclear lamina
c. Microfilaments
d. Microtubules
e. Cell wall

75. The surface area of some eukaryotic cells is greatly increased by
a. microtubules.
b. pili.
c. thylakoid membranes.
d. myosin.
e. microvilli.

76. Microvilli are created by projections of
a. microtubules.
b. actin.
c. myosin.
d. intermediate filaments.
e. None of the above

77. Hair and intermediate filaments are composed of
a. microtubules.
b. microfilaments.
c. collagen.
d. hydroxyapatite.
e. keratin.

78. Microtubules are composed of subunits of
a. α and β tubulin.
b. δ and λ actin.

c. ρ and σ myosin.
d. kappa tubules.
e. kappa actinomin.

79. The usefulness of microvilli to cells that have them is
a. to aid in their locomotion.
b. to help concentrate food particles.
c. for intracellular trafficking of molecules.
d. to greatly increase their surface area.
e. for intercellular communications.

80. The ability for red blood cells to squeeze through capillaries without breaking apart is due to
a. actin.
b. ankyrin.
c. spectrin.
d. All of the above
e. None of the above

81. This cellular component can be found at the base of each cilium.
a. Centriole
b. Basal body
c. Nucleolus
d. Flagellum
e. Microvillus

82. The cellular structures that are most like centrioles are
a. basal bodies.
b. microbodies.
c. chromoplasts.
d. microfilaments.
e. centromeres.

83. What would you expect would happen if you removed a plant cell's wall and placed it into a drop of water?
a. The cell would begin to grow.
b. The cell would shrink in size.
c. The cell would burst.
d. The cell would first swell and then shrink.
e. The cell would first shrink and then swell.

5 Cellular Membranes

Fill in the Blank

1. **Lipids** within membranes act as barriers to the passage of many materials, and serve to maintain the membrane's physical integrity.

2. In a complex solution, the **diffusion** of each substance is independent of that of the other substances.

3. **Receptor-mediated endocytosis** is the movement of specified macromolecules into a cell; it involves coated pits, clathrin, and coated vesicles.

4. Components like hormones, growth factors, and parts of viruses and bacteria that bind to specific cell surface receptors are termed **ligands.**

5. Some materials move through biological membranes more readily than others. This characteristic of biological membranes is called **selective permeability**.

6. The major lipids in biological membranes are called **phospholipids**.

7. The cells of the intestinal epithelium are joined to one another by **tight junctions** that prevent substances from passing between the cells of this tissue.

8. The coupled transport system by which glucose and sodium ions enter intestinal epithelial cells is called a **symport**.

9. Insulin is a hormone that binds to specific proteins, called receptors, on the plasma membrane. Insulin and other molecules that bind to specific receptors are called **ligands**.

10. Mammalian embryos have protein complexes that couple cells together, allowing communication by small molecules between cells. These complexes of proteins are called **gap junctions**.

11. Diffusion occurs **down** a concentration gradient.

12. The pressure that increases inside a plant cell when it is placed in water, which finally prevents further net movement of water molecules into the cell, is called **turgor**.

13. The sodium–potassium pump of cell membranes is a(n) **antiport** active transport system.

14. G protein is a protein that is involved with the transfer of information from the outside of the cell to the inside in a process generally referred to as **signal transduction.**

Multiple Choice

1. The chemical makeup, physical organization, and function of a biological membrane depend on which of the following classes of biochemical compounds?
   a. Proteins
   b. Lipids
   c. Fats
   d. Carbohydrates
   e. *a, b,* **and** *d*

2. Integral membrane proteins have
   a. hydrophobic regions within the lipid portion of the bilayer.
   b. hydrophilic regions that protrude in aqueous environments on either side of the membrane.
   c. lateral but not vertical movement within the bilayer.
   d. *a* and *b*
   e. *a, b,* **and** *c*

3. Specialized cell junctions include
   a. gap junctions.
   b. tight junctions.
   c. desmosomes.
   d. *a, b,* **and** *c*
   e. *a* and *b*

4. Substances move through biological membranes via
   a. simple osmosis.
   b. **active transport.**
   c. reverse osmosis.
   d. *a* and *b*
   e. None of the above

5. Because the sodium–potassium pump imports K^+ ions while exporting Na^+ ions, it is a coupled transport system termed a(n)
   a. symport.
   b. **antiport.**
   c. secondary active transporter.
   d. facilitated transport.
   e. diffusion mechanism.

6. Whether a membrane protein is integral, traverses the entire membrane, and/or remains in the membrane is principally a function of its
   a. primary structure.
   b. secondary structure.
   c. **tertiary structure.**
   d. quaternary structure.
   e. None of the above

7. Houseplants adapted to indoor temperatures may die when accidentally left outdoors because
   a. their DNA cannot function.
   b. **their membrane lipid composition cannot adapt.**
   c. their photosynthesis is impaired.

    d. their chloroplasts malfunction.

    e. None of the above

8. The compounds in biological membranes that form a barrier to the movement of materials across the membrane are
   - *a.* integral membrane proteins.
   - *b.* carbohydrates.
   - ***c.* lipids.**
   - *d.* nucleic acids.
   - *e.* peripheral membrane proteins.

9. The interior of the phospholipid bilayer is
   - *a.* hydrophilic.
   - ***b.* hydrophobic.**
   - *c.* aqueous.
   - *d.* solid.
   - *e.* charged.

10. In biological membranes, the phospholipids are arranged in
    - ***a.* a bilayer with the fatty acids pointing toward each other.**
    - *b.* a bilayer with the fatty acids facing outward.
    - *c.* a single layer with the fatty acids facing the interior of the cell.
    - *d.* a single layer with the phosphorus-containing region facing the interior of the cell.
    - *e.* a bilayer with the phosphorus groups in the interior of the membrane.

11. A protein that forms an ion channel through a membrane is most likely to be
    - *a.* a peripheral protein.
    - ***b.* an integral protein.**
    - *c.* a phospholipid.
    - *d.* an enzyme.
    - *e.* entirely outside the phospholipid bilayer.

12. When a mouse cell and a human cell are fused, the membrane proteins of the two cells become uniformly distributed over the surface of the hybrid cell. This occurs because
    - ***a.* many proteins can move around within the bilayer.**
    - *b.* all proteins are anchored within the membrane.
    - *c.* proteins are asymmetrically distributed within the membrane.
    - *d.* all proteins in the plasma membrane are peripheral.
    - *e.* different membranes contain different proteins.

13. The hydrophilic regions of a membrane protein are most likely to be found
    - *a.* only in muscle cell membranes.
    - *b.* associated with the fatty acid region of the lipids.
    - *c.* in the interior of the membrane.
    - ***d.* exposed on the surface of the membrane.**
    - *e.* either on the surface or inserted into the interior of the membrane.

14. Biological membranes are composed of
    - *a.* nucleotides and nucleosides.
    - *b.* enzymes, electron acceptors, and electron donors.
    - *c.* fatty acids.
    - *d.* monosaccharides.
    - ***e.* lipids, proteins, and carbohydrates.**

15. The LDL receptor is an integral protein that crosses the plasma membrane, with portions of the protein extending both outside and into the interior of the cell. The amino acid side chains in the region of the protein that crosses the membrane are most likely to be
    - *a.* charged.
    - *b.* hydrophilic.
    - ***c.* hydrophobic.**
    - *d.* carbohydrates.
    - *e.* lipids.

16. Which of the following compounds functions as recognition signals between cells?
    - *a.* RNA
    - *b.* Phospholipids
    - *c.* Cholesterol
    - *d.* Fatty acids
    - ***e.* Glycolipids**

17. When a membrane is prepared by freeze-fracture and examined under the electron microscope, the exposed interior of the membrane bilayer appears to be covered with bumps. These bumps are
    - ***a.* integral membrane proteins.**
    - *b.* ice crystals.
    - *c.* platinum.
    - *d.* organelles.
    - *e.* vesicles.

18. Structures that contain networks of keratin fibers and hold adjacent cells together are called
    - *a.* extracellular matrices.
    - *b.* glycoproteins.
    - *c.* gap junctions.
    - ***d.* desmosomes.**
    - *e.* phospholipid bilayers.

19. The electric signal for contraction passes rapidly from one muscle cell to the next by way of
    - *a.* tight junctions.
    - *b.* desmosomes.
    - ***c.* gap junctions.**
    - *d.* integral membrane proteins.
    - *e.* freeze fractures.

20. Tight junctions serve an important function in epithelial cell layers by
    - ***a.* restricting the extracellular movement of molecules between the adjacent cells.**
    - *b.* allowing the movement of nerve impulses from one cell to the next.
    - *c.* providing cytoplasmic channels between adjacent cells.
    - *d.* providing channels between the cytoplasm and the extracellular environment.
    - *e.* acting as recognition sites for foreign substances.

21. You fill a shallow pan with water and place a drop of red ink in one end of the pan and a drop of green ink in the other end. Which of the following is true at equilibrium?
    - *a.* The red ink is uniformly distributed in one half of the pan, and the green ink is uniformly distributed in the other half of the pan.
    - ***b.* The red and green inks are both uniformly distributed throughout the pan.**
    - *c.* Each ink is moving down its concentration gradient.
    - *d.* The concentration of each ink is higher at one end of the pan than at the other end.
    - *e.* No predictions can be made without knowing the molecular weights of the pigment molecules.

22. Which of the following does *not* affect the rate of diffusion of a substance?
    a. The temperature
    b. The concentration gradient
    c. The electrical charge of the diffusing material
    d. **The presence of other substances in the solution**
    e. The molecular diameter of the diffusing material

23. For cells where carbon dioxide crosses the plasma membrane by simple diffusion, what determines the rate at which carbon dioxide enters the cell?
    a. **The concentration of carbon dioxide on each side of the membrane**
    b. The amount of ATP being produced by the cell
    c. The amount of carrier protein in the membrane
    d. The amount of energy available
    e. The concentration of hydrogen ions on each side of the membrane

24. Plant cells transport sucrose across the vacuole membrane against its concentration gradient by a process known as
    a. simple diffusion.
    b. **active transport.**
    c. passive transport.
    d. facilitated diffusion.
    e. cellular respiration.

25. You place cells in a solution of glucose and measure the rate at which glucose enters the cells. As you increase the concentration of the glucose solution, the rate at which glucose enters the cells increases. However, when the glucose concentration of the solution is increased above 10 *M*, the rate at which glucose enters the cells no longer increases. Which of the following is the most likely mechanism for glucose transport into the cell?
    a. Simple diffusion
    b. **Facilitated diffusion**
    c. Endocytosis
    d. Connexons
    e. A symport

26. Transporting substances across a membrane from an area of lower concentration to an area of higher concentration requires
    a. phospholipids.
    b. diffusion.
    c. gap junctions.
    d. facilitated diffusion.
    e. **energy.**

27. In the parietal cells of the stomach, the uptake of chloride ions is coupled to the transport of bicarbonate ions out of the cell. This type of transport system is called
    a. a uniport.
    b. a symport.
    c. an exchange channel.
    d. diffusion.
    e. **an antiport.**

28. When a red blood cell is placed in an isotonic solution, which of the following will occur?
    a. The cell will shrivel.
    b. The cell will swell and burst.
    c. The cell will shrivel, and then return to normal.
    d. The cell will swell, and then return to normal.
    e. **Nothing.**

29. When a plant cell is placed in a hypotonic solution, which of the following occurs?
    a. **The cell takes up water until the osmotic potential equals the pressure potential of the cell wall.**
    b. The cell takes up water and eventually bursts.
    c. The cell shrinks away from the cell wall.
    d. There is no movement of water into or out of the cell.
    e. Water moves out of the cell.

30. When vesicles from the Golgi apparatus deliver their contents to the exterior of the cell, they add their membranes to the plasma membrane. Why doesn't the plasma membrane increase in size?
    a. Some vesicles from the Golgi apparatus fuse with the lysosomes.
    b. Membrane vesicles carry proteins from the endoplasmic reticulum to the Golgi apparatus.
    c. **Membrane is continually being lost from the plasma membrane by endocytosis.**
    d. New phospholipids are synthesized in the endoplasmic reticulum.
    e. The phospholipids become more tightly packed together in the membrane.

31. Human growth hormone binds to a specific protein on the plasma membrane. This protein is called
    a. a ligand.
    b. clathrin.
    c. **a receptor.**
    d. spectrin.
    e. a cell adhesion molecule.

32. Receptor-mediated endocytosis is the mechanism for transport of
    a. clathrin.
    b. all macromolecules.
    c. ions.
    d. **specific macromolecules.**
    e. integral membrane proteins.

33. During the formation of muscle, the association of individual muscle cells with one another to form a tissue requires specific membrane proteins. These proteins are called
    a. coated vesicles.
    b. **cell adhesion molecules.**
    c. glycolipids.
    d. carrier molecules.
    e. transport proteins.

34. The neurotransmitter acetylcholine can activate a muscle cell and cause it to contract, even though the acetylcholine molecule never enters the cell. How is this possible?
    a. The acetylcholine receptor protein is a peripheral protein.
    b. Acetylcholine can bind to all proteins in the plasma membrane.
    c. **The acetylcholine receptor protein spans the plasma membrane.**
    d. Acetylcholine is hydrophobic.
    e. Acetylcholine enters the cell by receptor-mediated endocytosis.

35. Insulin is a protein secreted by cells of the pancreas. What is the pathway for the synthesis and secretion of insulin?
    a. **Rough ER, Golgi apparatus, vesicle, plasma membrane**
    b. Golgi apparatus, rough ER, lysosome
    c. Lysosome, vesicle, plasma membrane
    d. Plasma membrane, coated vesicle, lysosome
    e. Rough ER, cytoplasm, plasma membrane

36. You are studying how low-density lipoproteins (LDL) enter cells. When you examine cells that have taken up LDL, you find that the LDL is inside clathrin-coated vesicles. What is the most likely mechanism for the uptake of LDL?
    a. Facilitated diffusion
    b. Proton antiport
    c. **Receptor-mediated endocytosis**
    d. Gap junctions
    e. Ion channels

37 If you compare the proteins of the plasma membrane and the proteins of the inner mitochondrial membrane, which of the following will be true?
    a. Both membranes will have only peripheral proteins.
    b. Only the mitochondrial membrane will have integral proteins.
    c. Only the mitochondrial membrane will have peripheral proteins.
    d. All of the proteins from both membranes will be hydrophilic.
    e. **The proteins from the two membranes will be different.**

38. The molecules in a membrane that limit its permeability are the
    a. carbohydrates.
    b. **phospholipids.**
    c. proteins.
    d. negative ions.
    e. water.

39. The plasma membrane of animals contains carbohydrates
    a. on the side of the membrane facing the cytosol.
    b. **on the side of the membrane facing away from the cell.**
    c. on both sides of the membrane.
    d. on neither side of the membrane.
    e. within the membrane.

40. Integral membrane proteins tend to
    a. be of small molecular weight.
    b. be very soluble in water.
    c. **have hydrophobic amino acids on much of their intramembrane surface.**
    d. be rare.
    e. contain much cholesterol.

41. An important function of integral proteins of a eukaryotic cell membrane is
    a. movement of the cell.
    b. **binding with hormones in the cell's environment.**
    c. usage of genetic information.
    d. digestion of food molecules.
    e. generation of ATP.

42. Cholesterol molecules act to
    a. help hold a membrane together.
    b. transport ions across membranes.
    c. attach to carbohydrates.
    d. disrupt membrane function.
    e. **increase the fluidity of the membrane.**

43. The rate of facilitated diffusion of a molecule across a membrane does *not* continue to increase as the concentration difference of the molecule across the membrane increases because
    a. facilitated diffusion requires ATP energy.
    b. as the concentration difference increases, molecules interfere with one another.
    c. **there are a limited number of carrier proteins in the membrane.**
    d. increased concentration difference causes a situation far from equilibrium.
    e. the diffusion constant depends on the concentration difference.

44. Active transport is important because it can move molecules
    a. from their high concentration to a lower concentration.
    b. **from their low concentration to a higher concentration.**
    c. that resist osmosis across the membrane.
    d. with less ATP than might otherwise be used to move the molecules.
    e. by increasing their diffusion coefficient.

45. Active transport usually moves molecules
    a. in the same direction as does diffusion.
    b. **in the opposite direction as does diffusion.**
    c. in a direction that tends to bring about equilibrium.
    d. toward higher pH.
    e. toward higher osmotic potential.

46. Osmosis is a specific form of
    a. **diffusion.**
    b. facilitated transport.
    c. active transport.
    d. secondary active transport.
    e. movement of water by carrier proteins.

47. Osmosis moves water from a region of
    a. high concentration of dissolved material to a region of low concentration.
    b. **low concentration of dissolved material to a region of high concentration.**
    c. hypertonic solution to a region of hypotonic solution.
    d. negative osmotic potential to a region of positive osmotic potential.
    e. low concentration of water to a region of high concentration of water.

48. Clathrin-coated pits are structures associated with
    a. active transport.
    b. phagocytosis.
    c. flagellar movement.
    d. **receptor-mediated endocytosis.**
    e. secretory vesicles.

49. A general feature of a ligand is that it
    a. diffuses across a membrane.
    b. is part of the sodium–potassium pump.
    c. **binds specifically with a cell surface molecule.**
    d. is part of the carbohydrate on the surface of a cell.
    e. is a cell adhesion molecule.

50. Which of the following molecules is probably the most likely to diffuse across a cell membrane?
    a. Glucose
    b. **Na+**
    c. A steroid
    d. A protein common to blood
    e. A peripheral protein

51. Cell growth could involve movement of membrane material from
    a. the cell membrane to the vesicles.
    b. **the Golgi apparatus to the cell membrane.**
    c. the smooth ER to the rough ER.
    d. coated pits to the inside of the cell.
    e. lysosomes to the cell membrane.

52. An important function of specialized membranes found in certain organelles is to
    a. help the organelles move.
    b. protect the organelles from increased temperatures.
    c. **store energy by holding positive and negative charges separated from each other.**
    d. use their internal genetic information.
    e. destroy cellular waste products.

53. Membrane molecules that help individual cells organize themselves into tissues are known as
    a. peripheral proteins.
    b. **cell adhesion molecules.**
    c. clathrins.
    d. secondary active transport proteins.
    e. symports.

54. Materials can be limited from moving through the spaces between cells by
    a. coated pits on the surfaces of the cells.
    b. desmosomes between the cells.
    c. carbohydrates on the surfaces of the cells.
    d. the extracellular matrix around the cells.
    e. **tight junctions between the cells.**

55. Keratin is a protein found in
    a. plasmodesmata.
    b. **desmosomes.**
    c. gap junctions.
    d. clathrin pits.
    e. microfilaments.

56. A concentration gradient of glucose across a membrane means
    a. **there are more moles of glucose on one side of the membrane than the other.**
    b. glucose molecules are more crowded on one side of the membrane than the other.
    c. there is less water on one side of the membrane than the other.
    d. the glucose molecules are chemically more tightly bonded together on one side than the other.
    e. there are more glucose molecules within the membrane than outside of the membrane.

57. Connexons occur in
    a. the cytoskeleton.
    b. tight junctions.
    c. desmosomes.
    d. plasmodesmata.
    e. **gap junctions.**

58. When placed in a hypertonic solution, plant cells
    a. **shrink.**
    b. swell.
    c. burst.
    d. transport water out.
    e. concentrate.

59. Transport proteins that simultaneously move two molecules across a membrane in the same direction are called
    a. uniports.
    b. **symports.**
    c. antiports.
    d. active transporters.
    e. diffusive ports.

60. When placed in water, wilted plants lose their limpness because of
    a. active transport of salts from the water into the plant.
    b. active transport of salts into the water from the plant.
    c. **osmosis of water into the plant cells.**
    d. osmosis of water from the plant cells.
    e. diffusion of water from the plant cells.

61. The only process that could possibly bring glucose molecules into cells that does *not* involve the metabolic energy of ATP is _____.
    a. phagocytosis
    b. pinocytosis
    c. active transport
    d. **diffusion**
    e. osmosis

62. The functional roles for different proteins found in membranes include all except which of the following?
    a. Allowing movement of molecules that would otherwise be excluded by the lipid components of the membrane
    b. Transferring signals from outside the cell to the inside of the cell
    c. Maintaining the shape of the cell
    d. Facilitating the transport of macromolecules across the membrane
    e. **Stabilizing the lipid bilayer**

63. Carbohydrates associated with cellular membranes are expected to be found associated with
    a. protein inside cells.
    b. lipids inside cells.
    c. **proteins outside cells.**
    d. proteins of the internal organelles.
    e. the nuclear membrane.

64. The site where transmembrane proteins are embedded into membranes is the
    a. smooth ER.
    b. **rough ER.**
    c. Golgi.
    d. nuclear membrane.
    e. lysosomes.

65. The site where carbohydrates are initially added to membrane proteins is the
    a. **ER.**
    b. mitochondria.
    c. Golgi.
    d. nuclear membrane.
    e. lysosomes.

66. Red blood cells' membranes are
    a. **more protein than lipid.**
    b. more lipid than protein.
    c. more carbohydrate than lipid.
    d. more carbohydrate than protein.
    e. more steroid than lipid.

67. The tensile strength of connections between adjacent cells in tissues come from
    a. gap junctions.
    b. tight junctions.
    c. **desmosomes.**
    d. slip junctions.
    e. weld junctions.

68. Desmosomes include or associate with all except
    a. dense plaque-like regions.
    b. keratin fibers.
    c. external cell adhesion molecules.
    d. **internal channel proteins.**
    e. All of the above

69. For each molecule of ATP consumed during active transport of sodium and potassium,
    a. 2 sodium ions are imported and 3 potassium ions are exported.
    b. 2 sodium ions are imported and 1 potassium ion is exported.
    c. 1 potassium ion is imported and 3 sodium ions are exported.
    d. **2 potassium ions are imported and 3 sodium ions are exported.**
    e. 3 potassium ions are imported and 2 sodium ions are exported.

70. Secondary active transport involves all the following except
    a. **the direct use of ATP.**
    b. coupling to another transport system.
    c. use of regained energy from an existing gradient.
    d. the requirement for energy.
    e. the ability to concentrate the transported molecule.

6 Energy, Enzymes, and Metabolism

Fill in the Blank

1. A **spontaneous** reaction is one that, given enough time, goes largely to completion by itself without the addition of energy.

2. Variations of enzymes that allow organisms to adapt to changing environments are termed **isozymes or isoenzymes**.

3. Although some enzymes consist entirely of one or more polypeptide chains, others possess a tightly bound nonprotein portion called a **prosthetic group**.

4. **Coupled** reactions, displayed by succinate dehydrogenase and other enzymes, are the major means of carrying out energy-requiring reactions within cells.

5. The second law of thermodynamics states that the **entropy**, or disorder, of the universe is constantly increasing.

6. When a drop of ink is added to a beaker of water, the dye molecules become randomly dispersed throughout the water. This is an example of an increase in **entropy**.

7. For a reaction to be spontaneous, the change in free energy of the reaction, ΔG, must be **negative**.

8. The enzyme phosphoglucoisomerase catalyzes the conversion of glucose 6-phosphate to fructose 6-phosphate. The region on phosphoglucoisomerase where glucose 6-phosphate binds is called **the active site**.

9. The ΔG of a spontaneous reaction is negative, indicating that the reaction releases free energy. Such a reaction is **exergonic**.

10. The zinc ion in the active site of the enzyme thermolysin is called a **prosthetic group**.

11. When an enzyme is heated until its three-dimensional structure is destroyed, the enzyme is said to be **denatured**.

12. Temperature of water above a waterfall is probably **colder** than the temperature where the water falls.

13. **Metabolism** is the term used for all chemical activity of a living organism.

14. Heat, light, electricity, and motion are all examples of **kinetic** energy.

15. The energy in a system that exists due to position is **potential** energy.

16. The building up of molecules in a living system is **anabolism**, while the breaking down is **catabolism**.

17. The primary directional flow of energy in and among earthly life forms is **light** to **chemical** to **heat**.

18. The first law of thermodynamics is that **energy is neither created nor destroyed**.

Multiple Choice

1. The change in free energy is related to
   a. change in heat.
   b. change in entropy.
   c. change in pressure.
   d. a and b
   e. a, b, and c

2. Enzymes are sensitive to
   a. temperature.
   b. pH.
   c. irreversible inhibitors such as DIPF.
   d. allosteric effectors.
   e. All of the above

3. End products of biosynthetic pathways often act to block the initial step in that pathway. This phenomenon is called
   a. allosteric inhibition.
   b. denaturation.
   c. branch pathway inhibition.
   d. feedback inhibition.
   e. binary inhibition.

4. Which of the following identifies a group of enzymes that is important in fine-tuning the metabolic activities of cells?
   a. Isozymes
   b. Alloenzymes
   c. Allosteric enzymes
   d. a and c
   e. a, b, and c

5. Competitive and noncompetitive enzyme inhibitors differ with respect to
   a. the precise location on the enzyme to which they bind.
   b. their pH.
   c. their binding affinities.
   d. their energies of activation.
   e. None of the above

6. During photosynthesis, plants use light energy to synthesize glucose from carbon dioxide. However, plants do not use up energy during photosynthesis; they merely convert it from light energy to chemical energy. This is an illustration of
   a. increasing entropy.
   b. chemical equilibrium.
   c. the first law of thermodynamics.
   d. the second law of thermodynamics.
   e. a spontaneous reaction.

7. The standard free energy change for the hydrolysis of ATP to ADP + phosphate is –7.3 kcal/mol. What can you conclude from this information?
   a. The reaction will never reach equilibrium.
   b. The free energy of ADP and phosphate is higher than the free energy of ATP.
   c. The reaction requires energy.
   d. The reaction is endergonic.
   e. The reaction is exergonic.

8. The hydrolysis of maltose to glucose is an exergonic reaction. Which of the following statements is true?
   a. The reaction requires the input of free energy.
   b. The free energy of glucose is larger than the free energy of maltose.
   c. The reaction is not spontaneous.
   d. The reaction releases free energy.
   e. At equilibrium, the concentration of maltose is higher than the concentration of glucose.

9. The first law of thermodynamics states that the total energy in the universe is
   a. decreasing.
   b. increasing.
   c. constant.
   d. being converted to free energy.
   e. being converted to matter.

10. If the enzyme phosphohexosisomerase is added to a 0.3 M solution of fructose 6-phosphate, and the reaction is allowed to proceed to equilibrium, the final concentrations are 0.2 M glucose 6-phosphate and 0.1 M fructose 6-phosphate. This data gives an equilibrium constant of 2. What is the equilibrium constant if the initial concentration of fructose 6-phosphate is 3 M?
   a. 2
   b. 3
   c. 5
   d. 10
   e. 20

11. You are studying the effects of temperature on the rate of a particular enzyme-catalyzed reaction. When you increase the temperature from 40°C to 70°C, what effect will this have on the rate of the reaction?
   a. It will increase.
   b. It will decrease.
   c. It will decrease to zero because the enzyme denatures.
   d. It will increase and then decrease.
   e. This cannot be answered without more information.

12. If the change in heat of a chemical reaction is negative and the change in entropy is positive, what can you conclude about the reaction?
   a. It requires energy.
   b. It is endergonic.
   c. It is spontaneous.
   d. It will not reach equilibrium.
   e. It decreases the disorder in the system.

13. Which of the following determines the rate of a reaction?
   a. ΔS
   b. ΔG
   c. ΔH
   d. The activation energy
   e. The overall change in free energy

14. In a spontaneous chemical reaction, transition-state species have free energies
   a. lower than either the reactants or the products.
   b. higher than either the reactants or the products.
   c. lower than the reactants, but higher than the products.
   d. higher than the reactants, but lower than the products.
   e. lower than the reactants, but the same as the products.

15. The hydrolysis of sucrose to glucose and fructose is a spontaneous reaction. However, if you dissolve sucrose in water and keep the solution overnight at room temperature, there is no detectable conversion to glucose and fructose. Why?
   a. The change in free energy of the reaction is positive.
   b. The activation energy of the reaction is high.
   c. The change in free energy of the reaction is negative.
   d. The reaction is endergonic.
   e. The free energy of the products is higher than the free energy of the reactants.

16. The enzyme α-amylase increases the rate at which starch is broken down into smaller oligosaccharides. It does this by
   a. decreasing the equilibrium constant of the reaction.
   b. increasing the change in free energy of the reaction.
   c. decreasing the change in free energy of the reaction.
   d. increasing the change in entropy of the reaction.
   e. lowering the activation energy of the reaction.

17. The enzyme glyceraldehyde 3-phosphate dehydrogenase catalyzes the reaction glyceraldehyde 3-phosphate → 1,3-diphosphoglycerate. The region of the enzyme where glyceraldehyde 3-phosphate binds is called
   a. the transition state.
   b. the groove.
   c. the catalyst.
   d. the active site.
   e. the energy barrier.

18. The enzyme glucose oxidase binds the six-carbon sugar glucose and catalyzes its conversion to glucono-1,4-actone. Mannose is also a six-carbon sugar, but glucose oxidase cannot bind mannose. The specificity of glucose oxidase is based on
   a. the free energy of the transition state.
   b. the activation energy of the reaction.
   c. the change in free energy of the reaction.
   d. the tertiary structure of the enzyme.
   e. the rate constant of the reaction.

19. In the presence of alcohol dehydrogenase, the rate of reduction of acetaldehyde to ethanol increases as you increase the concentration of acetaldehyde. Eventually the rate of the reaction reaches a maximum, where further increases in the concentration of acetaldehyde have no effect. Why?
    a. **All of the alcohol dehydrogenase molecules are bound to acetaldehyde molecules.**
    b. At high concentrations of acetaldehyde, the activation energy of the reaction increases.
    c. At high concentrations of acetaldehyde, the activation energy of the reaction decreases.
    d. The enzyme is no longer specific for acetaldehyde.
    e. At high concentrations of acetaldehyde, the change in free energy of the reaction decreases.

20. When an enzyme catalyzes both a spontaneous reaction and a nonspontaneous reaction, the two reactions are said to be
    a. substrates.
    b. endergonic.
    c. kinetic.
    d. activated.
    e. **coupled.**

21. In glycolysis, the exergonic reaction 1,3-diphosphoglycerate → 3-phosphoglycerate is coupled to the reaction $ADP + P_i → ATP$. Which of the following is most likely to be true about the reaction $ADP + P_i → ATP$?
    a. The reaction never reaches equilibrium.
    b. The reaction is spontaneous.
    c. There is a large decrease in free energy.
    d. **The reaction is endergonic.**
    e. Temperature will not affect the rate constant of the reaction.

22. X-ray crystallography can be used to obtain which type of information?
    a. The rate constant for a reaction
    b. $ΔG$ of a reaction
    c. $ΔS$ of a reaction
    d. The amount of enzyme activation energy
    e. **The shape of an enzyme**

23. Trypsin and elastase are both enzymes that catalyze hydrolysis of peptide bonds. But trypsin only cuts next to lysine and elastase only cuts next to alanine. Why?
    a. Trypsin is a protein and elastase is not.
    b. $ΔG$ for the two reactions is different.
    c. **The shape of the active site for the two enzymes is different.**
    d. One of the reactions is endergonic, and the other is exergonic.
    e. Hydrolysis of lysine bonds requires water; hydrolysis of alanine bonds does not.

24. The enzyme catalase has a ferric ion tightly bound to the active site. The ferric ion is called
    a. a side chain.
    b. an enzyme.
    c. a coupled reaction.
    d. **a prosthetic group.**
    e. a substrate.

25. The addition of the competitive inhibitor mevinolin slows the reaction HMG-CoA → mevalonate, which is catalyzed by the enzyme HMG-CoA reductase. How could you overcome the effects of mevinolin and increase the rate of the reaction?
    a. Add more mevalonate.
    b. **Add more HMG-CoA.**
    c. Lower the temperature of the reaction.
    d. Add a prosthetic group.
    e. Lower the rate constant of the reaction.

26. How does a noncompetitive inhibitor inhibit binding of a substrate to an enzyme?
    a. It binds to the substrate.
    b. It binds to the active site.
    c. It lowers the activation energy.
    d. It increases the $ΔG$ of the reaction.
    e. **It changes the shape of the active site.**

27. Which type of inhibitor can be overcome completely by the addition of more substrate?
    a. Irreversible
    b. Noncompetitive
    c. **Competitive**
    d. Prosthetic
    e. Isotonic

28. Binding of substrate to the active site of an enzyme is
    a. **reversible.**
    b. irreversible.
    c. noncompetitive.
    d. coupled.
    e. allosteric.

29–30. Consider the following metabolic pathway. Reactants and products are designated by capital letters; enzymes are designated by numbers.

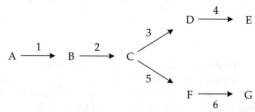

29. Which enzyme is end product G most likely to inhibit?
    a. 1
    b. 2
    c. 3
    d. **5**
    e. 6

30. Assume that end product E is a negative feedback regulator of enzyme 1. What happens to a cell when it is grown in the presence of large amounts of E?
    a. **The cell can't make G.**
    b. The cell can't make A.
    c. The cell makes too much G.
    d. The cell makes too much E.
    e. The cell makes too much D.

31. An allosteric inhibitor
    a. decreases the concentration of inactive enzyme.
    b. **decreases the concentration of active enzyme.**
    c. increases the concentration of product.
    d. decreases the concentration of substrate.
    e. increases the concentration of enzyme–substrate complex.

32. An RNA molecule that catalyzes the cleavage of another RNA chain at a specific site is called
    a. an abzyme.
    b. an isozyme.
    c. an allosteric enzyme.
    d. a regulatory enzyme.
    e. a ribozyme.

33. Energy present in a system that is unusable to do work relates to the system's
    a. temperature.
    b. entropy.
    c. work.
    d. thermodynamics.
    e. equilibrium.

34. The molecules that are acted upon by an enzyme are called
    a. products.
    b. substrates.
    c. carriers.
    d. prosthetics.
    e. effectors.

35. The sum total of all the chemical reactions in a living structure is called its
    a. energetics.
    b. activity.
    c. digestive power.
    d. entropy.
    e. metabolism.

36. The rate of a chemical reaction in a cell is how
    a. often the reaction occurs.
    b. quickly it reaches equilibrium.
    c. much energy must be added to have the reaction occur.
    d. much activation energy is required to have the reaction occur.
    e. easily the reaction is inhibited.

37. The statement "Enzymes are highly specific" means certain
    a. enzymes are found in certain cells.
    b. reactions involving certain substrates are catalyzed by certain enzymes.
    c. enzymes require certain concentrations of substrates.
    d. reactions with certain activation energies are catalyzed by certain enzymes.
    e. concentrations of substrates work with certain enzymes.

38. An active site is
    a. the part of the substrate that binds with an enzyme.
    b. the part of the enzyme that binds with a substrate.
    c. where energy is added to an enzyme catalyst.
    d. where enzymes are found in cells.
    e. None of the above

39. Enzymatic reactions can become saturated as substrate concentration increases because
    a. enzymes have the maximum possible number of hydrogen atoms attached to them.
    b. the concentration of substrate cannot increase any higher.

    c. substrates are inhibitors of enzymes.
    d. the activation energy of the reaction cannot be further lowered.
    e. there are a limited number of the enzyme molecules present.

40. Trypsin is an enzyme whose substrate is
    a. polysaccharide chains of bacteria.
    b. sucrose.
    c. another protein.
    d. $FADH_2$.
    e. carboxyl groups.

41. Competitive inhibitors of enzymes work by
    a. fitting into the active site.
    b. fitting into a site other than the active site.
    c. altering the shape of the enzyme.
    d. changing the enzyme into an inactive form.
    e. increasing the activation energy of the enzyme-catalyzed reaction.

42. Allosteric inhibitors act by
    a. decreasing the amount of enzyme molecules present.
    b. increasing the amount of enzyme molecules present.
    c. decreasing the amounts of inactive form of the enzyme present.
    d. decreasing the amounts of active form of the enzyme present.
    e. increasing the amounts of substrate present.

43. Negative feedback in a sequence of chemical reactions involves a chemical occurring
    a. late in the sequence inhibiting an earlier reaction.
    b. early in the sequence inhibiting a later reaction.
    c. early in the sequence activating a later reaction.
    d. late in the sequence activating an earlier reaction.
    e. late in the sequence inhibiting a later reaction.

44. Denatured enzymes are the same as
    a. ribozymes.
    b. abzymes.
    c. isozymes.
    d. destroyed enzymes.
    e. coenzymes.

45. The inhibition of enzyme activity by competitive inhibitors can be reduced by
    a. decreasing the concentration of allosteric enzymes.
    b. increasing the concentration of noncompetitive inhibitors.
    c. increasing the inhibitor concentration.
    d. increasing the concentration of substrate.
    e. decreasing the concentration of substrate.

46. The concentration of a substrate present in a reaction at equilibrium depends most strongly on the
    a. concentration of enzyme present.
    b. concentration of active form of the enzyme present.
    c. concentration of activator present.
    d. concentrations of other substrates and products present.
    e. concentrations of products present.

47. An allosteric site of an enzyme is where
    a. a competitive inhibitor may bind.
    b. a substrate may bind.
    c. a prosthetic may group-bind.
    d. an activator may bind.
    e. a coenzyme may bind.

48. The rate of a chemical reaction is strongly related to its
    a. equilibrium constant.
    b. change in free energy.
    c. change in entropy.
    d. activation energy.
    e. standard free energy change.

49. Factors that can either activate or inhibit allosteric enzymes are called
    a. proteins.
    b. coenzymes.
    c. sites.
    d. effectors.
    e. competitors.

50. The diversity of chemical reactions occurring in a cell depends mostly on certain molecules present in the cell termed
    a. isozymes.
    b. coenzymes.
    c. ribozymes.
    d. abzymes.
    e. enzymes.

51. When organisms move from one environment to another, they sometimes synthesize variations of existing enzymes termed
    a. coenzymes.
    b. abzymes.
    c. isozymes.
    d. effectors.
    e. activators.

52. A type of inhibitor of an enzyme, which binds within the enzyme's active site, is termed
    a. allosteric.
    b. noncompetitive.
    c. competitive.
    d. extracompetitive.
    e. None of the above

53. The process that involves an end product acting as an inhibitor of an earlier step in a metabolic pathway is called
    a. feedback activation.
    b. feedback inhibition.
    c. positive feedback.
    d. concerted activation.
    e. competitive inhibition.

54. What can never be created or destroyed?
    a. Entropy
    b. Energy
    c. Free energy
    d. Thermal energy
    e. Potential energy

55. The maximum possible rate of an enzyme reaction influenced by a competitive inhibitor depends on the concentration of _____ present.
    a. inhibitor
    b. substrate
    c. product
    d. enzyme
    e. free energy

56. Enzymes of the acid–base catalysis type contain a
    a. metal ion bound to a side chain.
    b. prosthetic group.
    c. coenzyme.
    d. acid or base amino acid residue in the active site.
    e. covalent activated active site.

7 Cellular Pathways That Harvest Chemical Energy

Fill in the Blank

1. The breakdown (hydrolysis) of ATP, which yields ADP and an inorganic phosphate ion, is an exergonic reaction yielding approximately **12** kcal of free energy per mole of ATP under biological conditions.

2. Part of the unusually large amount of free energy that results from the hydrolysis of ATP derives from the large number of **negative** charges near each other on neighboring phosphate groups.

3. Oxidation and **reduction** occur together.

4. Thanks to its ability to carry electrons and free energy, **NAD** is the major and universal energy intermediary in cells.

5. The chemiosmotic formation of ATP during the operation of the respiratory chain is called **oxidative phosphorylation**.

6. The loss of an electron by a ferrous ion (Fe^{2+}) to give a ferric ion (Fe^{3+}) is called **oxidation**.

7. In a redox reaction, the reactant that becomes oxidized is called a **reducing agent**.

8. A chemical reaction resulting in the transfer of electrons or hydrogen atoms is called a **redox** reaction.

9. The pathway for the oxidation of glucose to pyruvate is called **glycolysis**.

10. The conversion of glucose to lactic acid is a form of **fermentation**.

11. During the citric acid cycle, malate is constantly being turned over, but the concentration of malate doesn't change. Thus the concentration of malate is said to be in a **steady state**.

12. Fatty acids must be converted to **acetyl CoA** before they can be used for respiratory ATP production.

13. During alcoholic fermentation, NAD^+ is regenerated by the reduction of acetaldehyde to **ethanol**.

14. An increase in the concentration of acetyl CoA results in the synthesis of oxaloacetate from **pyruvate**.

15. NAD is an abbreviation for **nicotinamide adenine dinucleotide**.

16. An enzyme that transfers phosphorous from ATP to another protein is called a **kinase**.

Multiple Choice

1. ATP is
   a. a short-term, energy-storage compound.
   b. the cell's principle compound for energy transfers.
   c. synthesized within mitochondria.
   d. the molecule all living cells rely on to do work.
   e. All of the above

2. When a molecule loses hydrogen atoms (*not* hydrogen ions), it becomes
   a. reduced.
   b. oxidized.
   c. redoxed.
   d. hydrogenated.
   e. hydrolyzed.

3. The end product of glycolysis is
   a. pyruvate.
   b. the starting point for the citric acid cycle.
   c. the starting point for the fermentation pathway.
   d. *a* and *b*
   e. *a*, *b*, and *c*

4. In the conversion of succinate to fumarate, hydrogen atoms are transferred to FAD. The conversion of succinate and FAD to fumarate and $FADH_2$ is an example of
   a. hydrolysis.
   b. an allosteric reaction.
   c. a metabolic pathway.
   d. an aerobic reaction.
   e. a redox reaction.

5. The oxidation of malate to oxaloacetate is coupled to the reduction of NAD^+ to $NADH + H^+$. NAD^+ is
   a. a reducing agent.
   b. an oxidizing agent.
   c. a vitamin.
   d. a phosphate ester.
   e. a phosphorylating agent.

6. During respiration, NADH donates two electrons to ubiquinone. When this happens, ubiquinone is
   a. reduced.
   b. oxidized.
   c. phosphorylated.
   d. aerobic.
   e. hydrolyzed.

7. Which of the following oxidizes other compounds by gaining free energy and hydrogen atoms and reduces other compounds by giving up free energy and hydrogen atoms?
   a. Vitamins
   b. Adenine
   c. ATP
   d. NAD
   e. Riboflavin

8. Isocitrate dehydrogenase is an enzyme of the citric acid cycle. Where in the cell is this enzyme located?
   a. In the thylakoids
   b. In the cytoplasm
   c. In the chloroplast
   d. In the mitochondrial matrix
   e. In the plasma membrane

9. In the first reaction of glycolysis, glucose receives a phosphate group from ATP. This reaction is
   a. respiration.
   b. a redox reaction.
   c. exergonic.
   d. endergonic.
   e. fermentation.

10. The reduction of pyruvate to lactic acid during fermentation allows glycolysis to continue in the absence of oxygen. Why?
    a. Water is formed during this reaction.
    b. This reaction is a kinase reaction.
    c. This reaction is coupled to the oxidation of NADH to NAD⁺.
    d. This reaction is coupled to the formation of ATP.
    e. This reaction is coupled to the reduction of NAD⁺ to NADH.

11. During glycolysis, for each mole of glucose oxidized to pyruvate
    a. 6 moles of ATP are produced.
    b. 4 moles of ATP are used, and 2 moles of ATP are produced.
    c. 2 moles of ATP are used, and 4 moles of ATP are produced.
    d. 2 moles of NAD⁺ are produced.
    e. no ATP is produced.

12. In steps 6 through 10 of glycolysis, the conversion of one mole of glyceraldehyde 3-phosphate to pyruvate yields 2 moles of ATP. But the oxidation of glucose to pyruvate produces a total of 4 moles of ATP. Where do the remaining 2 moles of ATP come from?
    a. One mole of glucose gives 2 moles of glyceraldehyde 3-phosphate.
    b. Two moles of ATP are used during the conversion of glucose to glyceraldehyde 3-phosphate.
    c. Glycolysis produces 2 moles of NADH.
    d. Fermentation of pyruvate to lactic acid yields 2 moles of ATP.
    e. Fermentation of pyruvate to lactic acid yields 2 moles of NAD⁺.

13. For glycolysis to continue, all cells require
    a. a respiratory chain.
    b. oxygen.
    c. mitochondria.
    d. chloroplasts.
    e. NAD⁺.

14. The free energy released during the oxidation of glyceraldehyde 3-phosphate to 1,3 bisphosphoglycerate is
    a. used to oxidize NADH.
    b. lost as heat.
    c. used to synthesize ATP.
    d. used to reduce NAD⁺.
    e. stored in lactic acid.

15. The oxidation of pyruvate to carbon dioxide is called
    a. fermentation.
    b. the citric acid cycle.
    c. glycolysis.
    d. oxidative phosphorylation.
    e. the respiratory chain.

16. Which of the following is produced during the citric acid cycle?
    a. FAD
    b. Pyruvate
    c. Reduced electron carriers
    d. Lactic acid
    e. Water

17. Some of the free energy released by oxidation of pyruvate to acetate is stored in acetyl CoA. How does acetyl CoA store free energy?
    a. Acetyl CoA has a higher free energy than acetate.
    b. Acetyl CoA is an electron carrier.
    c. Acetyl CoA is a phosphate donor.
    d. Acetate + CoA→acetyl CoA is an exergonic reaction.
    e. Reduction of acetyl CoA is coupled to ATP synthesis.

18. The oxidizing agent at the end of the respiratory chain is
    a. O₂.
    b. NAD.
    c. ATP.
    d. FAD.
    e. ubiquinone.

19. During the citric acid cycle, energy stored in acetyl CoA is used to
    a. create a proton gradient.
    b. drive the reaction ADP + Pᵢ – ATP.
    c. reduce NAD⁺ to NADH.
    d. drive the reaction oxaloacetate→citric acid.
    e. reduce FAD to FADH₂.

20. During the citric acid cycle, oxidative steps are coupled to
    a. oxidative phosphorylation.
    b. the oxidation of water.
    c. the oxidation of electron carriers.
    d. the hydrolysis of ATP.
    e. the reduction of electron carriers.

21. The drug 2,4-dinitrophenol (DNP) destroys the proton gradient across the inner mitochondrial membrane. What would you expect to be the effect of incubating isolated mitochondria in a solution of DNP?
    a. Oxygen would no longer be reduced to water.
    b. No ATP would be made during transport of electrons down the respiratory chain.
    c. Mitochondria would show a burst of increased ATP synthesis.
    d. Glycolysis would stop.
    e. Mitochondria would switch from glycolysis to fermentation.

22. Electron transport within NADH-Q reductase, cytochrome reductase, and cytochrome oxidase can be coupled to proton transport from the mitochondrial matrix to the space between the inner and outer mitochondrial membranes because those protein complexes are located
    a. in the mitochondrial matrix.
    b. **within the inner mitochondrial membrane.**
    c. in the space between the inner and outer mitochondrial membranes.
    d. in the cytoplasm.
    e. loosely attached to the inner mitochondrial membrane.

23. According to the chemiosmotic theory, the energy for the synthesis of ATP during the flow of electrons down the respiratory chain is provided directly by
    a. the hydrolysis of GTP.
    b. the reduction of NAD^+.
    c. **the diffusion of protons.**
    d. the reduction of FAD.
    e. the hydrolysis of ATP.

24. In the absence of oxygen, cells capable of fermentation
    a. accumulate glucose.
    b. no longer produce ATP.
    c. accumulate pyruvate.
    d. oxidize FAD.
    e. **oxidize NADH to produce NAD^+.**

25. For bacteria to continue growing rapidly when they are shifted from an environment containing oxygen to an anaerobic environment, they must
    a. increase the rate of the citric acid cycle.
    b. produce more ATP per mole of glucose during glycolysis.
    c. produce ATP during the oxidation of NADH.
    d. increase the rate of transport of electrons down the respiratory chain.
    e. **increase the rate of the glycolytic reactions.**

26. In alcoholic fermentation, NAD^+ is produced during
    a. the oxidation of pyruvate to acetyl CoA.
    b. the reduction of pyruvate to lactic acid.
    c. **the reduction of acetaldehyde to ethanol.**
    d. the hydrolysis of ATP to ADP.
    e. the oxidation of glucose.

27. During the fermentation of 1 molecule of glucose, the net production of ATP is
    a. 1 molecule.
    b. **2 molecules.**
    c. 3 molecules.
    d. 6 molecules.
    e. 8 molecules.

28. The portion of aerobic respiration that produces the most ATP per mole of glucose is
    a. **oxidative phosphorylation.**
    b. the citric acid cycle.
    c. glycolysis.
    d. lactic acid fermentation.
    e. alcoholic fermentation.

29. More free energy is released during the citric acid cycle than during glycolysis, but only 1 mole of ATP is produced for each mole of acetyl CoA that enters the cycle. What happens to most of the remaining free energy that is produced during the citric acid cycle?
    a. It is used to synthesize GTP.
    b. **It is used to reduce electron carriers.**
    c. It is lost as heat.
    d. It is used to reduce pyruvate.
    e. It is converted to kinetic energy.

30. When the supply of acetyl CoA being produced exceeds the demands of the citric acid cycle, some of the acetyl CoA is diverted to the synthesis of
    a. pyruvate.
    b. NAD.
    c. proteins.
    d. **fatty acids.**
    e. lactic acid.

31. Before starch can be used for respiratory ATP production, it must be hydrolyzed to
    a. pyruvate.
    b. fatty acids.
    c. amino acids.
    d. **glucose.**
    e. oxaloacetate.

32. When yeast cells are switched from aerobic to anaerobic growth conditions, the rate of glycolysis increases. The rate of glycolysis is regulated by the concentration of _____ in the cell.
    a. **ATP**
    b. acetyl CoA
    c. oxaloacetate
    d. FAD
    e. protein

33. When acetyl CoA builds up in the cell, it increases the activity of the enzyme that synthesizes oxaloacetate from pyruvate and carbon dioxide. Acetyl CoA is acting as
    a. an electron carrier.
    b. a substrate.
    c. **an allosteric activator.**
    d. an acetate donor.
    e. a proton pump.

34. In yeast, if the citric acid cycle is shut down because of a lack of oxygen, glycolysis will probably
    a. shut down.
    b. **increase.**
    c. produce more ATP per mole of glucose.
    d. produce more NADH per mole of glucose.
    e. produce acetyl CoA for fatty acid synthesis.

35. Which of the following lowers the activation energy of a reaction, thus allowing equilibrium to be reached more quickly?
    a. Carbonic acid
    b. Free energy from other reactions
    c. **Enzymes**
    d. Inorganic catalysts
    e. None of the above

36. The function of NAD⁺ is to
    a. cause the release of energy to adjacent cells when energy is needed in aerobic conditions.
    b. hasten the release of energy when the cell has been deprived of oxygen.
    c. carry hydrogen atoms and free energy from compounds being oxidized and to give hydrogen atoms and free energy to compounds being reduced.
    d. block the release of energy to adjacent cells.
    e. None of the above

37. The end result of glycolysis is that
    a. 38 ATP molecules are created.
    b. eight NAD molecules get reduced.
    c. two molecules of pyruvate are formed.
    d. one molecule of glucose is converted to lactic acid.
    e. None of the above

38. The results of the first five reactions of the glycolytic pathway are
    a. adding phosphates; modifying sugars; forming G3P.
    b. oxidative steps; proton pumping; reactions with oxygen.
    c. oxidation of pyruvate; formation of acetyl CoA.
    d. removal of hydrogen and protons from glucose.
    e. None of the above

39. For the citric acid cycle to proceed, it is necessary for
    a. pyruvate to bind to oxaloacetate.
    b. carbon dioxide to bind to oxaloacetate.
    c. an acetyl group to bind to oxaloacetate.
    d. water to be oxidized.
    e. None of the above

40. Which of the following events occurs in the respiratory chain?
    a. Carbon dioxide is released.
    b. Carbon dioxide is reduced.
    c. Cytochromes, FADH, and NADH are oxidized.
    d. Only NAD⁺ is reduced.
    e. None of the above

41. The respiratory chain contains three large enzymes: NADH-Q reductase, cytochrome reductase, and cytochrome oxidase. The function of these enzymes is to
    a. allow electrons to be transported.
    b. insure the production of water and oxygen.
    c. regulate the passage of water through the chain.
    d. oxidize NADH.
    e. None of the above

42. Water is a by-product of cellular respiration. The water is produced as a result of
    a. combining carbon dioxide with protons.
    b. the conversion of pyruvate to acetyl CoA.
    c. the degradation of glucose to pyruvate.
    d. the reduction of oxygen at the end of the electron transport chain.
    e. None of the above

43. The formation of ethanol from pyruvate is an example of
    a. an exergonic reaction.
    b. providing an extra source of energy from glycolysis.
    c. a fermentation process that takes place in the absence of oxygen.
    d. cellular respiration.
    e. None of the above

44. Regardless of the electron or hydrogen acceptor employed, fermentation always produces
    a. AMP.
    b. DNA.
    c. P_i.
    d. NAD⁺.
    e. None of the above

45. Yeast cells tend to create anaerobic conditions because they use oxygen faster than it can be replaced by diffusion through the cell membrane. For this reason, yeast cells
    a. exhibit a red pigment.
    b. exhibit a green pigment.
    c. die.
    d. produce ethanol.
    e. None of the above

46. In human cells (muscle cells), the fermentation process produces
    a. lactic acid.
    b. 12 moles of ATP.
    c. pyruvic acid.
    d. an excessive amount of energy.
    e. None of the above

47. If a cell has an abundant supply of ATP, acetyl CoA may be used to
    a. enhance fermentation.
    b. enhance oxidative metabolism.
    c. promote fatty acid synthesis.
    d. convert glucose to glycogen.
    e. None of the above

48. In order for glucose to be used as an energy source, it is necessary that
    a. glucose be formed from fructose.
    b. glucose phosphate be formed from fructose phosphate.
    c. glucose be degraded to carbon dioxide.
    d. two ATP molecules be invested in the system.
    e. None of the above

49. Many species derive their energy from fermentation. The function of fermentation is to
    a. reduce NAD⁺.
    b. oxidize carbon dioxide.
    c. oxidize NADH + H⁺, ensuring a continued supply of ATP.
    d. produce acetyl CoA.
    e. None of the above

50. The chemiosmotic generation of ATP is driven by
    a. osmotic movement of water into an area of high solute concentration.
    b. the addition of protons to ADP and phosphate via enzymes
    c. oxidative phosphorylation.
    d. a difference in H⁺ ion concentration of both sides of a membrane.
    e. None of the above

51. When a cell needs energy, cellular respiration is regulated by the citric acid cycle enzyme isocitrate dehydrogenase, which is stimulated by
    a. H$^+$ ions.
    b. heat.
    c. oxygen.
    d. ADP.
    e. None of the above

52. Substrate-level phosphorylation is
    a. transfer of a phosphate to a protein.
    b. transfer of a phosphate to a substrate.
    c. transfer of a phosphate to an ADP.
    d. transfer of an ATP to a protein.
    e. transfer of a phosphate from ATP to a substrate.

53. The proton-motive force is
    a. the force a proton has on the motive.
    b. the proton concentration gradient and electric charge difference.
    c. a metabolic pathway.
    d. ATP synthase.
    e. a redox reaction.

54. Most ATP made in our bodies is made
    a. by glycolysis.
    b. in the citric acid cycle.
    c. using ATP synthase.
    d. from photosynthesis.
    e. burning fat.

55. Brown fat is "burned" to raise the body temperature of some small mammals by
    a. thermogenin uncoupling respiration.
    b. increasing the rate of glycolysis.
    c. shivering.
    d. hydrogen ions leaking across the cell's plasma membrane.
    e. cytochrome reductase.

56. In eukaryotic cells, some glycolytic enzymes are found to be associated with
    a. the mitochondrial membrane.
    b. the mitochondrial matrix.
    c. the nucleus.
    d. the cytoskeleton.
    e. None of the above

57. ATP can be used to drive nonspontaneous reactions because
    a. nonspontaneous reactions are exergonic.
    b. the breakdown of ATP to ADP is exergonic.
    c. the breakdown of ATP to ADP is endergonic.
    d. when ATP is broken down to ADP, P$_i$ is released.
    e. ADP possesses more free energy than ATP.

8 Photosynthesis: Energy from the Sun

Fill in the Blank

1. During the light reactions of photosynthesis, the synthesis of **ATP** is coupled to the diffusion of protons.

2. Atmospheric carbon dioxide enters plant leaves through openings called **stomata**.

3. In noncyclic photophosphorylation, the electrons for the reduction of chlorophyll in photosystem II come from **water**.

4. When a mixture of radioactive compounds is separated by paper chromatography, the location of the individual compounds on the paper can be determined by a technique known as **autoradiography**.

5. When **C₄ plants** are exposed to light and $^{14}CO_2$, four-carbon acids are the first ^{14}C-containing products.

6. During the process of **photorespiration**, rubisco catalyzes the reaction of RuBP with oxygen.

7. A group of scientists led by **Melvin Calvin** conducted experiments demonstrating that RuBP is the CO_2 acceptor in the dark reactions of photosynthesis.

8. When isolated chloroplasts are shifted from a low-pH solution to a more alkaline solution, ATP synthesis occurs, even in the absence of light. This experiment was used to support the **chemiosmotic** mechanism of ATP formation in chloroplasts.

9. During cyclic photophosphorylation, the energy of photons is converted to the chemical energy of the product, **ATP**.

10. In C₃ plants, the Calvin–Benson cycle occurs in the chloroplasts of **mesophyll** cells, but in C₄ plants the cycle occurs in the **bundle sheath** cells.

11. In both photosynthesis and respiration, **ATP** synthesis is coupled to the diffusion of protons across a membrane.

12. The dark reactions take place in the **light**.

13. NADP is the abbreviation for **nicotinamide adenine dinucleotide phosphate**.

14. The Calvin–Benson cycle is sometimes called the **carbon reduction cycle or dark reactions**.

15. The oxygen found in Earth's atmosphere is generated from the photosystem **II** of noncyclic photophosphorylation.

16. **Plastoquinone** instead of $NADP^+$ receives the electron from ferredoxin during cyclic photophosphorylation.

17. The most abundant enzyme in the biosphere is **rubisco or RuBP carboxylase**.

Multiple Choice

1. To obtain free energy, heterotrophs require a source of
   a. **partially reduced organic compounds.**
   b. light energy.
   c. kinetic energy.
   d. carbon dioxide.
   e. water.

2. How do red and blue light differ from one another?
   a. They differ in intensity.
   b. They have a different number of photons in each quantum.
   c. **Their wavelengths are different.**
   d. They differ in duration.
   e. Red is radiant, and blue is electromagnetic.

3. The wavelength of X rays is shorter than the wavelength of infrared rays. Which of the following is true?
   a. **X rays have more energy per photon than infrared rays.**
   b. X rays have a smaller value for Planck's constant than infrared waves.
   c. X rays have a different absorption spectrum than infrared waves.
   d. X rays and infrared waves have the same frequency.
   e. Infrared waves are in the ground state, and X rays are in the excited state.

4. A molecule has an absorption spectrum that shows maximum absorption within the wavelengths of visible light. This molecule is
   a. a reducing agent.
   b. a quantum.
   c. a photon.
   d. electromagnetic radiation.
   e. **a pigment.**

5. When white light strikes a blue pigment, blue light is
   a. reduced.
   b. absorbed.
   c. converted to chemical energy.
   d. **reflected or transmitted.**
   e. used to synthesize ATP.

6. A graph that plots the rate at which carbon dioxide is converted to glucose versus the wavelength of light illuminating a leaf is called
   a. a Planck equation.
   b. an absorption spectrum.
   c. enzyme kinetics.
   d. an electromagnetic spectrum.
   e. **an action spectrum.**

7. The photosynthetic pigment chlorophyll *a* absorbs
   a. infrared light.
   b. red and blue light.
   c. X rays.
   d. gamma rays.
   e. white light.

8. Accessory pigments
   a. play no role in photosynthesis.
   b. transfer energy from chlorophyll to the electron transport chain.
   c. absorb only in the red wavelengths.
   d. allow plants to harvest visible light of intermediate wavelengths.
   e. transfer electrons to NADP.

9. Why are the absorption spectrum of chlorophyll *a* and the action spectrum of photosynthesis *not* identical?
   a. Accessory pigments contribute energy to drive photosynthesis.
   b. Chlorophyll *a* absorbs both red and blue light.
   c. Chlorophyll *a* reflects green light.
   d. Different wavelengths of light have different energies.
   e. Chlorophyll *a* can be activated by absorbing a photon of light.

10. Excited chlorophyll is a better reducing agent than ground-state chlorophyll because
    a. excited chlorophyll can release energy by fluorescence.
    b. one of the electrons is farther from the atomic nucleus.
    c. excited chlorophyll is reduced by NADPH.
    d. excited chlorophyll absorbs light in the green wavelengths.
    e. only excited chlorophyll contains a porphyrin ring.

11. In cyclic photophosphorylation, chlorophyll is reduced by
    a. NADPH.
    b. a chemiosmotic mechanism.
    c. plastoquinone.
    d. ATP.
    e. hydrogens liberated by the splitting of a water molecule.

12. Free energy is released in cyclic photophosphorylation
    a. by the formation of ATP.
    b. during the excitation of chlorophyll.
    c. during the fluorescence of chlorophyll.
    d. during each of the redox reactions of the electron transport chain.
    e. when electrons are transferred from photosystem I to photosystem II.

13. During cyclic photophosphorylation, the energy to produce ATP is provided by
    a. heat.
    b. NADPH.
    c. ground state chlorophyll.
    d. the redox reactions of the electron transport chain.
    e. the Calvin–Benson cycle.

14. In noncyclic photophosphorylation, water is used for
    a. the hydrolysis of ATP.
    b. the excitation of chlorophyll.

c. **the reduction of chlorophyll.**
d. the oxidation of NADPH.
e. the synthesis of chlorophyll.

15. Photophosphorylation provides the Calvin–Benson cycle with
    a. protons and electrons.
    b. CO_2 and glucose.
    c. water and photons.
    d. light and chlorophyll.
    e. ATP and NADPH.

16. In noncyclic photophosphorylation, the chlorophyll in photosystem I is reduced by
    a. water.
    b. an electron from the transport chain of photosystem II.
    c. two photons of light.
    d. NADPH.
    e. ATP.

17. The enzyme ATP synthase couples the synthesis of ATP to
    a. the diffusion of protons.
    b. the reduction of $NADP^+$.
    c. the excitation of chlorophyll.
    d. the reduction of chlorophyll.
    e. carbon dioxide fixation.

18–19. You incubate a suspension of algae in a flask in the presence of light and carbon dioxide. When you transfer the flask from the light to the dark, you block the reduction of 3-phosphoglycerate to glyceraldehyde 3-phosphate.

18. Why does this reaction stop when the algae are placed in the dark?
    a. It requires carbon dioxide.
    b. It is an exergonic reaction.
    c. It requires ATP and NADPH + H^+.
    d. It requires oxygen.
    e. Chlorophyll is *not* synthesized in the dark.

19. When the reduction of 3-phosphoglycerate to glyceraldehyde 3-phosphate is blocked, the concentration of ribulose bisphosphate declines. Why?
    a. Ribulose bisphosphate is synthesized from glyceraldehyde 3-phosphate.
    b. Glyceraldehyde 3-phosphate is converted to glucose.
    c. Ribulose bisphosphate is used to synthesize 3-phosphoglycerate.
    d. a and c
    e. a and b

20. The enzyme rubisco is found in
    a. chloroplasts.
    b. mitochondria.
    c. the cytoplasm.
    d. the nucleus.
    e. yeast.

21. During carbon dioxide fixation, carbon dioxide combines with
    a. NADPH.
    b. 3PG.
    c. G3P.
    d. water.
    e. ribulose bisphosphate.

22. _____ mole(s) of carbon dioxide must enter the Calvin–Benson cycle for the synthesis of one mole of glucose.
    a. One
    b. Two
    c. Three
    d. Six
    e. Twelve

23. In the experiments conducted to identify the first compound that is formed during CO_2 fixation, why weren't all of the compounds of the Calvin–Benson cycle labeled with ^{14}C?
    a. The cells were incubated with $^{14}CO_2$ for a very short time.
    b. The cells were incubated in the dark.
    c. The cells were incubated in the absence of CO_2.
    d. The cells were incubated at very low concentrations of CO_2.
    e. The cells were incubated in the absence of water.

24. The NADPH required for the reduction of 3PG to G3P comes from
    a. the dark reactions.
    b. the light reactions.
    c. the synthesis of ATP.
    d. the Calvin–Benson cycle.
    e. oxidative phosphorylation.

25. In C_4 plants, the function of the four-carbon acids that are synthesized in the mesophyll cells is to
    a. reduce $NADP^+$.
    b. combine with CO_2 to produce glucose.
    c. carry CO_2 to the bundle sheath cells.
    d. drive the synthesis of ATP.
    e. close the stomata.

26. In the mesophyll layer of C_4 plants, light energy is used to synthesize
    a. O_2.
    b. G3P.
    c. 3PG from RuBP.
    d. CO_2.
    e. PEP.

27. In C_4 plants, starch grains are found in the chloroplasts of
    a. the thylakoids.
    b. mesophyll cells.
    c. the intracellular space.
    d. the stroma.
    e. bundle sheath cells.

28. During photorespiration, rubisco uses _____ as a substrate.
    a. carbon dioxide
    b. oxygen
    c. glyceraldehyde 3-phosphate
    d. 3-phosphoglycerate
    e. NADPH

29. Photorespiration starts
    a. in mitochondria.
    b. in chloroplasts.
    c. only in C_4 plants.
    d. in the microbodies.
    e. in the cytoplasm.

30. In plants, the reactions of glycolysis occur
    a. only in C_3 plants.
    b. in the mitochondria.
    c. in the chloroplasts.
    d. only in the presence of light.
    e. in the cytosol.

31. In both photosynthesis and respiration, protons are pumped across a membrane during
    a. electron transport.
    b. photolysis.
    c. CO_2 fixation.
    d. reduction of oxygen.
    e. glycolysis.

32. The enzyme PEP carboxylase
    a. can trap CO_2 even at relatively low CO_2 concentrations.
    b. catalyzes the synthesis of RuBP.
    c. catalyzes the synthesis of 3PG.
    d. is found in the chloroplasts of bundle sheath cells.
    e. couples the synthesis of ATP to the diffusion of protons.

33. The function of photorespiration is
    a. CO_2 fixation.
    b. unknown.
    c. ATP production.
    d. to generate a proton gradient.
    e. to synthesize glucose.

34. The NADPH required for CO_2 fixation is formed
    a. by the reduction of oxygen.
    b. by the hydrolysis of ATP.
    c. during the light reactions.
    d. only in C_4 plants.
    e. in the mitochondria.

35. The O_2 gas produced during photosynthesis is derived from
    a. carbon dioxide.
    b. glucose.
    c. water.
    d. carbon monoxide.
    e. bicarbonate ions.

36. Photosynthesis and respiration have which of the following in common?
    a. In eukaryotes, both processes reside in specialized organelles.
    b. ATP synthesis in both processes relies on the chemiosmotic mechanism.
    c. Electron transport
    d. Both require light.
    e. a, b, and c

37. The expression "We are creatures of the chloroplasts" means that
    a. all life possesses chloroplasts.
    b. all life depends ultimately on photosynthesis.
    c. chloroplasts are models of all organelles.
    d. a and c
    e. None of the above

38. When a photon interacts with molecules such as those within chloroplasts,
    a. the photons may bounce off the molecules, having no effect.

b. the photons may pass through the molecules, having no effect.

c. the photons may be absorbed by the molecules.

d. *a* and *c*

e. *a*, *b*, and *c*

39. Which of the following scientific tools "cracked" the Calvin–Benson cycle?

a. Isotopes

b. Paper partition chromatography

c. Autoradiography

d. Centrifugation and electron microscopy

e. *a*, *b*, and *c*

40. In noncyclic photophosphorylation, electrons from which source replenish chlorophyll molecules that have given up electrons?

a. Carbon dioxide

b. Water

c. NADPH + H$^+$

d. Gaseous oxygen

e. None of the above

41. What was the substance that turned out to be the mystery compound of the Calvin–Benson cycle (i.e., the previously unknown "compound X")?

a. G3P

b. NADP

c. RuBP

d. Diacylglycerol

e. None of the above

42. In our past evolutionary history, noncyclic photophosphorylation by cyanobacteria, algae, and plants poured enough gas into the atmosphere to make possible the evolution of cellular respiration. That gas was

a. oxygen.

b. methane.

c. hydrogen.

d. nitrogen.

e. chlorine.

43. After World War II, the Berkeley group (Calvin et al.) made progress in understanding the dark reactions when these researchers devised a way to grow dense quantities of algae in a flattened flask called a

a. plant tissue culture.

b. *Chlorella* suspension.

c. pipette.

d. lollipop.

e. sustained phytoculture.

44. Which of the following biological groups is dependent on photosynthesis for its survival?

a. Vertebrates

b. Class Mammalia

c. Fish

d. *a* and *b*

e. *a*, *b*, and *c*

45. Photosynthesis is divided into two main phases, the first of which is a series of reactions that requires the absorption of photons. This phase is referred to as the

a. reduction phase.

b. dark reactions phase.

c. carbon fixation phase.

d. light reactions phase, or photophosphorylation.

e. None of the above

46. Based on its electronic structure, a molecule has a range of photon energies. These photon energies are known as the molecule's

a. chloroplasts.

b. light reactions.

c. absorption spectrum.

d. photosystems I and II.

e. None of the above

47. Heterotrophs are dependent on autotrophs for their food supply. Autotrophs can make their own food by

a. feeding on bacteria and converting the nutrients into usable energy.

b. using light and an inorganic carbon source to make reduced carbon compounds.

c. synthesizing it from water and carbon dioxide.

d. All of the above

e. None of the above

48. Photosynthesis is the process that uses light energy to extract H atoms from which of the following sources?

a. Glucose

b. Chlorophyll

c. Carbon dioxide

d. Water

e. None of the above

49. Which of the following occurs during the dark reactions of photosynthesis?

a. Water is converted into hydrogen and water.

b. Carbon dioxide is converted into sugars.

c. Chlorophyll acts as an enzyme only in the dark.

d. Nothing occurs, the plant rests in the dark.

e. None of the above

50. In the Calvin–Benson cycle, NADPH is necessary to make glucose. How many molecules of NADPH are used to make one molecule of glucose?

a. 18

b. 36

c. 3

d. 6

e. None of the above

51. In bright light, the pH of the thylakoid space can become

a. more acidic.

b. more alkaline.

c. Nothing happens; the pH of thylakoid spaces never changes.

d. neutral.

e. None of the above

52. When a photon is absorbed by a molecule, what exactly happens to the photon?

a. It loses its ability to generate any energy.

b. It raises the molecule from a ground state of low energy to an excited state.

c. The exact relationship of the photon to the molecule is not clearly understood.

d. It causes a change in the velocity of the wavelengths.

e. None of the above

53. A range of energy that cannot be seen by human eyes, but has slightly more energy per photon than visible light, is
    a. adaptive radiation.
    b. solar radiation.
    c. gamma radiation.
    d. ultraviolet radiation.
    e. None of the above

54. The main photosynthetic pigments in plants are
    a. chlorophyll *s* and chlorophyll *a*.
    b. chlorophyll *x* and chlorophyll *y*.
    c. retinal pigment and accessory pigment.
    d. chlorophyll *a* and chlorophyll *b*.
    e. None of the above

55. If one were to compare long wavelengths to short wavelengths, it would be evident that short-wavelength photons have
    a. an insignificant amount of energy.
    b. more energy.
    c. energy not available to plant cells.
    d. a ladder of energy.
    e. None of the above

56. The oxygen produced during photosynthesis comes from
    a. the air the plant takes in.
    b. carbon dioxide.
    c. water.
    d. carbohydrates.
    e. None of the above

57. The chemiosmotic hypothesis states that the energy for the production of ATP comes from
    a. the transfer of phosphate from intermediate compounds.
    b. the reduction of NADP.
    c. a proton gradient set up across the thylakoid membrane.
    d. the oxidation of carbon dioxide.
    e. None of the above

58. When carbon dioxide is added to RuBP, the first stable product synthesized is
    a. pyruvate.
    b. glyceraldehyde 3-phosphate.
    c. phosphoglycerate.

    d. ATP.
    e. None of the above

59. Photosynthesis takes place in plants only in the light. Respiration takes place
    a. in the dark only.
    b. in the light only.
    c. in all organisms except for plants.
    d. both with and without light.
    e. None of the above

60. The revised, balanced equation for the generation of sugar from sunlight, water and carbon dioxide is
    a. $6 CO_2 + 6 H_2O \rightarrow C_6H_{12}O_6 + O_2$.
    b. $6 CO_2 + 12 H_2O \rightarrow C_6H_{12}O_6 + 6 O_2 + 6 H_2O$.
    c. $6 CO_2 + 6 H_2O \rightarrow C_6H_{12}O_6 + 6 O_2$.
    d. $12 CO_2 + 12 H_2O \rightarrow 2 C_6H_{12}O_6 + 2 O_2$.
    e. None of the above

61. After removal of the carbon, the oxygen in the carbon dioxide ends up in
    a. the air.
    b. the sugar.
    c. water molecules.
    d. the air and sugar.
    e. the sugar and water molecules.

62. The net energy outcome of cyclic photophosphorylation is
    a. ATP.
    b. ATP and NADH.
    c. ATP and NADPH.
    d. sugar.
    e. NADPH.

63. In C_4 plants, CO_2 is first fixed into a compound called
    a. pyruvate.
    b. glucose.
    c. oxaloacetate.
    d. ribulose bisphosphate.
    e. 3-phosphoglycerate.

64. In cacti, CO_2 is stored for use in the Calvin–Benson cycle
    a. in the stems, roots, and leaves.
    b. during the evening.
    c. in glucose molecules.
    d. in the stroma.
    e. None of the above

9 Chromosomes, the Cell Cycle, and Cell Division

Fill in the Blank

1. When a DNA molecule doubles, a chromosome is then comprised of two joined **chromatids.**

2. In general, the division of the cell, called **cytokinesis,** follows immediately upon mitosis.

3. During prophase I of meiosis a unique event occurs that results in the formation of recombinant chromosomes. This event is termed **crossing over.**

4. In humans, **cri-du-chat** syndrome results from a deletion of a part of chromosome and takes its name from the cry of a cat.

5. The structure present during mitosis composed of two identical DNA molecules complexed with proteins and joined at the centromere is called a **chromosome.**

6. After the DNA has replicated, the chromosome is made up of two **chromatids.**

7. The stage of the cell cycle during which DNA replicates is called the **S phase.**

8. **Fertilization** is the fusion of two gametes.

9. During prometaphase, microtubules associate with specialized structures in the centromere regions called **kinetochores.**

10. Occasionally, a homologous chromosome pair will fail to separate during anaphase I of meiosis. The resulting cells, one of which lacks a copy of this chromosome while the other contains both members of the homologous pair, are called **aneuploid** cells.

11. During a process known as **translocation,** a piece of one chromosome can break off and become joined to a different chromosome.

12. The orderly distribution of genetic information occurs in prokaryotic cells by a process known as **fission.**

13. A cell with three homologous sets of chromosomes is called a **triploid** cell.

14. The heritable information of the cell is **DNA.**

15. The process that ensures the genetic information is passed on to a cell's daughter cells is **mitosis.**

16. The process that ensures that only one of each pair of chromosomes is included in a gamete is **meiosis.**

17. The main role of nucleosomes in eukaryotic cells is to **package** the DNA.

18. The G2 phase always follows **S.**

19. The G in G1 and G2 is short for **gap.**

20. Components of maturation promoting factor are **Cdk's and cyclins.**

21. The chromatin **condenses** during prophase.

22. The milestone event that defines entry into prometaphase is **disintegration of nuclear envelope.**

23. **Cytokinesis** refers to the division of the cell.

24. In plants a **cell plate** forms at the equatorial region of the cell.

25. The cell plate is derived from the **Golgi apparatus** of the cell.

26. The "invisible thread" that pinches cells into two during cell division is made of **actin and myosin.**

27. A zygote usually has **two** copies of each chromosome.

28. A **karyotype** is the number, form, and type of chromosomes found in a cell, usually similar within a species and different between species.

29. A **homolog** is one of a pair of chromosomes having the same overall genetic composition and sequence.

30. Nondisjunction causes the production of **aneuploid** cells.

31. Down syndrome can be caused by having an extra chromosome **21.**

32. Of a tetraploid or a triploid plant, more problems are expected from a **triploid** during meiosis.

Multiple Choice

1. Chromosomes contain large amounts of five different interacting proteins, all of which are known as
   a. pentanes.
   b. hexosamines.
   c. histones.
   d. protein hormones.
   e. histamines.

2. The appropriate decisions to enter the S phase and the M phase of the cell cycle depend on a pair of biochemicals called
   a. actin and myosin.
   b. Cdk's and cyclin.
   c. ligand and receptor.
   d. MSH and MSH-receptor.
   e. ATP and ATPase.

3. Sex and reproduction
   a. **are not the same thing**.
   b. are identical.
   c. both involve combining genetic material from two cells.
   d. both involve the formation of new individuals.
   e. None of the above

4. Circular chromosomes are found in
   a. prokaryotes.
   b. some viruses.
   c. chloroplasts.
   d. mitochondria.
   e. **All of the above**

5. The molecules that make up a chromosome are
   a. DNA and RNA.
   b. **DNA and proteins.**
   c. proteins and lipids.
   d. nucleotides and nucleosides.
   e. proteins and phospholipids.

6. During mitosis and meiosis the chromatin compacts. Which of the following processes is made easier by this compaction?
   a. **The orderly distribution of genetic material to two new nuclei**
   b. The replication of the DNA
   c. Exposing the genetic information on the DNA
   d. The unwinding of DNA from around the histones
   e. The disappearance of the nuclear membrane

7. The basic structure of chromatin has sometimes been referred to as beads on a string of DNA. Those beads are called
   a. chromosomes.
   b. chromatids.
   c. supercoils.
   d. interphases.
   e. **nucleosomes.**

8. DNA replication occurs
   a. during both mitosis and meiosis.
   b. only during mitosis.
   c. only during meiosis.
   d. **during the S phase.**
   e. during G2.

9. Mature nerve cells are incapable of cell division. These cells are probably in
   a. **G1.**
   b. the S phase.
   c. G2.
   d. mitosis.
   e. meiosis.

10. A cell cycle consists of
    a. mitosis and meiosis.
    b. G1, the S phase, and G2.
    c. prophase, metaphase, anaphase, and telophase.
    d. **interphase and mitosis.**
    e. meiosis and fertilization.

11. The cells of the intestinal epithelium are continually dividing, replacing dead cells lost from the surface of the intestinal lining. If you examined a population of intestinal epithelial cells under the microscope, most of the cells would
    a. be in meiosis.
    b. be in mitosis.

c. be in interphase.
d. have condensed chromatin.
e. b and d

12. The products of mitosis are
    a. one nucleus containing twice as much DNA as the parent nucleus.
    b. two genetically identical cells.
    c. four nuclei containing half as much DNA as the parent nucleus.
    d. four genetically identical nuclei.
    e. **two genetically identical nuclei.**

13. The mitotic spindle is composed of
    a. chromosomes.
    b. chromatids.
    c. **microtubules.**
    d. chromatin.
    e. centrosomes.

14. When dividing cells are examined under the light microscope, chromosomes first become visible during
    a. interphase.
    b. the S phase.
    c. **prophase.**
    d. G1.
    e. G2.

15. The structures that line up the chromosomes on the equatorial plate during metaphase are called
    a. asters.
    b. **polar and kinetochore microtubules.**
    c. centrosomes.
    d. centrioles.
    e. histones.

16. The microtubules of the mitotic spindle attach to a specialized structure in the centromere region of each chromosome, called the
    a. **kinetochore.**
    b. nucleosome.
    c. equatorial plate.
    d. aster.
    e. centrosome.

17. After the centromere divides during mitosis, the chromatids, now called _____, move toward opposite poles of the spindle.
    a. centrosomes
    b. kinetochores
    c. half spindles
    d. asters
    e. **daughter chromosomes**

18. In plant cells, cytokinesis is accomplished by the formation of
    a. an aster.
    b. a membrane furrow.
    c. an equatorial plate.
    d. **a cell plate.**
    e. a spindle.

19. The distribution of mitochondria between the daughter cells during cytokinesis
    a. **is random.**
    b. is directed by the mitotic spindle.
    c. is directed by the centrioles.
    d. results in all of the mitochondria remaining in the parent cell.
    e. only occurs during meiosis.

20. Genetically diverse offspring result from
    a. mitosis.
    b. cloning.
    c. sexual reproduction.
    d. cytokinesis.
    e. anaphase.

21. During asexual reproduction, the genetic material of the parent is passed on to the offspring by
    a. homologous pairing.
    b. meiosis and fertilization.
    c. mitosis and cytokinesis.
    d. karyotyping.
    e. chiasmata.

22. Meiosis can occur
    a. in all organisms.
    b. only when an organism is diploid.
    c. only in multicellular organisms.
    d. only in haploid organisms.
    e. only in single-celled organisms.

23. All zygotes are
    a. multicellular.
    b. diploid.
    c. animals.
    d. clones.
    e. gametes.

24. In all sexually reproducing organisms, the diploid phase of the life cycle begins at
    a. spore formation.
    b. gamete formation.
    c. meiosis.
    d. mitosis.
    e. fertilization.

25. The members of a homologous pair of chromosomes
    a. are identical in size and appearance.
    b. contain identical genetic information.
    c. separate to opposite poles of the cell during mitosis.
    d. are only found in haploid cells.
    e. are only present after the S phase.

26. The diagnosis of Down syndrome is made by examining the individual's
    a. spores.
    b. karyotype.
    c. chromatin.
    d. nucleosomes.
    e. kinetochores.

27. During meiosis, the sister chromatids separate during
    a. anaphase II.
    b. anaphase I.
    c. the S phase.
    d. synapsis.
    e. telophase II.

28. The exchange of genetic material between chromatids on homologous chromosomes occurs during
    a. interphase.
    b. mitosis and meiosis.
    c. prophase I.
    d. anaphase I.
    e. anaphase II.

29. At the end of the first meiotic division, each chromosome consists of
    a. chiasmata.
    b. a homologous chromosome pair.
    c. four copies of each DNA molecule.
    d. two chromatids.
    e. a pair of polar microtubules.

30. The four haploid nuclei found at the end of meiosis differ from one another in their exact genetic composition. Some of this difference is the result of
    a. cytokinesis.
    b. replication of DNA during the S phase.
    c. separation of sister chromatids at anaphase II.
    d. spindle formation.
    e. crossing over during prophase I.

31. Diploid cells of the fruit fly *Drosophila* have 10 chromosomes. How many chromosomes does a *Drosophila* gamete have?
    a. One
    b. Two
    c. Five
    d. 10
    e. 20

32. During meiosis I in humans, one of the daughter cells receives
    a. only maternal chromosomes.
    b. a mixture of maternal and paternal chromosomes.
    c. the same number of chromosomes as a diploid cell.
    d. a sister chromatid from each chromosome.
    e. one-fourth the amount of DNA in the parent nucleus.

33. The fact that most monosomies and trisomies are lethal to human embryos illustrates
    a. the importance of the orderly distribution of genetic material during meiosis.
    b. the exchange of genetic information during crossing over.
    c. the advantage of sexual reproduction to the survival of a population.
    d. that each chromosome contains a single molecule of DNA.
    e. that meiosis results in the formation of haploid gametes.

34. A triploid nucleus *cannot* undergo meiosis because
    a. the DNA cannot replicate.
    b. not all of the chromosomes can form homologous pairs.
    c. the sister chromatids cannot separate.
    d. cytokinesis cannot occur.
    e. a cell plate cannot form.

35. A bacterial cell gives rise to two genetically identical daughter cells by a process known as
    a. nondisjunction.
    b. mitosis.
    c. meiosis.
    d. fission.
    e. fertilization.

36. Which of the following is *not* part of sexual reproduction?
    a. The segregation of homologous chromosomes during gamete formation
    b. The fusion of sister chromatids during fertilization
    c. The fusion of haploid cells from a diploid zygote
    d. The reduction in chromosome number during meiosis
    e. The production of genetically distinct gametes during meiosis

37. Chromatin condenses to form discrete, visible chromosomes
    a. early in G1.
    b. during S.
    c. during telophase.
    d. during prophase.
    e. at the end of cytokinesis.

38. Chromosomes "decondense" into diffuse chromatin
    a. at the end of telophase.
    b. at the beginning of prophase.
    c. at the end of interphase.
    d. at the end of metaphase.
    e. only in dying cells.

39. Microtubules that form the mitotic spindle tend to originate from or terminate in
    a. centromeres and telomeres.
    b. euchromatin.
    c. centrioles and telomeres.
    d. the nuclear envelope.
    e. centrioles and centromeres (kinetochores).

40. Asexual reproduction produces genetically identical individuals because
    a. chromosomes do not have to replicate.
    b. it involves chromosome replication without cytokinesis.
    c. no meiosis or fertilization takes place.
    d. the only cell division that occurs is meiosis.
    e. the mitotic spindle prevents nondisjunction.

41. One difference between mitosis and meiosis I is that
    a. homologous chromosome pairs synapse during mitosis.
    b. chromosomes do not replicate in the interphase preceding meiosis.
    c. homologous chromosome pairs synapse during meiosis but not mitosis.
    d. spindles composed of microtubules are not required during meiosis.
    e. sister chromatids separate during meiosis but not mitosis.

42. Genetic recombination occurs during
    a. prophase of meiosis I.
    b. interphase preceding meiosis II.
    c. mitotic telophase.
    d. fertilization.
    e. formation of somatic cells.

43. The number of chromosomes is reduced to half during
    a. anaphase of mitosis and meiosis.
    b. meiosis II.

c. meiosis I.
d. fertilization.
e. interphase.

44. The total DNA content of each daughter cell is reduced during meiosis because
    a. chromosomes do not replicate during the interphase preceding meiosis I.
    b. chromosomes do not replicate between meiosis I and II.
    c. half of the chromosomes from each gamete are lost during fertilization.
    d. sister chromatids separate during anaphase of meiosis I.
    e. chromosome arms are lost during crossing over.

45. Many chromosome abnormalities (trisomies and monosomies) are *not* observed in the human population because
    a. they are lethal and cause spontaneous abortion of the embryo early in development.
    b. all trisomies and monosomies are lethal early in embryonic development.
    c. meiosis distributes chromosomes to daughter cells with great precision.
    d. they occur but are not observed because human chromosomes are so difficult to count.
    e. the human meiotic spindle is self-correcting.

46. In a haploid organism, most mitosis occurs
    a. after fertilization and before meiosis.
    b. after meiosis and before fertilization.
    c. between meiosis I and II.
    d. during G1.
    e. in diploid cells.

47. The event in the cell division process that clearly involves microfilaments rather than microtubules is
    a. chromosome separation during anaphase.
    b. movement of chromosomes to the metaphase plate.
    c. chromosome condensation during prophase.
    d. disappearance of the nuclear envelope during prophase.
    e. cytokinesis in animal cells.

48. Trisomies and monosomies can result from accidents that occur during meiosis called
    a. nondisjunctions.
    b. inversions.
    c. reciprocal translocations.
    d. recombinations.
    e. acrocentricities.

49. Chromosome number is reduced during meiosis because the process consists of
    a. two cell divisions without any chromosome replication.
    b. a single cell division without any chromosome replication.
    c. two cell divisions in which half of the chromosomes are destroyed.
    d. two cell divisions and only a single round of chromosome replication.
    e. four cell divisions with no chromosome replication.

50. Which of the following phases of the cell cycle is *not* part of interphase?
    ***a.* M**
    b. S
    c. G1
    d. G2
    e. G0

51. During mitotic anaphase, chromatids migrate
    a. from the poles of the cell toward the metaphase plate.
    ***b.* from the metaphase plate toward the poles.**
    c. toward the nuclear envelope.
    d. along with their sister chromatids toward one pole.
    e. along with the other member of the homologous pair toward the metaphase plate.

52. Which of the following does *not* occur during mitotic prophase?
    a. Disappearance of the nuclear envelope
    b. Chromosome condensation
    c. Migration of centrioles toward the cell poles
    ***d.* Synapsis of homologous chromosomes**
    e. Formation of the mitotic spindle

53. Which of the following is *not* true of homologous chromosome pairs?
    ***a.* They come from only one of the individual's parents.**
    b. They usually contain slightly different versions of the same genetic information.
    c. They segregate from each other during meiosis I.
    d. They synapse during meiosis I.
    e. Each contains two sister chromatids at the beginning of meiosis I.

54. Which of the following is *not* true of sister chromatids?
    a. They arise by replication during S phase.
    b. They segregate from each other during each mitotic anaphase.
    c. They usually contain identical versions of the same genetic information.
    ***d.* They segregate from each other during meiosis I.**
    e. They are joined during prophase and metaphase at their common centromere.

55. Chromatin consists of
    a. DNA and histones.
    ***b.* DNA, histones, and many other nonhistone proteins.**
    c. mostly RNA and DNA.
    d. RNA, DNA, and nonhistone proteins.
    e. DNA alone.

56. The DNA of an eukaryotic cell is
    ***a.* double stranded.**
    b. single stranded.
    c. circular.
    d. complex inverted.
    e. conservative.

57. Nucleosomes are made of
    a. RNA.
    b. lipid.
    ***c.* histones.**
    d. chromatin.
    e. All of the above

58. A cell that is postreproductive will remain in
    a. S.
    ***b.* G1.**
    c. G2.
    d. M.
    e. prophase.

59. The kinetochore is made of
    a. protein.
    b. microtubules.
    c. DNA.
    ***d.* a and b.**
    e. All of the above

60. When chromosomes segregate incorrectly during meiosis, it is called
    ***a.* nondisjunction.**
    b. mitosis.
    c. meiosis.
    d. fission.
    e. fertilization.

61. The milestone that defines metaphase is when the chromosomes
    a. separate.
    b. come together.
    c. are at opposite poles.
    ***d.* line up.**
    e. cross over.

62. The milestone that defines anaphase is when the chromosomes
    ***a.* separate.**
    b. come together.
    c. are at opposite poles.
    d. line up.
    e. cross over.

63. The milestone that defines telophase is when the chromosomes
    a. separate.
    b. come together.
    ***c.* are at opposite poles.**
    d. line up.
    e. cross over.

64. The major drawback of asexual reproduction is
    a. that it takes too little time.
    b. the variation generated.
    c. it prevents change.
    d. it requires cytokinesis.
    ***e.* the lack of variation among the progeny.**

65. Haploid means
    a. the genes are arranged haphazardly.
    ***b.* containing only one copy of each chromosome.**
    c. the process of meiosis.
    d. half again the number of chromosomes.
    e. All of the above

66. A reduction step during meiosis is important because
    a. it returns the chromosome number to normal before fertilization.
    b. there is a mechanism for doing this.
    c. only one copy of each chromosome is necessary.
    ***d.* otherwise chromosome copies would double each fertilization.**
    e. fertilization uses this.

67. Each diploid cell of a human female contains
    a. one of each type of chromosome.
    b. two of each type of chromosome.
    c. four of each type of chromosome.
    d. a total of 23 of each type of chromosome.
    e. a total of 46 of each type of chromosome.

68. The importance of synapsis and the formation of chiasmata is that
    a. reciprocal exchange of chromosomal sections occur.
    b. the DNA on homologous chromosomes mix.
    c. as a result an increase in the variation of progeny occurs.
    d. overt evidence is provided for maternal and paternal chromosomes mixing.
    e. All of the above

69. A triploid plant has
    a. one extra chromosome.
    b. one extra set of chromosomes.
    c. three ploids.
    d. three times the chance of a monoploid.
    e. three attempts at ploid.

70. Nucleosomes are found
    a. only in interphase cells.
    b. only during mitosis.
    c. only during meiosis.
    d. in all chromatin.
    e. in the cytoplasm.

71. The energy to move chromosomes during mitosis is provided by
    a. centrioles.
    b. DNA polymerization.
    c. migration of the centrosomes.
    d. formation of the cell plate.
    e. ATP.

72. During bacterial cell division, the two DNA molecules are separated by
    a. centrosomes.
    b. spindle fibers.
    c. nucleosomes.
    d. cell elongation.
    e. aneuploidy.

10 Transmission Genetics: Mendel and Beyond

Fill in the Blank

1. A **heritable** trait is one that can be passed from one generation to another.

2. A **gene** is a portion of DNA that resides at a particular locus or site on a chromosome and encodes a particular function.

3. To determine the overall probability of independent events, **multiply** the probabilities of the individual events.

4. One particular allele of a gene may be defined as **wild type**, or standard, because it is present in most individuals and gives rise to an expected trait, or phenotype.

5. Geneticists make use of **recombinant** frequencies to map chromosomes, that is, to locate genetic loci on the chromosome.

6. A cross between two parents that differs by a single trait is a **monohybrid** cross.

7. The physical appearance of a character is the **phenotype** while the genetic constitution is the **genotype**.

8. A cross between two parents that differs by two independent traits is a **dihybrid** cross.

9. For genes that fail to independently assort, phenotypes that appear in combinations that are not present in either parent are **recombinant**.

10. The region of the chromosome occupied by a gene is called a **locus**.

11. When a cross is made and a trait disappears in the F_1 generation, only to reappear in the F_2, the trait is probably **recessive**.

12. A female who is heterozygous for a recessive, sex-linked character is a **carrier**.

13. When the expression of one gene depends on the expression of another gene, the genes demonstrate **epistasis**.

14. When many genes contribute to the phenotype, variation is said to be **continuous**.

15. A **character** is a feature such as a flower color; a **trait** is a particular form of a character, such as a white flower.

Multiple Choice

1. Gregor Mendel concluded that each pea has two units for each trait, and each gamete contains one unit. Mendel's "units" are now referred to as
   a. **genes.**
   b. characters.
   c. allelic pairs.
   d. transcription factors.
   e. None of the above

2. A particular genetic cross in which the individual in question is crossed with an individual known to be homozygous for a recessive trait is referred to as a
   a. parental cross.
   b. dihybrid cross.
   c. filial generation mating.
   d. reciprocal cross.
   e. **test cross.**

3. Incomplete dominance occurs when
   a. chromosomes are deleted.
   b. **heterozygotes synthesize a reduced amount of an enzyme, producing an intermediate phenotype.**
   c. the genes fail to segregate.
   d. the law of independent assortment is upheld.
   e. one gene is epistatic to the other.

4. Although the law of independent assortment is generally applicable, when two loci are on the same chromosome the phenotypes of the progeny sometimes do not fit the phenotypes predicted. This is due to
   a. translocation.
   b. inversions.
   c. chromatid affinities.
   d. **linkage.**
   e. reciprocal chromosomal exchanges.

5. When a given trait is the result of multigene action, one of the genes may mask the expression of one or all other genes. This phenomenon is termed
   a. **epistasis.**
   b. epigenesis.
   c. dominance.
   d. incomplete dominance.
   e. None of the above

6. Which of the following is *not* a characteristic that makes an organism suitable for genetic studies?
   a. **A small number of chromosomes**
   b. A short generation time
   c. Ease of cultivation
   d. The ability to control crosses
   e. The availability of a variation for traits

7. A key factor that allowed Mendel to interpret his breeding experiments was that
    a. the varieties of peas he started with were "true-breeding."
    b. peas naturally self-pollinate.
    c. peas can reproduce asexually.
    d. pollination could be controlled.
    e. a and d

8. Crossing spherical-seeded pea plants with dented-seeded pea plants resulted in progeny that all had spherical seeds. This indicates that the dented-seed trait is
    a. codominant.
    b. dominant.
    c. recessive.
    d. a and b
    e. a and c

9–11. Schmoos live in widely separated groups and rarely interbreed, so geneticists know very little about them. On one occasion, two Schmoos from different groups did mate. A big-footed white Schmoo mated with a small-footed brown Schmoo. Three offspring resulted: one big-footed brown and two small-footed brown.

9. Which statement about the inheritance of color in Schmoos is most likely to be correct?
    a. Brown is dominant to white.
    b. White is dominant to brown.
    c. White and brown are codominant.
    d. a and c
    e. You cannot reach any conclusions.

10. Which statement about the inheritance of footedness in Schmoos is most likely to be correct?
    a. Big is dominant to small.
    b. Small is dominant to big.
    c. Big and small are codominant.
    d. a and c
    e. You cannot reach any conclusions.

11. If big feet (B) in Schmoos is dominant to small feet (b), what is the genotype of the big-footed white parent Schmoo with respect to the foot gene?
    a. bb
    b. BB
    c. Bb
    d. a and b
    e. a and c

12. When reciprocal crosses produce identical results,
    a. the trait is sex-linked.
    b. the trait is not sex-linked.
    c. the trait is not autosomally inherited.
    d. a and b
    e. b and c

13. The physical appearance of a character is called
    a. the genotype.
    b. the phenotype.
    c. an allele.
    d. a trait.
    e. a gene.

14. Different forms of a gene are called
    a. traits.
    b. phenotypes.

c. genotypes.
    d. alleles.
    e. None of the above

15. When genes for two different characters segregate in a cross, what type of cross is it?
    a. Monohybrid
    b. Dihybrid
    c. Trihybrid
    d. F_1
    e. F_2

16. In Mendel's experiments, if the allele for tall (T) plants was incompletely dominant over the allele for short (t) plants, what would be the result of crossing two Tt plants?
    a. 1/4 would be tall; 1/2 intermediate height; 1/4 short.
    b. 1/2 would be tall; 1/4 intermediate height; 1/4 short.
    c. 1/4 would be tall; 1/4 intermediate height; 1/2 short.
    d. All the offspring would be tall.
    e. All the offspring would be intermediate.

17. The region of the chromosome occupied by a gene is called a(n)
    a. allele.
    b. region.
    c. locus.
    d. type.
    e. phenotype.

18. In Mendel's experiments, the spherical seed character (SS) is completely dominant over the dented seed character (ss). If the characters for height were incompletely dominant, such that TT are tall, Tt are intermediate, and tt are short, what would be the result of crossing a spherical-seeded, short (SStt) plant to a dented-seeded, tall (ssTT) plant?
    a. 1/2 would be smooth-seeded and intermediate height; 1/2 would be smooth-seeded and tall.
    b. All the progeny would be smooth-seeded and tall.
    c. All the progeny would be smooth-seeded and short.
    d. All the progeny would be smooth-seeded and intermediate height.
    e. You cannot predict the outcome.

19. If Mendel's crosses between spherical-seeded, tall and dented-seeded, short plants had produced many more than 1/16 dented-seeded, short plants in the F_2 generation, he might have concluded that
    a. the spherical seed and tall traits are linked.
    b. the dented seed and short traits are unlinked.
    c. all traits in peas assort independently of each other.
    d. all traits in peas are linked.
    e. He would not have concluded any of the above.

20. An organism that produces either male gametes or female gametes, but not both, is called
    a. monoecious.
    b. dioecious.
    c. heterozygous.
    d. homozygous.
    e. parthenogenic.

21. Why would you predict that half of the human babies born will be males and half will be females?
    a. **Because of the segregation of the X and Y chromosomes during male meiosis**
    b. Because of the segregation of the X chromosomes during female meiosis
    c. Because all eggs contain an X chromosome
    d. *a* and *b*
    e. *a* and *c*

22. A human male carrying an allele for a trait on the X chromosome is
    a. heterozygous.
    b. homozygous.
    c. **hemizygous.**
    d. monozygous.
    e. holozygous.

23. Cleft chin is a sex-linked dominant trait. A man with a cleft chin marries a woman with a round chin. What proportion of their female progeny will show the trait?
    a. 0%
    b. 25%
    c. 50%
    d. 75%
    e. **100%**

24–25. A female carrier of the recessive, sex-linked trait red color blindness marries a normal male.

24. What proportion of their male progeny will show the trait?
    a. 0%
    b. 25%
    c. **50%**
    d. 75%
    e. 100%

25. What proportion of their female progeny will show the trait?
    a. **0%**
    b. 25%
    c. 50%
    d. 75%
    e. 100%

26. A linkage group corresponds to
    a. a group of genes on different chromosomes.
    b. the linear order of chromomeres on a chromosome.
    c. the length of a chromosome.
    d. **a group of genes on the same chromosome.**
    e. None of the above

27. It would have been very difficult for Mendel to draw conclusions about the patterns of inheritance if he had used cattle instead of peas. Why?
    a. Cattle reproduce asexually.
    b. **Cattle have small numbers of offspring.**
    c. Cattle do not have observable phenotypes.
    d. Cattle do not have genotypes.
    e. Cattle do not have autosomes.

28. Epistasis refers to
    a. a group of genes that are close together.
    b. the interaction of two genes so that a new phenotype is produced.
    c. the expression of two genes in the same individual.
    d. the linear order of genes on a chromosome.
    e. **the expression of one gene masking the expression of another.**

29–31. An agouti mouse that is heterozygous at the albino and agouti loci (*AaBb*) is mated to an albino mouse that is heterozygous at the agouti locus (*aaBb*).

29. What proportion of the progeny do you expect to be albino?
    a. 0%
    b. 12.5%
    c. 37.5%
    d. **50%**
    e. 100%

30. What proportion of the progeny do you expect to be agouti?
    a. 0%
    b. 12.5%
    c. **37.5%**
    d. 50%
    e. 100%

31. What proportion of the progeny do you expect to be black? Black is the alternate phenotype to agouti.
    a. 0%
    b. **12.5%**
    c. 37.5%
    d. 50%
    e. 100%

32. The complete phenotype of an organism is dependent on
    a. genotype.
    b. penetrance.
    c. expressivity.
    d. polygenes.
    e. **All of the above**

33. When a black, straight-winged fly is crossed to a brown, curly-winged fly, the frequency at which black curly-winged and brown straight-winged flies are seen in the progeny is called
    a. the mutation frequency.
    b. the mitotic frequency.
    c. the meiotic frequency.
    d. the allele frequency.
    e. **the recombinant frequency.**

34. Alleles for genes located on mitochondrial DNA are said to be maternally inherited. What is the reason for this pattern of inheritance?
    a. The egg and sperm contribute equal numbers of cytoplasmic organelles to the zygote.
    b. **The egg contributes virtually all of the cytoplasmic organelles to the zygote.**
    c. Half of the nuclear chromosomes in the zygote come from the father.
    d. Half of the nuclear chromosomes in the zygote come from the mother.
    e. All of the nuclear chromosomes in the zygote come from the mother.

35. Which of the following methods was *not* used by Mendel in his study of the genetics of the garden pea?
    a. Maintenance of true-breeding lines
    b. Cross-pollination
    c. Microscopy
    d. Production of hybrid plants
    e. Quantitative analysis of results

36. In Kölreuter's studies, reciprocal crosses
    a. always gave identical results.
    b. only involved heterozygous individuals.
    c. supported the blending hypothesis of inheritance.
    d. could only be done with homozygous individuals.
    e. consist of an F_1 and an F_2 generation.

37. Which of the following statements about Mendelian genetics is false?
    a. Alternate forms of genes are called alleles.
    b. A locus is a gene's location on its chromosome.
    c. Only two alleles can exist for a given gene.
    d. A genotype is a description of the alleles that represent an individual's genes.
    e. Individuals with the same phenotype can have different genotypes.

38. Segregation of alleles occurs
    a. during gamete formation.
    b. at fertilization.
    c. during mitosis.
    d. during the random combination of gametes to produce the F_2 generation.
    e. only in monohybrid crosses.

39. A pea plant with red flowers is test crossed, and one half of the resulting progeny have red flowers, while the other half have white flowers. You know that the genotype of the test-crossed parent was
    a. *RR*.
    b. *Rr*.
    c. *rr*.
    d. either *RR* or *Rr*.
    e. Cannot tell unless the genotypes of both parents are known

40. Mendel performed a cross between individuals heterozygous for three different traits: yellow versus green seeds (green is dominant), red versus white flowers (red is dominant), and green versus yellow pods (green is dominant). What fraction of the offspring are expected to have green seeds, red flowers, and green pods?
    a. 27/64
    b. 12/64
    c. 9/64
    d. 6/64
    e. 3/64

41. Classical albinism results from a recessive allele. Which of the following is expected of the offspring of a normally pigmented male with an albino father and an albino wife?
    a. 3/4 normal; 1/4 albino
    b. 3/4 albino; 1/4 normal
    c. 1/2 normal; 1/2 albino
    d. All normal
    e. All albino

42. In humans, a widow's peak is caused by a dominant allele *W*, and a continuous hairline, by a recessive allele *w*. Short fingers are caused by a dominant allele *S*, and long fingers, by a recessive allele *s*. Suppose a woman with a continuous hairline and short fingers and a man with a widow's peak and long fingers have three children. One child has short fingers and a widow's peak, one has long fingers and a widow's peak, and one has long fingers and a continuous hairline. What are the genotypes of the parents?
    a. Female *wwSS*; male *WWss*
    b. Female *wwSs*; male *Wwss*
    c. Female *wwSs*; male *WWss*
    d. Female *WwSs*; male *WwSs*
    e. None of the above

43. In garden peas, the allele for tall plants is dominant over the allele for short plants. A true-breeding tall plant is crossed with a short plant, and one of their offspring is test crossed. Out of 20 offspring resulting from the test cross, about _____ should be tall.
    a. 0
    b. 5
    c. 10
    d. 15
    e. 20

44. Which of the following phenomena *cannot* be observed using only dihybrid crosses?
    a. Crossing over
    b. Segregation of alleles
    c. Independent assortment of alleles
    d. Recessive lethal alleles
    e. None of the above

45. Separation of the alleles of a single gene into different gametes is called
    a. synapsis.
    b. segregation.
    c. independent assortment.
    d. heterozygous separation.
    e. recombination.

46. In mice, short hair is dominant to long hair. If a short-hair individual is crossed with a long-hair individual and both long- and short-hair offspring result, you can conclude that
    a. the short-hair individual was homozygous.
    b. the short-hair individual was heterozygous.
    c. the long-hair individual was homozygous.
    d. the long-hair individual was heterozygous.
    e. more offspring are required in order to decide.

47. In dogs, erect ears and barking while following a scent are due to dominant alleles; droopy ears and silence while following a scent are due to recessive alleles. A dog homozygous for both traits is mated to a droopy-eared, silent follower. If the two genes are unlinked, the expected F_1 phenotypic ratios should be
    a. 9:3:3:1.
    b. 1:1.
    c. 100% of one phenotype.
    d. 1:2:1.
    e. None of the above

48. In cocker spaniels, black color (B) is dominant over red (b), and solid color (S) is dominant over spotted (s). If the offspring between BBss and bbss individuals are mated with each other, what fraction of their offspring will be expected to be black and spotted? Assume the genes are unlinked.
    a. 1/16
    b. 9/16
    c. 1/9
    d. 3/16
    e. **3/4**

49. If the same allele has two or more phenotypic effects, it is said to be
    a. codominant.
    b. a marker.
    c. linked.
    d. **pleiotropic.**
    e. hemizygous.

50. In Netherlands dwarf rabbits, a gene showing intermediate inheritance produces three phenotypes. Rabbits that are homozygous for one allele are small rabbits; individuals homozygous for the other allele are deformed and die; heterozygous individuals are dwarf. If two dwarf rabbits are mated, what proportion of their surviving offspring should be dwarf?
    a. 1/4
    b. 1/3
    c. 1/2
    d. **2/3**
    e. 3/4

51. In tomatoes, tall is dominant to short, and smooth fruits are dominant to hairy fruits. A plant homozygous for both dominant traits is crossed with a plant homozygous for both recessive traits. The F_1 progeny are tested and crossed with the following results: 78 tall, smooth fruits; 82 dwarf, hairy fruits; 22 tall, hairy fruits; and 18 dwarf, smooth fruits. These data indicate that
    a. the genes are on different chromosomes.
    b. the genes are linked, but do *not* cross over.
    c. the genes are linked and show 10% recombination.
    d. **the genes are linked and show 20% recombination.**
    e. the genes are linked and show 40% recombination.

52. Remembering that white eyes is a recessive, sex-linked trait, if a white-eyed female fruit fly is mated to a red-eyed male, their offspring should be
    a. 50% red-eyed, 50% white-eyed for both sexes.
    b. all white-eyed for both sexes.
    c. **all white-eyed males, all red-eyed females.**
    d. all white-eyed females, all red-eyed males.
    e. 50% red-eyed males, 50% white-eyed males, all red-eyed females.

53. A dominant allele K is necessary for normal hearing. A dominant allele M of a different locus results in deafness no matter what other alleles are present. If a kkMm individual is crossed with a Kkmm individual, what percentage of the offspring will be deaf?
    a. 0%
    b. 25%
    c. 50%
    d. **75%**
    e. None of the above

54. The genetic disease blue sclera is determined by an autosomal dominant allele. The eyes of individuals with this allele have bluish sclera. These same individuals may also suffer from fragile bones and deafness. This is an example of
    a. incomplete dominance.
    b. **pleiotropy.**
    c. epistasis.
    d. codominance.
    e. linkage.

55. The blue sclera allele has 90% penetrance for producing blue sclera, 60% penetrance for fragile bone, and 40% penetrance for deafness. If these probabilities of penetrance are independent, what is the probability that an individual with the blue sclera allele will be deaf, have blue sclera, and fragile bones?
    a. **22%**
    b. 40%
    c. 60%
    d. 90%
    e. None of the above

56. Y-linked genes include a gene that produces hairy pinna (the external ear). A male with hairy pinna should pass this trait
    a. usually to his sons, but rarely also to a daughter.
    b. **only to his sons.**
    c. only to his daughters.
    d. only to his grandsons.
    e. to all his children if the mother is a carrier.

57. Which of the following organelles contain DNA?
    a. Nucleus
    b. Chloroplast
    c. Mitochondria
    d. Ribosome
    e. **a, b, and c**

58. Two strains of true-breeding plants that have different alleles for a certain character are crossed. Their progeny are called
    a. P generation.
    b. **F_1 generation.**
    c. F_2 generation.
    d. F_1 crosses.
    e. F_2 progeny.

59. A mutation at a single locus causes a change in many different characters. This an example of a
    a. polygene effect.
    b. epigenetic effect.
    c. cytoplasmic effect.
    d. multiple negativity effect.
    e. **pleiotropic effect.**

60. In guppies, fan tail is dominant to flesh tail, and rainbow color is dominant to pink. F_1 female guppies are crossed to flesh-tailed, pink-colored males, and 401 fan-tailed, pink-colored; 399 flesh-tailed, rainbow-colored; 98 flesh-tailed, pink-colored; and 102 fan-tailed, rainbow-colored guppies are observed in the progeny. The map distance between these two genes is
    a. 80 cM.
    b. 25 cM.
    c. .8 cM.
    d. **20 cM.**
    e. None of the above

61. How many linkage groups are present in a female human?
    a. 46
    b. Thousands
    c. 23
    d. 2
    e. 496

62. Tall pea plants are crossed to short and the progeny are medium height. The F_1 plants are crossed together, but the progeny observed among the F_2 have nine different size classes. This character's mode of inheritance is

    a. pleiotropic.
    b. epistasis.
    c. multiallelic
    d. polygenic.
    e. hypostatic.

63. Sex in humans is determined by
    a. a gene called *SRY* found on the Y chromosome.
    b. a gene called *SRY* found on the X chromosome.
    c. a gene found on an autosomal chromosome called *SDG*.
    d. the simple presence of a Y chromosome.
    e. a gene called *SDG* found on the Y chromosome.

11 *DNA and Its Role in Heredity*

Fill in the Blank

1. The X-ray crystallographs of the English chemist **Rosalind Franklin** were essential for the discovery of the structure of the DNA molecule.

2. Since the DNA molecule runs on and on, without kinks or bulges, nucleotide pair after nucleotide pair, information must lie in the **linear** sequence of the nitrogenous bases.

3. Arthur Kornberg showed that DNA could replicate in the test tube if it contained intact DNA for a template, a mixture of the four precursors—the four nucleoside triphosphates, and **DNA polymerase**.

4. The material that changed R strain pneumococcus into the virulent S strain was originally called the **transforming principle**.

5. The basic units of DNA and RNA molecules are the **nucleotides** (or **nitrogenous bases**).

6. The experiments of Meselson and Stahl established the **semiconservative replication** of DNA.

7. The nitrogenous bases classified as purines are **adenine** and **guanine**.

8. The nitrogenous bases classified as pyrimidines are **cytosine** and **thymine**.

9. The nitrogenous base is covalently bonded to the **1'** carbon of the deoxyribose.

10. Using Meselson and Stahl's experimental system for studying the mode of replication of DNA, the genetic information from a life form from Mars was analyzed. It was found that after the first round of replication, two distinct bands appeared in the cesium chloride gradient. This would be consistent with **conservative** replication.

11. The enzyme that replicates the lagging strand is **DNA polymerase III**.

12. The enzyme that replicates the leading strand is **DNA polymerase III**.

13. The region of DNA where DNA begins replication is an **origin of replication**.

14. The fragments of RNA–DNA found on the lagging strand of DNA prior to RNA removal and ligation are called **Okazaki fragments**..

15. The purines take up **more** (more/less) space in the center of a DNA molecule.

Multiple Choice

1. Why was it thought that genes were made of proteins and *not* nucleic acids?
   a. Nucleic acids are made up of 20 different bases, while proteins are made up of only 5 amino acids.
   b. Protein subunits can combine to form larger proteins.
   c. **Proteins seemed to be much more diverse chemically.**
   d. Proteins can be enzymes.
   e. None of the above

2. Why is it that DNA, made up of only 4 different bases, can encode the information necessary to specify the workings of an entire organism?
   a. **DNA molecules are extremely long.**
   b. DNA molecules are found in the nucleus.
   c. DNA is transcribed into RNA and then into proteins with specific functions.
   d. DNA is eventually translated into proteins, which are made up of 20 different amino acids.
   e. None of the above

3. In Griffith's experiments, what happened when heat-killed S strain pneumococci were injected into a mouse along with live R strain pneumococci?
   a. DNA from the live R was taken up by the heat-killed S, converting them to R and killing the mouse.
   b. **DNA from the heat-killed S was taken up by the live R, converting them to S and killing the mouse.**
   c. Proteins released from the heat-killed S killed the mouse.
   d. RNA from the heat-killed S was translated into proteins that killed the mouse.
   e. Nothing

4. Experiments designed to identify the transforming principle were based on
   a. purifying each of the macromolecule types from a cell-free extract.
   b. removing each of the macromolecules from a cell, then testing its type.
   c. **selectively destroying the different macromolecules in a cell-free extract.**
   d. *a* and *b*
   e. *a* and *c*

5. The Hershey–Chase experiment determined that
   a. protein and DNA are the hereditary materials of viruses.
   b. protein, not DNA, is the hereditary material of viruses.
   c. viruses do not contain hereditary material.
   d. **DNA, not protein, is the hereditary material of viruses.**
   e. the blender would be useful in the kitchen.

6. The rules formulated by Erwin Chargaff state that
   a. **A = T and G = C in any molecule of DNA.**
   b. A = C and G = T in any molecule of DNA.
   c. A = G and C = T in any molecule of DNA.
   d. A = U and G = C in any molecule of RNA.
   e. DNA and RNA are made up of the same four nitrogenous bases.

7. Purines include
   a. cytosine, uracil, and thymine.
   b. adenine and cytosine.
   c. adenine and thymine.
   d. cytosine and thymine.
   e. **adenine and guanine.**

8. The structure of DNA is characterized by
   a. a double, right- or left-handed helix and antiparallel strands.
   b. **a double, right-handed helix and antiparallel strands.**
   c. a single, right-handed helix.
   d. a double, right-handed helix and parallel strands.
   e. All of the above

9. The nitrogenous bases (and the two strands of the DNA double helix) are held together by
   a. weak van der Waals forces.
   b. covalent bonds.
   c. **hydrogen bonds.**
   d. a and b
   e. a and c

10. The base-paired structure of DNA implies that
    a. it can replicate to form identical molecules.
    b. it can be used as a template to make RNA.
    c. it is the hereditary material.
    d. **a and b**
    e. a and c

11. What are the three major properties of genes that are explained by the structure of DNA?
    a. They contain information, direct the synthesis of proteins, and are contained in the cell nucleus.
    b. They contain nitrogenous bases, direct the synthesis of RNA, and are contained in the cell nucleus.
    c. They replicate exactly, are contained in the cell nucleus, and direct the synthesis of cellular proteins.
    d. They encode the organism's phenotype, are passed on from one generation to the next, and contain nitrogenous bases.
    e. **They contain information, replicate exactly, and change to produce a mutation.**

12. Semiconservative replication of DNA involves
    a. **each of the original strands acting as a template for a new strand.**
    b. only one of the original strands acting as a template for a new strand.

    c. the complete separation of the original strands, the synthesis of new strands, and the reassembly of double-stranded molecules.
    d. the use of the original double-stranded molecule as a template, without unwinding.
    e. None of the above

13. The molecules that function to replicate DNA in the cell are
    a. DNA nucleoside triphosphatases.
    b. **DNA polymerases.**
    c. nucleoside polymerases.
    d. DNAses.
    e. ribonucleases.

14. During replication, the new DNA strand is synthesized
    a. in the 3′ to 5′ direction.
    b. **in the 5′ to 3′ direction.**
    c. in both the 3′ to 5′ and 5′ to 3′ directions from the replication fork.
    d. from one end to the other, in the 3′ to 5′ or the 5′ to 3′ directions.
    e. None of the above

15. DNA replication in eukaryotes differs from replication in bacteria because
    a. synthesis of the new DNA strand is from 3′ to 5′ in eukaryotes and from 5′ to 3′ in bacteria.
    b. synthesis of the new DNA strand is from 5′ to 3′ in eukaryotes and from 3′ to 5′ in bacteria.
    c. **there are many replication forks in each eukaryotic chromosome and only one in bacterial DNA.**
    d. synthesis of the new DNA strand is from 5′ to 3′ in eukaryotes and is random in prokaryotes.
    e. Okazaki fragments are produced in eukaryotic DNA replication but not in prokaryotic DNA replication.

16. Which of the following molecules functions to transfer information from one generation to the next?
    a. **DNA**
    b. mRNA
    c. tRNA
    d. Proteins
    e. Lipids

17. Mutations are
    a. heritable changes in the sequence of DNA bases that produce an observable phenotype.
    b. **heritable changes in the sequence of DNA bases.**
    c. mistakes in the incorporation of amino acids into proteins.
    d. heritable changes in the mRNA of an organism.
    e. None of the above

18. In order to show that DNA is the "transforming principle," Avery and MacLeod showed that DNA could transform avirulent strains of pneumococcus. This hypothesis was strengthened by their demonstration that
    a. enzymes that destroyed proteins also destroyed transforming activity.
    b. **enzymes that destroyed nucleic acids also destroyed transforming activity.**
    c. enzymes that destroyed complex carbohydrates also destroyed transforming activity.
    d. the transformation activity was destroyed by boiling.
    e. other strains of bacteria could also be successfully transformed.

19. During infection of *E. coli* cells by bacteriophage T2,
    a. proteins are the only phage components which actually enter the infected cell.
    b. both proteins and nucleic acids enter the cell during infection.
    c. only protein from the infecting phage can also be detected in progeny phage.
    d. only nucleic acids enter the cell during infection.
    e. more than one infecting phage particle is required to produce infection.

20. Bacteriophage nucleic acids were labeled by carrying out an infection of *E. coli* cells growing in
    a. ^{14}C-labeled CO_2.
    b. ^{3}H-labeled water.
    c. ^{32}P-labeled phosphate.
    d. ^{35}S-labeled sulfate.
    e. ^{18}O-labeled water.

21. Information used by Watson and Crick to determine the structure of DNA included
    a. electron micrographs of individual DNA molecules.
    b. light micrographs of bacteriophage particles.
    c. light micrographs of individual bacterial chromosomes.
    d. nuclear magnetic resonance analysis of DNA.
    e. X-ray crystallography images of double-stranded DNA.

22. Double-stranded DNA looks a little like a ladder that has been twisted into a helix, or spiral. The ladder supports are
    a. individual nitrogenous bases.
    b. alternating bases and sugars.
    c. alternating bases and phosphate groups.
    d. alternating sugars and phosphates.
    e. alternating bases, sugars, and phosphates.

23. The steps of the ladder are
    a. individual nitrogenous bases.
    b. pairs of bases.
    c. alternating bases and phosphate groups.
    d. alternating sugars and bases.
    e. alternating bases, sugars, and phosphates.

24. If a double-stranded DNA molecule contains 30% T, how much G does it contain?
    a. 20%
    b. 30%
    c. 40%
    d. 50%
    e. 60%

25. You have analyzed the DNA isolated from a newly discovered virus and found that its base composition is 32% A, 17% C, 32% G, and 19% T. What would be a reasonable explanation of this observation?
    a. The virus must be extraterrestrial.
    b. In some viruses, double-stranded DNA is made up of base pairs containing two purines or two pyrimidines.
    c. Some of the T was converted to C during the isolation procedure.
    d. The genome of the phage is single-stranded, not double-stranded.
    e. The genome of the phage must be circular, not linear.

26. The base composition of DNA that is complementary to the viral DNA described in the previous question would be
    a. 32% A, 17% C, 32% G, 19% T.
    b. 32% T, 17% G, 32% C, 19% A.
    c. 32% C, 17% A, 32% G, 19% T.
    d. 25% A, 25% G, 25% C, 25% T.
    e. 32% A, 32% T, 18% C, 18% G.

27. In the Meselson–Stahl experiment, the conservative model of DNA replication is ruled out by which of the following observations?
    a. No completely heavy DNA is observed after the first round of replication.
    b. No completely light DNA ever appears, even after several replications.
    c. The product which accumulates after two rounds of replication is completely heavy.
    d. Completely heavy DNA is observed throughout the experiment.
    e. Three different DNA densities are observed after a single round of replication.

28. During DNA replication
    a. one parental strand must be degraded to allow the other strand to be copied.
    b. the parental strands must separate so that both can be copied.
    c. the parental strands come back together after the passage of the replication fork.
    d. origins of replication always give rise to single replication forks.
    e. two replication forks diverge from each origin but one always lags behind the other.

29. The enzyme DNA ligase is required continuously during DNA replication because
    a. fragments of the leading strand must be joined together.
    b. fragments of the lagging strand must be joined together.
    c. the parental strands must be joined back together.
    d. 3'-deoxynucleoside triphosphates must be converted to 5'-deoxynucleoside triphosphates.
    e. the complex of proteins that works together at the replication fork must be kept from falling apart.

30. In DNA replication, each newly made strand is
    a. identical in DNA sequence to the strand from which it was copied.
    b. complementary in sequence to the strand from which it was copied.
    c. oriented in the same 3' to 5' direction as the strand from which it was copied.
    d. an incomplete copy of one of the parental strands.
    e. a hybrid molecule consisting of both ribo- and deoxyribonucleotides.

31. The feature of the Watson–Crick model of DNA structure that explains its ability to function in replication and gene expression is
    a. that each strand contains all the information present in the double helix.
    b. the structural and functional similarities of DNA and RNA.
    c. that the double helix is right-handed and not left-handed.

d. that DNA replication does not require enzyme catalysts.

e. the exposure of the bases in the major groove of the double helix.

32. The Hershey and Chase experiment convinced most scientists that
    a. bacteria can be transformed.
    b. **DNA is indeed the carrier of hereditary information.**
    c. DNA replication is semiconservative.
    d. the transforming principle requires host factors.
    e. All of the above

33. Which of the following features summarizes the molecular architecture of DNA?
    a. The two strands run in opposite directions.
    b. The molecule twists in the same direction as the threads of most screws.
    c. The molecule is a double-stranded helix.
    d. DNA has a uniform diameter.
    e. **All of the above**

34. The fidelity of DNA replication is astounding. The error rate is on the order of one wrong nucleotide per
    a. 10,000.
    b. 100,000.
    c. **10^8–10^{12}.**
    d. 10^{13}–10^{16}.
    e. one trillion.

35. Chargaff's rule says
    a. the DNA must be replicated before a cell can divide.
    b. viruses enter cells without their protein coat.
    c. only protein from the infecting phage can also be detected in progeny phage.
    d. only nucleic acids enter the cell during infection.
    e. **the amount of cytosine equals the amount of guanine.**

36. The first scientist(s) to suggest a mode of replication for DNA was (were)
    a. Linus and Pauling.
    b. Hershey and Chase.
    c. Albert Leverman.
    d. **Watson and Crick.**
    e. Meselson and Stahl.

37. In eukaryotic cells, each chromosome has
    a. one origin of replication.
    b. two origins of replication.
    c. **many origins of replication.**
    d. only one origin of replication per nucleus.
    e. None of the above

38. When adding the next monomer to a growing DNA strand, the monomer is added to the
    a. 1' carbon of the deoxyribose.
    b. 2' carbon of the deoxyribose.
    c. **3' carbon of the deoxyribose.**
    d. 4' carbon of the deoxyribose.
    e. 5' carbon of the deoxyribose.

39. The energy necessary for making a DNA molecule comes directly from the
    a. sugar.
    b. ATP.

c. **release of phosphates.**
d. NADPH.
e. NADH.

40. A deoxyribose nucleotide is
    a. a deoxyribose plus a nitrogenous base.
    b. a sugar and a phosphate.
    c. **a deoxyribose plus a nitrogenous base and a phosphate.**
    d. a ribose plus a nitrogenous base.
    e. a nitrogenous base bonded 5' to a sugar–phosphate.

41. Synthesis of DNA is
    a. spontaneous.
    b. **endergonic.**
    c. exergonic.
    d. pseudogonic.
    e. quasigonic.

42. The enzyme that removes the RNA primers is called
    a. DNA ligase.
    b. primase.
    c. reverse transcriptase.
    d. helicase.
    e. **DNA polymerase I.**

43. The enzyme that restores the phosphodiester linkage between adjacent fragments in the lagging strand during DNA replication is
    a. **DNA ligase.**
    b. primase.
    c. reverse transcriptase.
    d. helicase.
    e. DNA polymerase I.

44. A deoxyribose nucleoside is
    a. **a deoxyribose plus a nitrogenous base.**
    b. a sugar and a phosphate.
    c. a deoxyribose plus a nitrogenous base and a phosphate.
    d. a ribose plus a nitrogenous base.
    e. a nitrogenous base bonded 5' to a sugar–phosphate.

45. The building blocks for a new DNA molecule are
    a. deoxyribose nucleoside monophosphates.
    b. deoxyribose nucleoside diphosphates.
    c. **deoxyribose nucleoside triphosphates.**
    d. deoxyribose nucleotide diphosphates.
    e. deoxyribose nucleotide triphosphates.

46. The force that holds DNA together in a double helix is
    a. the force of the twist.
    b. covalent bonds.
    c. ionic bonds.
    d. ionic interactions.
    e. **hydrogen bonds.**

47. The enzyme that unwinds the DNA prior to replication is called
    a. DNA polymerase III.
    b. DNA ligase.
    c. single-stranded DNA binding proteins.
    d. primase.
    e. **helicase.**

48. The first repair of mistakes made during DNA replication is made by
    a. mismatch repair system.
    b. DNA polymerase.
    c. excision repair.
    d. SOS repair.
    e. post-replication repair.

49. If no proofreading occurred during DNA replication, the error rate would climb as high as
    a. 1 in 10 bases.
    b. 1 in 100 bases.
    c. 1 in 1000 bases.
    d. 1 in 10,000 bases.
    e. 1 in 1,000,000 bases.

50. The difference between DNA and RNA is that
    a. DNA has thymine and RNA has uracil.
    b. DNA has no oxygen bonded to the 2′ carbon; RNA does.
    c. DNA is the genetic material; RNA is not.
    d. DNA is double stranded and RNA cannot hydrogen bond.
    e. DNA is older than RNA.

12 From DNA to Protein: Genotype to Phenotype

Fill in the Blank

1. The phrase "inborn errors of metabolism" is virtually synonymous with the name **Archibald Garrod**.

2. Prototrophs ("original eaters") grow on minimal media whereas auxotrophs ("increased eaters") require **specific additional nutrients**.

3. The basic units of DNA and RNA molecules are the **nucleotides** (or **nitrogenous bases**).

4. RNA differs from DNA in base composition because it contains **uracil** instead of thymine.

5. The strand of DNA that is transcribed into RNA is the **template strand**.

6. The part of a protein that determines whether translation will continue in the cytosol or at the endoplasmic reticulum is the **signal sequence**.

7. The excess of codons (64) over amino acids (20) indicates that the genetic code is **degenerate**.

8. Garrod hypothesized that urine was dark in people afflicted with alkaptonuria because of a defective **enzyme**.

9. The genetic molecule for tobacco mosaic virus is **RNA**.

10. The enzyme that can use RNA as a template to generate DNA is **reverse transcriptase**.

11. A mRNA molecule with several ribosomes attached at the same time is called a(n) **polysome**.

12. Small ribosomal subunits are dispersed into smaller components by placing them into a detergent solution. Upon removal of the detergent, the components interact to create new intact subunits by a process called **self-assembly**.

13. A mutation that causes a change in the nitrogenous base sequence of a DNA molecule but no change in the amino acid sequence it codes for is called a(n) **silent mutation**.

Multiple Choice

1. After irradiating *Neurospora*, Beadle and Tatum collected mutants that require arginine to grow. These mutants
   a. **will not grow on minimal media but will grow on minimal media with arginine.**
   b. will grow on minimal media and on minimal media with arginine.
   c. will not grow on minimal media and will not grow on minimal media with arginine.
   d. will grow on minimal media and will grow on minimal media with arginine.
   e. None of the above

2. Within a group of mutants with the same growth requirement, that is, the same overt phenotype, mapping studies determined that individual mutations were on different chromosomes. This indicates that
   a. the same gene governs all the steps in a particular biological pathway.
   b. **different genes can govern different individual steps in the same biological pathway.**
   c. different genes govern the same step in a particular biological pathway.
   d. all biological pathways are governed by different genes.
   e. genes do not govern steps in biological pathways.

3. The study of *Neurospora* mutants that grew on various supplemented media led to
   a. determining the steps in biological pathways.
   b. the "one-gene, one-enzyme" theory.
   c. the idea that genes are "on" chromosomes.
   d. **a and b**
   e. a and c

4. Genes code for
   a. enzymes.
   b. polypeptides.
   c. RNA.
   d. **All of the above**
   e. None of the above

5. Why is it that DNA, made up of only four different bases, can encode the information necessary to specify the workings of an entire organism?
   a. DNA molecules are extremely long.
   b. DNA molecules form codons of three bases that code for amino acids.
   c. The same DNA sequence can be used repeatedly.
   d. DNA can be replicated with low error rates.
   e. **All of the above**

6. RNA differs from DNA because
   a. **RNA contains uracil instead of thymine and it is (usually) single-stranded.**
   b. RNA contains uracil instead of thymine.
   c. RNA contains thymine instead of uracil and it is (usually) single-stranded.
   d. RNA contains uracil instead of cytosine.
   e. None of the above

7. What are the three major properties of genes that are explained by the structure of DNA?
   a. They contain information, direct the synthesis of proteins, and are contained in the cell nucleus.
   b. They contain nitrogenous bases, direct the synthesis of RNA, and are contained in the cell nucleus.
   c. They replicate exactly, are contained in the cell nucleus, and direct the synthesis of cellular proteins.
   d. They encode the organism's phenotype, are passed on from one generation to the next, and contain nitrogenous bases.
   e. They contain information, replicate exactly, and change to produce a mutation.

8. Which of the following statements about the flow of genetic information is correct?
   a. Proteins encode information that is used to produce other proteins of the same amino acid sequence.
   b. RNA encodes information that is translated into DNA, and DNA encodes information that is translated into proteins.
   c. Proteins encode information that can be translated into RNA, and RNA encodes information that can be transcribed into DNA.
   d. DNA encodes information that is translated into RNA, and RNA encodes information that is translated into proteins.
   e. None of the above

9. Which of the following molecules functions to transfer information from the nucleus to the cytoplasm?
   a. DNA
   b. mRNA
   c. tRNA
   d. Proteins
   e. Lipids

10. Which of the following molecules functions to transfer information from one generation to the next?
    a. DNA
    b. mRNA
    c. tRNA
    d. Proteins
    e. Lipids

11. Which of the following molecules functions to transfer information from mRNA to protein?
    a. DNA
    b. mRNA
    c. tRNA
    d. Proteins
    e. Lipids

12. A molecule of which type can function to transfer information from RNA to DNA?
    a. DNA
    b. mRNA
    c. tRNA
    d. Proteins
    e. Lipids

13. A sequence of three RNA bases can function as a
    a. codon.
    b. anticodon.
    c. gene.
    d. a and b
    e. a and c

14. The difference between mRNA and tRNA is that
    a. tRNA has a more elaborate three-dimensional structure due to extensive base pairing.
    b. tRNAs are usually very much smaller than mRNAs.
    c. mRNA has a more elaborate three-dimensional structure due to extensive base pairing.
    d. a and b
    e. None of the above

15. Ribosomes are a collection of
    a. small proteins that function in translation.
    b. proteins and small RNAs that function in translation.
    c. proteins and tRNAs that function in transcription.
    d. proteins and mRNAs that function in translation.
    e. mRNAs and tRNAs that function in translation.

16. The endoplasmic reticulum functions in protein synthesis
    a. as the site where mRNA attaches.
    b. as the site where all ribosomes bind.
    c. as the site of translation of membrane-bound and exported proteins.
    d. to produce tRNAs.
    e. to bring together mRNA and tRNA.

17. Retroviruses do *not* follow the "central dogma" of DNA→ RNA→ protein because
    a. they contain RNA that is used to make DNA.
    b. they contain DNA that is used to make more RNA.
    c. they contain DNA that is used to make tRNA.
    d. they contain only RNA as the genetic material.
    e. they do not contain either DNA or RNA as the genetic material.

18. Mutations are
    a. heritable changes in the sequence of DNA bases that produce an observable phenotype.
    b. heritable changes in the sequence of DNA bases.
    c. mistakes in the incorporation of amino acids into proteins.
    d. heritable changes in the mRNA of an organism.
    e. None of the above

19. The type of mutation that stops translation of a protein is a(n)
    a. missense mutation.
    b. nonsense mutation.
    c. frame-shift mutation.
    d. aberration.
    e. None of the above

20. The type of mutation that is an insertion or deletion of a single base is a(n)
    a. missense mutation.
    b. nonsense mutation.
    c. frame-shift mutation.
    d. aberration.
    e. None of the above

21. The "central dogma" of molecular biology states that
    a. information flow between DNA, RNA, and protein is reversible.
    b. information flow in the cell is from protein to RNA to DNA.
    c. information flow in the cell is unidirectional from DNA to RNA to protein.
    d. the DNA sequence of a gene can be predicted if we know the amino acid sequence of the protein it encodes.
    e. the genetic code is ambiguous but not degenerate.

22. Transcription is the process of
    a. synthesizing a DNA molecule from an RNA template.
    b. assembling ribonucleoside triphosphates into an RNA molecule without a template.
    c. synthesizing an RNA molecule using a DNA template.
    d. synthesizing a protein using information from a messenger RNA.
    e. replicating a single-stranded DNA molecule.

23. A transcription start signal is called
    a. an initiation codon.
    b. a promoter.
    c. an origin.
    d. an operator.
    e. a nonsense codon.

24. Initiation of transcription requires
    a. a temporary stoppage of DNA replication.
    b. a temporary separation of the strands in the template DNA.
    c. destruction of one of the strands of the template DNA.
    d. relaxation of positive supercoils in the DNA template.
    e. induction of positive supercoils in the DNA template.

25. The adapters that allow translation of the four-letter nucleic acid language into the 20-letter protein language are called
    a. aminoacyl tRNA synthetases.
    b. transfer RNAs.
    c. ribosomal RNAs.
    d. messenger RNAs.
    e. ribosomes.

26. The number of codons that actually specify amino acids is
    a. 20.
    b. 23.
    c. 45.
    d. 60.
    e. 61.

27. The genetic code is best described as
    a. degenerate but not ambiguous.
    b. ambiguous but not redundant.
    c. both ambiguous and redundant.
    d. neither ambiguous nor redundant.
    e. nonsense.

28. The three codons in the genetic code that do *not* specify amino acids are called
    a. missense codons.
    b. start codons.
    c. stop codons.
    d. promoters.
    e. initiator codons.

29. In eukaryotes, ribosomes become associated with endoplasmic reticulum membranes when
    a. a signal sequence on the mRNA interacts with a receptor protein on the membrane.
    b. a signal sequence on the ribosome interacts with a receptor protein on the membrane.

c. a signal sequence at the amino terminus of the protein being synthesized interacts with a receptor protein on the ribosome.
    d. a signal sequence on the protein being synthesized interacts with a membrane attachment site on the ribosome.
    e. the messenger RNA passes through a pore in the membrane.

30. Viruses that violate the "central dogma" through the use of an enzyme that makes DNA copies of an RNA molecule are called
    a. bacteriophage.
    b. retroviruses.
    c. RNA viruses.
    d. DNA viruses.
    e. enveloped viruses.

31. Sickle-cell disease is caused by a _____ mutation.
    a. nonsense
    b. missense
    c. frame-shift
    d. temperature sensitive
    e. silent

32. The classic work of Beadle and Tatum, later refined by others, provided evidence for
    a. the one-gene, one-enzyme hypothesis.
    b. the one-gene, one-polypeptide hypothesis.
    c. explaining how information in genes is translated into traits.
    d. explaining how some mutations affect organisms.
    e. All of the above

33. Two insightful predictions made by Francis Crick were the existence of
    a. transposons and retrotransposons.
    b. retroviruses and DNA viruses.
    c. promoters and enhancers.
    d. the vinegar yeast and collagenase.
    e. mRNA and tRNA.

34. RNA polymerase is a
    a. RNA-directed DNA polymerase.
    b. RNA-directed RNA polymerase.
    c. DNA-directed RNA polymerase.
    d. typical enzyme.
    e. form of RNA.

35. The direction of synthesis for a new mRNA molecule is
    a. 5′ to 3′ from a 5′ to 3′ template strand.
    b. 5′ to 3′ from a 3′ to 5′ template strand.
    c. 3′ to 5′ from a 5′ to 3′ template strand.
    d. 3′ to 5′ from a 3′ to 5′ template strand.
    e. 5′ to 5′ from a 3′ to 3′ template strand.

36. The region of DNA in prokaryotes that RNA polymerase binds to most tightly is the
    a. promoter.
    b. poly C center.
    c. enhancer.
    d. operator site.
    e. minor groove.

37. Promoters are made of
    a. protein.
    b. carbohydrate.
    c. lipid.
    d. **nucleic acids.**
    e. amino acids.

38. DNA is composed of two strands, and only one strand is typically used as a template for RNA synthesis. The correct strand is chosen because
    a. both are tried and the one that works is remembered.
    b. only one strand has the start codon.
    c. **the promoter acts to aim the RNA polymerase.**
    d. a start factor informs the system.
    e. The correct strand is picked half the time.

39. There are differences in the amount of transcription that takes place for different genes. One cause for this would be that
    a. **different promoters have different affinities for RNA polymerase.**
    b. longer genes take longer to transcribe.
    c. the outcome is influenced by random chance.
    d. ribosomes tend to attach to transcripts even before transcription is completed.
    e. None of the above answers are true.

40. There are _____ different RNA polymerases in eukaryotes.
    a. 1
    b. 2
    c. **3**
    d. 4
    e. 5

41. In eukaryotes, the RNA polymerase that synthesizes mRNA is
    a. I.
    b. **II.**
    c. III.
    d. IV.
    e. V.

42. The study of the Rous sarcoma virus by Howard Temin was particularly fruitful because it led to the discovery of
    a. viruses.
    b. cancer.
    c. chicken diseases.
    d. **reverse transcriptase.**
    e. polycluck syndrome.

43. Proteins are synthesized from the _____, in the _____ direction along the mRNA.
    a. **N terminus to C terminus; 5′ to 3′**
    b. C terminus to N terminus; 5′ to 3′
    c. C terminus to N terminus; 3′ to 5′
    d. N terminus to C terminus; 3′ to 5′
    e. N terminus to N terminus; 5′ to 5′

44. Two mutants, 1 and 2, are found that have defects in different genes that code for enzymes, which are part of the same biochemical pathway. Substance X makes it possible for both mutants to grow. Substance Z is only effective for mutant 2. Which mutant has the defective enzyme that is earliest in the biochemical pathway?
    a. Mutant 1
    b. **Mutant 2**
    c. Z is before X.
    d. X is before Z.
    e. None of the above

45. Imagine that a novel life form is found deep within Earth's crust. Evaluation of its DNA yields no surprises. However, it is found that a codon for this life form is just two bases in length. How many different amino acids could this organism be composed of?
    a. 4
    b. 8
    c. **16**
    d. 32
    e. 64

46. Poly uracil codes for
    a. three different amino acids.
    b. poly tryptophan.
    c. mRNA.
    d. a fatty acid.
    e. **poly phenylalanine.**

47. The enzyme that charges the tRNA molecules with appropriate amino acids is
    a. tRNA chargeatase.
    b. amino tRNA chargeatase.
    c. transcriptase.
    d. **aminoacyl-tRNA synthetase.**
    e. None of the above

48. It is currently believed that the enzyme that catalyses formation of the peptide bond during translation is composed of
    a. amino acids.
    b. protein.
    c. carbohydrate.
    d. **RNA.**
    e. DNA.

49. The termination of transcription is signaled by
    a. the stop codon.
    b. **a sequence of nitrogenous bases.**
    c. a protein bound to a certain region of DNA.
    d. rRNA.
    e. tRNA.

50. During translation initiation, the first site occupied by a charged tRNA is
    a. the A site.
    b. the B site.
    c. the large subunit.
    d. the T site.
    e. **the P site.**

51. During translation elongation, the existing polypeptide chain is transferred to
    a. **the tRNA occupying the A site.**
    b. the tRNA occupying the P site.
    c. the ribosomal rRNA.
    d. signal recognition particle.
    e. None of the above

52. The stop codons code for
    a. **no amino acid.**
    b. methionine.
    c. glycine.
    d. halt enzyme.
    e. DNA binding protein.

53. The Ames test detects the mutagenic activity of chemicals by
    a. **the number of bacteria that are killed.**
    b. the number of bacteria that live.
    c. the change in the color of the bacteria that are treated.
    d. the number of defective bacteria generated.
    e. None of the above

13 The Genetics of Viruses and Prokaryotes

Fill in the Blank

1. Bacteria that house bacteriophages that are not lytic are called **lysogenic**.

2. **Plasmids** are nonessential genetic elements that exist only as free, independently replicating cycles of DNA that cannot be incorporated into the bacterial chromosome.

3. When the synthesis of an enzyme is turned off in response to an external biochemical cue (such as an excess in tryptophan), the enzyme is said to be **repressible**.

4. A circular clearing in a lawn of bacteria caused by bacteriophages is called a **plaque**.

5. Bacteriophage DNA that is stably integrated into a bacterial chromosome is called a **prophage**.

6. Bacteriophages that can integrate their DNA into the host bacterial chromosome are called **temperate**.

7. DNA that can exist in the cytoplasm of a cell or be integrated into the bacterial chromosome is an **episome**.

8. Pieces of DNA that move from place to place in the bacterial chromosome are called **transposable elements**.

9. A gene that produces an mRNA containing information for more than one protein is a(n) **operon**.

10. The region of the gene that binds RNA polymerase is the **promoter**.

11. The site on the operon DNA where a repressor binds is the **operator** sequence.

12. **Regulatory genes** encode repressor proteins.

13. **Catabolite repression** is a positive control process that relies on increasing the affinity of promoters for RNA polymerase.

14. An infection that was acquired by a patient during a hospital stay is called a(n) **nocosomial** infection.

15. The first virus that was discovered was the **tobacco mosaic** virus.

16. Antibiotics are **ineffective** against viral infections.

17. A **virion** is the name of an individual viral particle when it is outside its host.

18. Phages are called **prophages** when their DNA is integrated into the host's chromosome.

Multiple Choice

1. Beginning with a single bacterium, how many cells would be present after 10 hours of growth if they can double every 20 minutes?
   a. 2,147,483,648
   b. 1,073,741,824
   c. 536,870,912
   d. 268,435,456
   e. 134,217,728

2. When a met^-bio^- strain of bacteria is mixed with a thr^-leu^- strain, wild-type bacteria result at a rate of one in every 10^7 cells. Why can't this be explained by mutation?
   a. The wild type would have to be a double mutation, which is too rare.
   b. The wild type would have to be a mutation in four genes, which is too rare.
   c. The wild type would have to be a deletion, which is extremely rare.
   d. Mutations can only occur in response to a mutagen.
   e. Bacterial genes do not mutate.

3. What is the result of mixing a few F^+ bacteria with many F^- bacteria?
   a. Half of the bacteria become F^+ and half become F^-.
   b. All the bacteria quickly become F^-.
   c. All the bacteria quickly become F^+.
   d. The culture will soon contain many F^+ bacteria and few F^- bacteria.
   e. Nothing happens.

4. The met^+leu^+ bacteria that result from a cross between met^+leu^- and met^-leu^+ bacteria are called
   a. Hfr strains.
   b. auxotrophs.
   c. parentals.
   d. recombinants.
   e. None of the above

5. Hfr strains are variants of F^+ strains that
   a. have the F plasmid integrated into the bacterial chromosome.
   b. transfer certain genes with a high frequency.
   c. are male.
   d. a and b
   e. a, b, and c

6. When an interrupted mating experiment is performed between a Hfr and an F⁻ strain, why do almost none of the F⁻ bacteria become Hfr?
   a. The recipient cells have the entire F plasmid but cannot make pili.
   b. The recipient cells have the entire F plasmid but cannot transfer genes.
   c. The F plasmid genes are the very last to transfer.
   d. The F⁻ cells do not receive a full chromosome compliment.
   e. None of the above

7. Which conclusion(s) can be drawn from the interrupted mating experiments of Jacob and Wollman?
   a. The E. coli chromosome is circular.
   b. Hfr strains have the F plasmid integrated into their chromosome.
   c. Different Hfr strains have the F plasmid integrated into their chromosomes at different sites.
   d. The transfer of the bacterial genes begins at the site of the F plasmid.
   e. All of the above

8. The transfer of a linear order of genes characterizes which type of gene transfer in bacteria?
   a. Hfr
   b. Sexduction
   c. Transduction
   d. Transformation
   e. None of the above

9. The transfer of genes by uptake of DNA from the medium characterizes which type of gene transfer in bacteria?
   a. Hfr
   b. Sexduction
   c. Transduction
   d. Transformation
   e. None of the above

10. The transfer of genes by a bacteriophage vector characterizes which type of gene transfer in bacteria?
    a. Hfr
    b. Sexduction
    c. Transduction
    d. Transformation
    e. None of the above

11. In generalized transduction,
    a. only a particular part of the bacterial chromosome can be transferred.
    b. any part of the bacterial chromosome might be transferred.
    c. only the F plasmid can be transferred.
    d. only the part of the bacterial chromosome near the F plasmid can be transferred.
    e. None of the above

12. Which of the following is an example of an episome?
    a. F plasmid
    b. Hfr element
    c. Viral prophage
    d. Viral DNA
    e. All of the above

13. An R factor
    a. is a plasmid that carries genes for antibiotic resistance.

b. is an episome that carries genes for antibiotic resistance.
   c. is a region of the bacterial chromosome that carries genes for antibiotic resistance.
   d. is a small portion of the F plasmid.
   e. is a measure of how well the bacterium is insulated from the cold.

14. For a plasmid to survive within the cytoplasm of a cell as it continually divides, it must
    a. be integrated in the bacterial chromosome.
    b. have an R factor.
    c. have an F plasmid.
    d. have an Hfr.
    e. have an origin of replication.

15. What is the difference between a plasmid and a transposable element?
    a. Plasmids exist in many copies in the cell, while a transposable element exists in a single copy.
    b. A plasmid replicates independently of the cell; a transposable element does not.
    c. A plasmid has an origin of replication; a transposable element does not.
    d. A plasmid exists independently in the cell cytoplasm, while a transposable element is integrated into another larger DNA molecule.
    e. None of the above

16. Transposable elements
    a. have contributed to the evolution of plasmids.
    b. were the original carriers of the resistance genes of the R factors.
    c. can move from one site to another.
    d. a and b
    e. a, b, and c

17. When E. coli are grown in a medium with little lactose,
    a. all of the enzymes of the lactose operon are present in small quantities.
    b. all of the enzymes of the lactose operon are present in large quantities.
    c. none of the enzymes of the lactose operon are present.
    d. β-galactosidase and permease are present in small quantities but transacetylase is present in large quantities.
    e. the mRNA of the lactose operon is not present at all.

18. A promoter is
    a. the region of a plasmid that binds the enzymes for replication.
    b. the region of the mRNA that binds to a ribosome.
    c. the region of DNA that binds RNA polymerase.
    d. the region of the mRNA that binds tRNAs.
    e. None of the above

19. The frequency of transcription of a particular bacterial gene is controlled by
    a. the DNA sequence of the particular promoter.
    b. the availability of the promoter to RNA polymerase.
    c. the number of ribosomes that are available in the cell.

d. a and *b*
e. a, b, and *c*

20. The three basic parts of an operon are
    a. the promoter, the operator, and the structural gene(s).
    b. the promoter, the structural gene(s), and the termination codons.
    c. the promoter, the mRNA, and the termination codons.
    d. the structural gene(s), the mRNA, and the tRNAs.
    e. None of the above

21. An inducer
    a. combines with a repressor and prevents it from binding the promoter.
    b. combines with a repressor and prevents it from binding the operator.
    c. binds to the promoter and prevents the repressor from binding to the operator.
    d. binds to the operator and prevents the repressor from binding at this site.
    e. binds to the termination codons and allows protein synthesis to continue.

22. The RNA transcribed from an operon
    a. is transcribed by the ribosomes to make more repressor.
    b. is translated by the ribosomes to make more inducer.
    c. is translated by the ribosomes to make the enzymes.
    d. is translated by the ribosomes to make more RNA polymerase.
    e. is usually not translated by the ribosomes at all.

23. The genes that encode repressor proteins are
    a. repressor genes.
    b. operons.
    c. inducer genes.
    d. regulatory genes.
    e. None of the above

24. When an operon is turned "off" in response to molecules present in the environment of the cell, it is
    a. repressible.
    b. suppressible.
    c. impressible.
    d. inducible.
    e. ostentatious.

25. In a repressible operon, the repressor molecule
    a. must first be activated by a corepressor.
    b. can repress the transcription of the operon on its own.
    c. is a molecule made from the operon.
    d. binds to the mRNA.
    e. must first be made negative to control the operon.

26. Catabolite repression refers to
    a. the increased transcription seen from many operons when glucose is present in the medium.
    b. the shutdown of transcription from many operons when glucose is present in the medium.
    c. the increased activity of inducers caused by glucose in the medium.
    d. a and *b*
    e. a and *c*

27. To be activated, the CRP must first bind
    a. the repressor molecule.
    b. the repressor protein.
    c. the activator protein.
    d. the corepressor molecule.
    e. cAMP.

28. The binding of the CRP–cAMP complex increases the binding of RNA polymerase at the promoter by
    a. twofold.
    b. tenfold.
    c. 50-fold.
    d. 100-fold.
    e. 1,000-fold.

29. When the concentration of glucose in the medium falls,
    a. the concentration of CRP rises.
    b. the concentration of cAMP rises.
    c. the concentration of repressors rises.
    d. the concentration of inducers rises.
    e. None of the above

30. The function of the F pili is
    a. to store the F plasmid.
    b. to make the initial contact between an F⁺ and an F⁻ cell that precedes conjugation.
    c. to uptake DNA during transformation.
    d. to transfer the DNA between mating partners during conjugation.
    e. the F pili have no function.

31. The difference between a plasmid and an episome is that
    a. a plasmid is a circular piece of DNA; an episome is a linear piece of DNA.
    b. a plasmid carries nonessential genetic elements, but an episome carries essential genetic elements.
    c. an episome carries nonessential genetic elements, but a plasmid carries essential genetic elements.
    d. a plasmid can replicate independently, but an episome cannot.
    e. a plasmid cannot incorporate into the bacterial chromosome, but an episome can.

32. The term lysogeny refers to
    a. a condition in which the bacteriophage DNA is stably integrated in the bacterial chromosome.
    b. a condition in which the bacteriophage DNA is excised from the bacterial chromosome.
    c. a condition in which a bacteriophage lyses a bacterium.
    d. a mutation induced by a bacteriophage.
    e. the exchange of genetic material between a bacteriophage and a bacteria.

33. Which of the following is *not* a mechanism of bacterial genetic recombination?
    a. Transformation
    b. Conjugation
    c. Transduction
    d. Catabolite repression
    e. None of the above

34. A population of genetically identical bacteria that arose from a single cell is known as a(n)
    a. plaque.
    b. lysogen.
    c. clone.
    d. hfr.
    e. K/4.

35. When a mixture of *E. coli* and bacteriophage T4 is poured over a laboratory plate containing solid media, circular clearing zones (plaques) appear because
    a. the bacteria killed the bacteriophage.
    b. the bacteriophage integrate the bacterial genome.
    c. **bacteria are lysed by the bacteriophage.**
    d. bacteria eat away the top surface of the solid media.
    e. the infected bacteria are auxotrophic and die.

36. The term auxotroph refers to
    a. **a mutant form of a bacteria that requires nutrient(s) *not* required by the wild-type bacteria.**
    b. a mutant form of a bacteria that requires no nutrients.
    c. a mutant form of a bacteria that can synthesize a nutrient which the wild-type bacteria cannot.
    d. a mutant form of a bacteria that can metabolize a nutrient which the wild-type bacteria cannot.
    e. a bacteria that can metabolize sugars.

37. Which of the following is *not* true of the F plasmid?
    a. The F plasmid can replicate independent of the bacterial chromosome.
    b. The F plasmid can be transferred from a donor to a recipient bacterium during mating.
    c. The F plasmid is involved in conjugation.
    d. The F plasmid contains genes that encode the pili.
    e. **The F plasmid contains genes that encode antibiotic resistance.**

38. One difference between Hfr bacteria and F⁺ bacteria is that
    a. F⁺ are males and Hfr are females.
    b. Hfr can participate in conjugation but F⁺ cannot.
    c. Hfr are auxotrophs and F⁺ are not.
    d. **F⁺ transfer their F plasmid to females during conjugation and Hfr generally do not.**
    e. There is no difference between Hfr and F⁺.

39. The low frequency of transfer of the F plasmid from an Hfr strain to an F⁻ strain during conjugation is because
    a. the F plasmid is not part of the chromosome.
    b. the F plasmid gets deleted from the chromosome prior to transfer.
    c. **the F plasmid is last to transfer.**
    d. the F⁻ bacteria already has an F plasmid.
    e. the F plasmid is transferred at high frequencies during Hfr conjugation.

40. Which of the following statements is true for both F plasmids and bacteriophages?
    a. They have a protein coat enclosing the DNA.
    b. They can lyse bacteria.
    c. They can participate in sexduction.
    d. **They can incorporate into the bacterial chromosome.**
    e. None of the above

41. The function of the promoter is
    a. to signal the RNA polymerase where to start transcribing the DNA.
    b. to signal the RNA polymerase which strand of the DNA to read.
    c. to signal the RNA polymerase where to stop transcribing the DNA.

    d. All of the above
    e. *a* and *b*

42. The mechanism by which the inducer causes the repressor to detach from the operator is an example of
    a. catabolite repression.
    b. transcription.
    c. transposition.
    d. **allosteric modification.**
    e. recombination.

43. _____ acts as a corepressor to block transcription of the tryptophan operon.
    a. cAMP
    b. Lactose
    c. **Tryptophan**
    d. Methionine
    e. CRP

44. The CRP–cAMP complex binds _____ of the operon.
    a. **close to the RNA polymerase binding site**
    b. close to the operator
    c. inside one of the structural genes
    d. at the termination
    e. None of the above

45. Which of the following is *not* true of conjugation between an F⁻ bacteria and an F⁺ bacteria?
    a. A conjugation tube is formed between the two bacteria.
    b. The F⁺ cell transfers a copy of the F plasmid to the F-cell.
    c. The F⁻ cell receives a copy of the F plasmid.
    d. The F⁻ cell becomes F⁺.
    e. **The F⁺ cell becomes F⁻.**

46. The _____ differ(s) among different Hfr strains.
    a. order of the genes along the circular chromosome
    b. F plasmid
    c. **insertion site of the F plasmid**
    d. genetic markers
    e. size of the chromosome

47. The brilliant "interrupted mating" experiments of Jacob and Wollman allowed them to conclude that
    a. the *E. coli* chromosome is circular.
    b. Hfr (high frequency) males have the F plasmid incorporated into their chromosomes.
    c. the location where the F plasmid is inserted varies, giving rise to different Hfr strains.
    d. one end—always the same end—of the opened chromosome leads the way into the female.
    e. **All of the above**

48. Episomes are
    a. viral prophages.
    b. nonessential genetic elements.
    c. elements that can exist independently or integrated into the main bacterial chromosome.
    d. *a* and *c*
    e. **a, b, and c**

49. R factors, or resistance factors, carried by plasmids or bacteria may be more widespread now than in the past because
    a. **the presence of antibiotics has stimulated the evolution of R factors.**

b. we are better at detecting them now than in the past.

c. current bacterial culture conditions favor the selection of R factors.

d. *a* and *c*

e. None of the above

50. All the following are advantages to working with bacteria over mice except
    a. bacteria have a smaller amount of DNA.
    b. bacteria are easy to contain.
    c. a large number of individuals can be produced quickly.
    d. bacteria are inexpensive to produce.
    e. bacteria take up very little space.

51. Viruses were first discovered in 1892 when
    a. a filtrate was found to be infectious.
    b. Albert Virus became ill from smoking infected tobacco.
    c. Stanley crystallized some and found that viruses contained protein and DNA.
    d. they were first observed using a light microscope.
    e. they were observed as plaques for the first time.

52. The genetic information of viruses is
    a. DNA.
    b. RNA.
    c. single-stranded.
    d. double-stranded.
    e. All of the above

53. A strain of virus that always lyses the cells it infects is called
    a. temperate.
    b. virulent.
    c. lytic.
    d. lysogenic.
    e. deadly.

54. Lysis of the host cell is caused by
    a. the cells simply bursting due to the amount of viral particles.
    b. the cells opening up in an attempt to dump out the viruses.
    c. a product of a viral gene attacking the cell wall.
    d. a yet unknown mechanism.
    e. None of the above

55. The HIV virus that causes AIDS is a(n)
    a. arbovirus.
    b. double-stranded DNA virus.
    c. single-stranded DNA virus.
    d. porcine virus.
    e. retrovirus.

56. The polio virus and the HIV virus differ because only the polio virus can
    a. produce and use reverse transcriptase.
    b. use RNA as its genetic material.
    c. use RNA as genetic information and mRNA, without generating a DNA molecule.
    d. infect both horses and humans.
    e. All of the above

57. Plants have a tough cell wall that viruses cannot pass through. Viruses enter plants
    a. through wounds.
    b. using insect vectors.
    c. by digesting the cell wall.
    d. *a* and *b*
    e. *b* and *c*

58. Once inside a plant cell, viruses spread by
    a. diffusion.
    b. Brownian motion.
    c. via plasmodesmata.
    d. attaching to shared endoplasmic reticulum.
    e. the movement of water each time it rains.

59. A patient is told that he has contracted a disease that is caused by an arbovirus. He most likely
    a. ate something he should not have.
    b. kissed someone he should not have.
    c. should have used insect repellent.
    d. got it because someone who already had it sneezed around him.
    e. will be contagious.

60. Prions are infectious particles composed of
    a. DNA and protein.
    b. protein.
    c. RNA.
    d. RNA and protein.
    e. DNA.

61. Viroids are composed of
    a. DNA and protein.
    b. protein.
    c. RNA.
    d. RNA and protein.
    e. DNA.

62. It is found that a certain enzyme is synthesized whenever the solution the cells are growing in lack substance X. This is most likely _____ gene regulation.
    a. inducible
    b. positive
    c. negative
    d. repressible
    e. positive–negative

14 The Eukaryotic Genome and Its Expression

Fill in the Blank

1. The association of single-stranded DNA with a complementary RNA molecule is called **nucleic acid hybridization**.

2. Segments of a gene that are transcribed but do *not* encode part of the protein product are called **introns**.

3. DNA sequences called **transposable elements** can replicate and insert themselves into different parts of the genome.

4. Members of a gene family that do not make a functional gene product are called **pseudogenes**.

5. snRNPs are complementary to **consensus** sequences located at the intron–exon boundaries.

6. When a cell begins to synthesize the proteins required for the specialized structure and function of a liver cell, it is undergoing the process of **differentiation**.

7. When you stain a preparation of epithelial cells from a female rat, you observe one highly condensed chromosome in interphase nuclei. This is the inactive X chromosome, or **Barr body**.

8. During the development of amphibian oocytes, the number of genes encoding the rRNA increases. This increase is called **gene amplification**.

9. Those DNA sequences that are *not* represented in the mRNA are referred to as intervening sequences or **introns**.

10. It is possible that some transposable elements arose from viruses called **retroviruses or RNA viruses**.

11. Most **pseudogenes** are probably duplicate genes that changed so much during evolution that they no longer function.

12. During RNA splicing, different snRNPs constitute a "splicing machine" called a **spliceosome**.

13. There are at least three ways in which **transcription** may be controlled: (1) genes may be inactivated, (2) specific genes may be amplified, and/or (3) specific genes may be selectively transcribed.

14. In both prokaryotes and eukaryotes, the **promoter** is a stretch of DNA to which RNA polymerase binds to initiate transcription.

Multiple Choice

1. When DNA is denatured by gentle heating
   a. **the hydrogen bonds between the base pairs break.**
   b. the DNA is broken down into nucleotides.
   c. the covalent bonds in the sugar–phosphate backbone are destroyed.
   d. the bases separate from the sugar–phosphate backbone.
   e. the DNA remains double-stranded.

2. mRNA will form hybrids only with the coding strand of DNA because
   a. DNA will not reanneal at high temperatures.
   b. the salt concentration will affect DNA reannealing.
   c. DNA will not reanneal at low temperatures.
   d. **RNA–DNA hybridization follows the base-pairing rules.**
   e. denatured DNA will not reanneal after it is diluted.

3. When a solution of denatured DNA is slowly cooled, the DNA will
   a. depolymerize.
   b. recombine.
   c. **reanneal.**
   d. hybridize to RNA.
   e. remain single-stranded.

4. When eukaryotic DNA is hybridized with mRNA, the hybrid molecules contain loops of double-stranded DNA. These regions of DNA are called
   a. retroviruses.
   b. **introns.**
   c. exons.
   d. transcripts.
   e. puffs.

5. The regions of DNA in a eukaryotic gene that encode a polypeptide product are called
   a. enhancers.
   b. mRNAs.
   c. hnRNAs.
   d. **exons.**
   e. leader sequences.

6. Exons are
   a. **translated.**
   b. found in most prokaryotic genes.
   c. removed during RNA processing.
   d. *a* and *b*
   e. *a* and *c*

7. What can be learned about a specific DNA sequence by measuring the rate at which it reanneals during liquid hybridization?
   a. Whether it encodes a protein product
   b. Whether it is located at the centromere
   c. **Whether it is a single-copy sequence**
   d. Whether it is an exon
   e. Whether it is an intron

8. Transposable elements can
   a. alter transcription of a gene.
   b. cause a mutation.
   c. cause gene duplication.
   d. **All of the above**
   e. None of the above

9. A DNA sequence that is not expressed but that hybridizes to globin mRNA is an example of
   a. a gene family.
   b. **a pseudogene.**
   c. a transposable element.
   d. a retrovirus.
   e. an exon.

10. The globin _____ includes a number of genes that encode similar proteins.
    a. pseudogene
    b. intron
    c. **gene family**
    d. retrovirus
    e. proto-oncogene

11. Most gene families probably arose by
    a. RNA processing.
    b. RNA splicing.
    c. DNA methylation.
    d. transcriptional regulation.
    e. **gene duplication.**

12. The modified G cap on eukaryotic mRNAs is found
    a. **at the 5′ end.**
    b. at the 3′ end.
    c. in the consensus sequence.
    d. in the poly A tail.
    e. in snRNA.

13. Poly A tails
    a. **are added after transcription.**
    b. are encoded by a sequence of thymines in the DNA.
    c. are found in all mRNAs.
    d. have no function.
    e. are removed during RNA processing.

14. Which of the following is *not* part of RNA processing in eukaryotes?
    a. Splicing of exons
    b. **Reverse transcription**
    c. Addition of a 5′ cap
    d. Addition of a poly A tail
    e. Intron removal

15. The binding of snRNPs to consensus sequences is necessary for
    a. gene duplication.
    b. addition of a poly A tail.
    c. capping an hnRNA.
    d. **RNA splicing.**
    e. transcription.

16. The expression of some genes can be regulated in part by the pattern of RNA splicing. This is an example of
    a. DNA methylation.
    b. transcriptional regulation.
    c. catalytic RNAs.
    d. **posttranscriptional control.**
    e. the endosymbiotic theory.

17. Heterochromatin
    a. contains poly A tails.
    b. **is usually not transcribed.**
    c. does not contain any DNA.
    d. is not replicated during the S phase.
    e. is found only in prokaryotes.

18. You are studying the expression of the globin gene. In neurons, this gene contains many methylated cytosines. What might this mean about the expression of the globin gene in neurons?
    a. It is expressed only in males.
    b. It is expressed only in females.
    c. **It is not expressed.**
    d. It is regulated by posttranscriptional control.
    e. It is regulated by posttranslational control.

19. You stain a preparation of normal rat epithelial cells and examine them under the microscope. Each interphase nucleus contains a single Barr body. What can you conclude about these cells?
    a. The cells are in meiotic prophase.
    b. All of the chromatin in these cells is inactive.
    c. The DNA in these cells has replicated.
    d. The cells are not transcribing any genes.
    e. **The cells came from a female rat.**

20. The cells in *Drosophila* that make eggshell (chorion) proteins increase the number of genes encoding these proteins. This is an example of
    a. **gene amplification.**
    b. DNA methylation.
    c. posttranscriptional control.
    d. RNA splicing.
    e. a gene family.

21. The region of a gene that binds RNA polymerase to initiate transcription is called
    a. an exon.
    b. the consensus sequence.
    c. heterochromatin.
    d. the cap.
    e. **the promoter.**

22. When an enhancer is bound, it
    a. increases the stability of a specific mRNA.
    b. **stimulates transcription of a specific gene.**
    c. stimulates transcription of all genes.
    d. stimulates splicing of a specific mRNA.
    e. stimulates splicing of all mRNAs.

23. Which of the following is *not* a feature of TATA boxes?
    a. They bind a specific transcription factor.
    b. They are found in the region of the promoter.
    c. **They are part of the intron consensus sequence.**
    d. They help specify the starting point for transcription.
    e. They contain thymine–adenine base pairs.

24. Transcription of eukaryotic genes requires
    a. binding of RNA polymerase to the promoter.
    b. binding of several transcription factors.
    c. capping of mRNA.
    d. *a* and *b*
    e. *a, b,* and *c*

25. Transferrin is a protein that transports iron into cells. The iron concentration in a cell regulates the amount of transferrin protein being made, but has no effect on the levels of transferrin mRNA. This is an example of
    a. translational control.
    b. an enhancer.
    c. transcriptional regulation.
    d. RNA processing.
    e. a promoter.

26. Some metabolic pathways are regulated in part by changing the rate of degradation of key enzymes. This is an example of
    a. an operon.
    b. transcriptional control.
    c. liquid hybridization.
    d. feedback inhibition.
    e. posttranslational control.

27. A DNA sequence, which can be distant to the gene, stimulates transcription, when bound by a protein. This sequence is a(n)
    a. TATA box.
    b. operon.
    c. enhancer.
    d. promoter.
    e. consensus sequence.

28. Expression of some eukaryotic genes can be regulated by translational control. What is an advantage of translational control?
    a. It provides a means for rapid change in protein concentrations.
    b. It prevents synthesis of excess RNA.
    c. It directs proteins to their proper subcellular location.
    d. It occurs only in zygotes.
    e. It degrades proteins that are no longer needed.

29. If double-helix DNA breaks down to a single strand when the hydrogen bonds have been disrupted by increasing the temperature, the DNA is then considered to be
    a. complementary.
    b. antiparallel.
    c. denatured.
    d. a template.
    e. a hybrid.

30. Transposable elements are found in both prokaryotes and eukaryotes. One of the major differences between the copying of transposable elements in some eukaryotes and prokaryotes is that
    a. in eukaryotes the elements must be denatured before they can be copied.
    b. in eukaryotes the elements all require an RNA intermediate.
    c. in eukaryotes the elements are not always copied.
    d. in eukaryotes the elements do not need an enzyme.
    e. None of the above

31. Which of the following is a posttranscriptional modification of mRNA found in eukaryotes?
    a. A 5′ cap
    b. A 3′ cap
    c. A poly T tail
    d. A polyadenylation at the 5′ end
    e. None of the above

32. A gene family is a set of genes that over time has changed slightly, extensively, or not at all. Which of the following statements is true for gene families?
    a. They must always be on the same chromosome.
    b. They must always code for the exact same proteins.
    c. One copy must retain the original function.
    d. They usually differ in their exons because these are coding segments.
    e. None of the above

33. Consensus sequences (short segments of DNA) appear in various genes. These sequences appear to be involved in
    a. directing the polymerases to the appropriate place on the DNA for transcription to begin.
    b. the splicing of introns out of the DNA.
    c. allowing the transcription to stop at the appropriate spot.
    d. catalyzing the synthesis of a protein.
    e. None of the above

34. A chemical modification that adds methyl groups to cytosine residues in some genes acts to
    a. enhance transcription.
    b. amplify the gene.
    c. inactivate the gene.
    d. stabilize the mRNA.
    e. None of the above

35. The interphase cells of normal female mammals have a stainable nuclear body called a Barr body. This body is
    a. an inactive X chromosome.
    b. made up of fat droplets.
    c. fragments of mRNA.
    d. extra chromosomal pieces.
    e. None of the above

36. Which of the following best describes the function of the addition of a methylated guanosine cap to the 5′ end of primary mRNA?
    a. It contains all of the coding and noncoding sequences of the DNA template.
    b. It provides the mRNA molecule with a poly A tail.
    c. It facilitates the binding of mRNA to ribosomes.
    d. It forms hydrogen bonds.
    e. It helps transfer amino acids to the ribosomes.

37. A cancerous cell is recognized by the fact that
    a. it is larger than a normal cell.
    b. its division is not regulated.
    c. its intracellular organelles multiply first.
    d. its cell membrane becomes fragile.
    e. None of the above

38. The class of DNA sequences that reanneals the fastest consists of
    a. single copy DNA.
    b. highly repetitive DNA.
    c. moderately repetitive DNA.
    d. a mixture of oligo-copy and single-copy DNA.
    e. None of the above

39. Moderately repetitive DNA is present in
    a. 2–10 copies per genome.
    b. 20–100 copies per genome.
    c. 200–1,000 copies per genome.
    d. a few hundred to 100,000 copies per genome.
    e. about 500 copies per genome.

40. RNA processing in eukaryotes involves
    a. the addition of a G cap.
    b. polyadenylation.
    c. removal of introns.
    d. splicing of exons.
    e. All of the above

41. Remote DNA sequences that bind activator proteins and stimulate specific promoters, thus facilitating the transcription of specific genes, are called
    a. enhancers.
    b. stimulators.
    c. inducers.
    d. allosteric modulators.
    e. derepressor elements.

42. The regulation of gene expression via differential degradation of proteins is an example of
    a. transcriptional control.
    b. translational control.
    c. posttranslational control.
    d. nuclear control.
    e. None of the above

43. DNA methylation
    a. is a mechanism of gene inactivation.
    b. adds methyl groups to cytosine residues in certain genes.
    c. inhibits transcription.
    d. a, b, and c
    e. a and b

44. The three RNA polymerases of eukaryotes catalyze
    a. rRNA synthesis.
    b. mRNA synthesis.
    c. tRNA synthesis.
    d. a, b, and c
    e. None of the above

45. The TATA box is
    a. a sequence common to the promoter region of many genes.
    b. a square-shaped, just-so sequence.
    c. an enhancer consensus sequence.
    d. an activator sequence necessary for proper translation.
    e. None of the above

46. RNA polymerase II by itself cannot bind to the chromosome and initiate transcription. It can bind and act only after the assembly of regulatory proteins called
    a. translation factors.
    b. posttranslation factors.
    c. initiation factors.

    d. transcription factors.
    e. None of the above

47. Mary Lyon, Liane Russell, and Ernest Beutler discovered
    a. the basis of hormone action.
    b. X chromosome inactivation.
    c. Barr bodies.
    d. melanosomes.
    e. heterochromatin.

48. The Barr body is evidence for
    a. X chromosome inactivation.
    b. cell death.
    c. ion pumps.
    d. posttranslational control of eukaryote gene expression.
    e. None of the above

49. The theoretical basis for DNA–DNA or DNA–RNA hybridization studies is
    a. base complementarity between nucleic acid strands.
    b. enzyme action.
    c. DNA and RNA looping.
    d. a and c
    e. None of the above

50. The DNA from a human cell, stretched out, would be ____ in length.
    a. 2 mm.
    b. 2 meters
    c. 2 km
    d. 2 μm
    e. None of the above

51. Human cells have approximately _____ the DNA found in prokaryotic cells.
    a. 10 times
    b. 100 times
    c. 1000 times
    d. 10,000 times
    e. 100,000 times

52. In bacteria, transcription
    a. is separate from translation.
    b. occurs in the nucleus.
    c. is not separated from translation.
    d. is the process of synthesizing proteins.
    e. is continuous for all genes.

53. Eukaryotic DNA is organized into chromosomes that must have
    a. DNA sequences that make up telomeres and centromeres.
    b. proteins that are centromeres and DNA that form telomeres.
    c. a 5' G cap.
    d. ubiquitin bound.
    e. an inactivation center.

54. In eukaryotic cells, transcription and translation
    a. are separated. Translation occurs in the nucleus and transcription in the cytoplasm.
    b. occur together in the cytosol.
    c. occur together in the nucleus.
    d. are separated. Translation occurs in the cytoplasm and transcription in the nucleus.
    e. are separated, except for proteins that bind to the DNA and ribosomes, which are translated in the nucleus.

55. "Beads on a string" describe the structural appearance of
    a. condensed chromosomes.
    b. lampbrush chromosomes.
    c. nucleosomes.
    d. the solenoid structure of DNA.
    e. the 30 nm DNA fibers.

56. Unpacked nucleosomes measure approximately _____ across.
    a. 1 nm
    b. 10 nm
    c. 30 nm
    d. 300 nm
    e. 3,000 nm

57. Highly repetitive DNA, when heated and cooled, will
    a. reanneal more slowly than single-copy DNA.
    b. reanneal more rapidly than single-copy DNA.
    c. reanneal with DNA that is not complementary.
    d. remain denatured until replicated.
    e. anneal to single-copy DNA.

58. Cancer cells are often found to have abnormally high amounts of
    a. nucleosomes.
    b. spliceosomes.
    c. centromereosomes.
    d. topoisomerase.
    e. telomerase.

59. Telomerase is important to eukaryotic cells because
    a. telomeres tend to get shorted with each cell division.
    b. telomeres tend to get longer with each cell division.
    c. telomerase digests telomeres to proper length.
    d. the leading strand of DNA causes the telomeres to shorten.
    e. it aids in making artificial chromosomes.

60. DNA transposons
    a. work by using reverse transcriptase.
    b. are transcribed but not translated.
    c. are also called retrotransposons.
    d. move to new locations from old locations.
    e. are circular DNA molecules.

61. In eukaryotic cells, promoters and terminators are
    a. transcribed.
    b. transcribed and translated.
    c. neither transcribed nor translated.
    d. transcribed and then removed.
    e. sequences of RNA that are spliced out.

62. A pseudogene is one that
    a. has gained a new function.
    b. is expressed in one type of cell but not another.
    c. is expressed in all types of cells.
    d. no longer functions.
    e. is a gene coding for the protein "pseudo."

63. Gene families
    a. always stay together.
    b. arise from duplication events.
    c. are genes that are inherited along family lines.
    d. are genetic pedigrees.
    e. are genes that are activated at the same time.

64. An interesting feature of the globin genes is
    a. different ones are expressed during different stages of prenatal development.
    b. only one copy is functional.
    c. the length of the mRNAs are very different.
    d. they are the result of differential posttranscriptional splicing.
    e. the transcripts are longer than the coding region.

65. Exons are
    a. spliced out of the original transcript.
    b. spliced together from the original transcript.
    c. spliced to introns to form the final transcript.
    d. much larger than introns.
    e. larger than the original coding region.

66. Three processes that must be completed before transcripts can be translated in eukaryotes are
    a. snurping, adding a poly A tail, and introning.
    b. snurping, transporting, and riboseolating.
    c. capping, transporting, and riboseolating.
    d. snurping, capping, and splicing.
    e. splicing, capping, and adding a poly A tail.

67. Snurps (snRNPs) are
    a. exon-intron boundary regions.
    b. small nuclear ribonucleoproteinparticles.
    c. protein fragments removed from the snRNA molecules.
    d. signal ribosomal nuclear proteins.
    e. glucose conjugated trapezoids.

68. Potential control points for regulating the amount of protein synthesized in eukaryotic cells include all of the following except
    a. transcription regulation.
    b. DNA amplification.
    c. transcript processing.
    d. breakdown of the synthesized protein.
    e. stabilization of the mRNA.

69. Although eukaryotic genes are solitary and not transcribed together using the operon system found in prokaryotic cells, coordinated gene expression occurs due to
    a. a single transcript containing the coding region for more than one protein.
    b. response elements that are similar among the genes that are expressed in a coordinated fashion.
    c. sigma and rho factors that switch banks of genes on and off.
    d. estrogen.
    e. TATA boxes.

70. The RNA polymerase that produces mRNA molecules in eukaryotic cells is
    a. RNA polymerase I.
    b. RNA polymerase II.
    c. RNA polymerase III.
    d. primase.
    e. reverse transcriptase.

71. Transcription factors are
    a. RNA sequences that bind to RNA polymerase.
    b. DNA sequences that up regulate transcription.
    c. protein that bind to DNA near the promoter sequence.

d. polysaccharides that bind to the transcripts.
e. factors that bind to enhancers.

72. Coordinated gene expression in eukaryotic cells can occur because
    a. one event often leads to another similar event.
    b. more than one protein is coded for by a transcript.
    c. **of similarities in promoters for different genes.**
    d. of the universal nature of the genetic code.
    e. all promoters are different.

73. Some genes only change transcription rates after DNA replication because
    a. changes occur in the cytoplasm due to cell division.
    b. this is when most proteins are synthesized.
    c. mRNA is generated mostly during this period.
    d. **the removal of histones allows proteins to bind to the DNA.**
    e. this is when new proteins are needed most.

74. Which of the following is a promoter?
    a. Protein
    b. RNA
    c. **DNA**
    d. Carbohydrate
    e. Enzyme

75. Which of the following is an enhancer?
    a. Protein
    b. RNA
    c. **DNA**
    d. Carbohydrate
    e. Enzyme

76. Which of the following is a transcription factor?
    a. **Protein**
    b. RNA
    c. DNA
    d. Carbohydrate
    e. Enzyme

77. In eukaryotic cells, a repressor
    a. is made of DNA.
    b. binds to the enhancer region to block transcription.
    c. is located both upstream and downstream from the promoter.
    d. binds to the operator to block RNA polymerase.
    e. **binds to a silencer to reduce transcription rates.**

78. The heat shock response is an example of
    a. a survival tactic.
    b. a survival strategy.
    c. a way to increase body temperature.
    d. **coordinated gene expression.**
    e. none of the above.

79. DNA binding motifs described in the chapter include all except
    a. **helix–straight–helix.**
    b. helix–turn–helix.
    c. helix–loop–helix.
    d. leucine zipper.
    e. zinc finger.

80. The DNA binding proteins that are associated with activation of genes by steroids have the _____ motif.
    a. helix–straight–helix.
    b. helix–turn–helix.
    c. helix–loop–helix.
    d. leucine zipper.
    e. **zinc finger.**

81. A cell is found that contains three Barr bodies. This cell has
    a. three Y chromosomes.
    b. three X chromosomes.
    c. four Y chromosomes
    d. **four X chromosomes.**
    e. three nucleoli.

82. DNA methylation in eukaryotic chromosomes involves adding a methyl to the
    a. 5′ position of G.
    b. **5′ position of C.**
    c. DNA binding proteins.
    d. RNA molecules.
    e. ribose.

83. One of the genes that is known to be transcribed from the inactive X chromosome is
    a. **Xist.**
    b. Zist.
    c. inactivation controller protein.
    d. lithozist.
    e. methyl-X

84. The interesting characteristic involving the control of tubulin production is that
    a. a translational repressor protein binds to the tubulin mRNA, which prevents ribosomes from attaching.
    b. alternative splicing of pre-mRNA results in the production of several different proteins.
    c. **tubulin can recognize and bind to tubulin mRNA causing an acceleration of its breakdown.**
    d. ubiquitin forms a complex with the tubulin, which causes its breakdown.
    e. the 5′ guanosine cap is added to the mRNA but is not modified so the mRNA is not expressed. When needed, it gets modified.

85. The interesting characteristic involving tropomyosin production is that
    a. a translational repressor protein binds to the tropomyosin mRNA, which prevents ribosomes from attaching.
    b. **alternative splicing of pre-mRNA results in the production of several different proteins.**
    c. tropomyosin can recognize and bind to tropomyosin RNA causing an acceleration of its breakdown.
    d. ubiquitin forms a complex with the tropomyosin, which causes its breakdown.
    e. the 5′ guanosine cap is added to the mRNA but is not modified so the mRNA is not expressed. When needed, it gets modified.

86. The interesting characteristic involving the control of protein synthesis in tobacco hornworms is that
    a. a translational repressor protein binds to the mRNA, which prevents ribosomes from attaching.
    b. alternative splicing of pre-mRNA results in the production of several different proteins.
    c. proteins can recognize and bind to mRNA causing an acceleration of its breakdown.
    d. ubiquitin forms a complex with the mRNA, which causes its breakdown.
    e. **the 5′ guanosine cap is added to the mRNA but is not modified so the mRNA is not expressed. When needed, it gets modified.**

87. The interesting characteristic involving the control of ferritin synthesis is that
    a. **a translational repressor protein binds to the ferritin mRNA, which prevents ribosomes from attaching.**
    b. alternative splicing of pre-mRNA results in the production of several different proteins.
    c. ferritin can recognize and bind to ferritin RNA causing an acceleration of its breakdown.
    d. ubiquitin forms a complex with the ferritin, which causes its breakdown.
    e. the 5′ guanosine cap is added to the mRNA but is not modified so the mRNA is not expressed. When needed, it gets modified.

15 Development: Differential Gene Expression

Fill in the Blank

1. Although we often stress the embryo in discussing animal development, it is a process that continues through all stages of life, ceasing only with **death**.

2. Presumed mechanisms of determination include determination by cytoplasm segregation and determination by **embryonic induction**, in which certain tissues induce the determinations of other tissues.

3. One important lesson to be learned from the two sequential steps of embryonic induction in the vulval cells of *Caenorhabditis elegans* is that much of development is controlled by switches that allow a cell to proceed down **alternative** tracks.

4. Related to mutant homeotic mutations is an important sequence of 180 base pairs of DNA called the **homeobox**, which has been found in both animals and plants, and which encodes a portion of some proteins called a homeodomain.

5. In the concluding paragraphs in our chapter on animal development, we learn how to "build a fly." One important lesson to learn from this exercise is that gene products like bicoid and nanos proteins (morphogens) of the sequentially expressed genes in a given developmental series function as **transcription factors**.

6. The process by which underlying mesoderm in chick embryos causes the differentiation of ectodermal cells into feathers is called **induction**.

7. Although some cells of the mesoderm are destined to become muscle cells, their **developmental potential**, or range of possible development, is greater and includes a number of different tissue types.

8. When the developmental fate of a cell does *not* change, even when the cell's surroundings are altered, the cell is said to be **determined**.

9. A sea urchin egg is said to have **polarity** because the distribution of cytoplasm components is different at one end of the egg than at the other end.

10. In *Caenorhabditis elegans*, germ line granules are present in the cytoplasm of the zygote. At the end of the early cell divisions, only those cells that contain germ line granules will give rise to egg and sperm in the adult. Germ line granules are an example of **cytoplasmic determinants**.

11. The number and polarity of segments formed during the development of insect larvae is determined by three classes of **segmentation** genes.

12. The **homeobox** is a small sequence of DNA that is found in some of the segmentation genes of *Drosophila*, as well as in some genes of all animals with segmented body plans.

13. One factor that regulates pattern formation is **positional** information, signals that indicate where one group of cells lies in relation to other cells.

14. Pattern formation in chicken wings is determined in part by positional information provided by a region called the **zone of polarizing activity**, or ZPA.

15. The process of cells becoming functionally distinct is called cellular **differentiation**.

16. *MyoD1* and homeotic genes are examples of **selector** genes.

17. Substances that are produced in one place and diffuse to another, and cause pattern formation are called **morphogens**.

18. The **anchor** cell controls the fate of six other cells, which are involved in the formation of the vulva in *Caenorhabditis elegans*.

Multiple Choice

1. Development occurs
   a. only during growth of the organism.
   b. **throughout the life of the organism.**
   c. only in nondividing cells.
   d. in ectoderm and endoderm, but not in mesoderm.
   e. only in animals.

2. During cleavage, the number of cells in a developing frog embryo increases. The cytoplasm in these new cells
   a. **comes from the egg cytoplasm.**
   b. is synthesized by the blastomeres.
   c. does not contain any yolk.
   d. is the vegetal pole.
   e. undergoes mitosis.

3. The fly larva is very different in appearance from the adult. The major changes that occur during the development from larva to adult are called
   a. gastrulation.
   b. involution.
   c. **complete metamorphosis.**
   d. discontinuous growth.
   e. neurulation.

4. A leg imaginal disc is removed from its normal position in a *Drosophila* larva and transplanted next to a wing imaginal disc. During metamorphosis, the transplanted disc will
   a. involute.
   b. form a wing.
   c. form a leg.
   d. remain undifferentiated.
   e. transdetermine.

5. A wing imaginal disc from a *Drosophila* larva is transplanted to a new location in the larva. During metamorphosis the transplanted disc still gives rise to a wing. What conclusion can you draw from this experiment?
   a. Imaginal disc cells are not yet determined.
   b. Imaginal disc cells are already differentiated.
   c. The developmental potential of the imaginal disc cells is more limited than their prospective fate.
   d. The developmental potential of the imaginal disc cells is greater than their prospective fate.
   e. The developmental potential of the imaginal disc cells is the same as their prospective fate.

6. Induction of ectoderm to form a lens placode requires
   a. competent ectoderm.
   b. an underlying optic vesicle.
   c. a signal from the optic vesicle.
   d. None of the above
   e. All of the above

7. The nuclear transplantation experiment conducted by Gurdon's group demonstrated that differentiation
   a. is due to changes in gene expression.
   b. in animals is irreversible.
   c. in plants is reversible.
   d. is due to loss of genetic material.
   e. is due to loss of nuclei.

8. Proteins in the egg cytoplasm that play a role in directing embryonic development are called
   a. instars.
   b. imaginal discs.
   c. mesenchyme proteins.
   d. polar proteins.
   e. cytoplasmic determinants.

9. When ectodermal cells are placed next to notochordal mesoderm, they form a neural tube. This is an example of
   a. induction.
   b. gastrulation.
   c. division.
   d. totipotency.
   e. cytoplasmic determination.

10. If cells from the neural tube of a frog embryo are transplanted onto the ventral surface of a second embryo, the transplanted tissue will still go on to develop into tissues of the nervous system. The transplanted cells are
    a. differentiated.
    b. totipotent.
    c. discontinuous.
    d. determined.
    e. endodermal.

11. The _____ genes determine the number and polarity of the segments in an insect larva.
    a. homeobox
    b. determinant
    c. segmentation
    d. positional
    e. involution

12. A mutation in a *Drosophila* homeotic gene might
    a. change the number of larval segments.
    b. change the prospective fate of an imaginal disc.
    c. produce a larva with two heads.
    d. prevent survival past the zygote stage.
    e. change the anterior–posterior organization of segments.

13. Homeobox DNA is found
    a. in all organisms except humans.
    b. only in tomatoes and sea urchins.
    c. only in organisms with segmented body plans.
    d. in both plants and animals.
    e. only in *Drosophila*.

14. The proteins made from genes containing a homeobox
    a. bind to DNA.
    b. are found only in the cytoplasm.
    c. regulate the transcription of other genes.
    d. *a* and *c*
    e. *b* and *c*

15. If the ZPA is removed from a wing bud in a developing chick embryo and retinoic acid is applied in the region of the missing ZPA, what will happen?
    a. The wing bud will develop normally.
    b. The distal part of the wing will be duplicated.
    c. The anterior part of the wing will be duplicated.
    d. The proximal part of the wing will be missing.
    e. Two normal wings will develop from the wing bud.

16. The roundworm *Caenorhabditis elegans* has been used for the detailed analysis of animal development. Which of the following is *not* a characteristic that makes this organism useful for such studies?
    a. The adult contains fewer than 1,000 cells.
    b. Development from the zygote to the adult takes only a few days.
    c. The body is relatively transparent.
    d. Symmetry in the adult body is the result of symmetrical cell divisions.
    e. Mutations in genes that control development have been identified.

17. The developmental fate of mesodermal cells in the chick wing bud is determined, at least in part, by the distance of those cells from the apical ectodermal ridge. This is an example of
    a. homeotic mutations.
    b. positional information.
    c. cytoplasm determinants.
    d. totipotency.
    e. irreversible differentiation.

18. A *Drosophila* in which the segment that normally gives rise to the halteres instead develops into a second pair of wings probably has a _____ mutation.
    a. chronogene
    b. gap gene
    c. homeotic
    d. segmentation gene
    e. cytoplasm

19. When the blastopore dorsal lip is grafted from one frog embryo onto a second embryo, the second dorsal lip will
    a. change the polarity of the adjacent segments.
    b. block gastrulation.
    c. change the developmental fate of the surrounding cells.
    d. change the prospective potency of the surrounding cells.
    e. cause rapid cell division.

20. Nuclear differentiation is characterized as
    a. irreversible in the frog nuclear transplant experiments.
    b. unimportant in the study of regeneration.
    c. a global or universal event in multicellular organisms.
    d. due to a major loss of DNA from chromosomes during cleavage.
    e. limited only to embryonic development through the gastrula stage.

21. P granules that determine which cells are destined to undergo meiosis
    a. are coded for by the gene called *bicoid*.
    b. have a position in the zygote determined by microfilaments.
    c. are responsible for embryonic induction.
    d. are pigment cells found in the developing eye.
    e. All of the above

22. Embryonic induction
    a. cannot explain the formation of the vertebrate eye.
    b. was first described by Briggs and King.
    c. initiates a sequence of differential gene expression.
    d. does not occur in the adult.
    e. is an example of a tissue inducing itself.

23. Imaginal discs
    a. have their fate determined prior to metamorphosis.
    b. cannot be transplanted between embryos.
    c. induce the dorsal blastopore lip.
    d. cannot be transplanted into an adult.
    e. are groups of cells that cannot undergo transdetermination.

24. The _____ is eight or more genes that control the development of the abdomen and posterior thorax of *Drosophila* sp.
    a. bithorax complex
    b. segment polarity
    c. antennapedia complex
    d. pair rule
    e. imaginal disc

25. Body segmentation in *Drosophila* sp. is controlled by which sequence of gene action?
    a. Segment polarity, pair rule, gap
    b. Gap rule, pair true, segment polarity
    c. Gap, pair rule, segment polarity
    d. Bithorax, pair rule, segment polarity
    e. None of the above

26. The homeobox gene complex
    a. codes for a transcriptional regulator.
    b. is found in segmented organisms.
    c. is present in some plants.
    d. is found in some genes that show differential expression in development.
    e. All of the above

27. Homeotic mutations
    a. do not create developmental abnormalities.
    b. affect the number of body segments.
    c. alter eye color in *Drosophila* sp.
    d. are easily studied in the adult.
    e. alter the developmental path of the imaginal disc.

28. Anteroposterior pattern formation in the chick wing is controlled in part by the
    a. bithorax mutant.
    b. homeobox domains.
    c. germ cell granules.
    d. imaginal discs.
    e. zone of polarizing activity.

29. Development of the vertebrate limb requires
    a. cellular differentiation.
    b. interactions between embryonic germ layers.
    c. coding for positional information.
    d. retinoic acid.
    e. All of the above

30. *Caenorhabditis elegans*
    a. has a fixed number of cells in the adult.
    b. has no mutations to help in its analysis.
    c. lacks a zygote.
    d. is a nematode worm of significant research value.
    e. *a* and *d*

31. Which of the following statements about *Caenorhabditis elegans* is *not* true?
    a. It is a worm.
    b. Genetic analysis has been productive.
    c. It has a fixed number of cells in the nervous system.
    d. It can reproduce asexually.
    e. Its life cycle is less than one week.

32. Which of the following statements about eye development is *not* true?
    a. The optic vesicle forms before the lens placode.
    b. The lens placode is lateral to the optic vesicle.
    c. The lens placode undergoes invagination.
    d. The optic cup connects with the optic nerve.
    e. The eye develops because of cytoplasmic determinants.

33. In a "global" sense, mutations in developmental biology
    a. can be explained by changes in DNA.
    b. can alter body segmentation.
    c. help explain congenital malformations in humans.
    d. can be experimentally studied with molecular biology techniques.
    e. All of the above

34. Transplantation has been done
    a. with nuclei from cells of an adult ewe's udder.
    b. with just very small amounts of cytoplasm as the graft.
    c. in mammals.
    d. in frogs.
    e. All of the above

35. Which of the following is *not* true of animal development?
    a. Genes regulate development.
    b. Development occurs by progressive loss of DNA.
    c. Blastomeres are early embryonic cells.
    d. The sea urchin blastopore forms the archenteron.
    e. Cells actively migrate during development.

36. Three major processes that reveal a great deal about animal development include
    a. determination, differentiation, and pattern formation.
    b. blastulation, gastrulation, and physiology.
    c. immunoregulation, neurulation, and histogenesis.
    d. oncogenesis, differentiation, and histogenesis.
    e. None of the above

37. Experiments by Briggs and King and by John Gurdon provided graphic evidence that
    a. neurulation is fixed.
    b. differentiation is not irreversible.
    c. amphibian embryos can cease development for long periods of time.
    d. a, b, and c
    e. a and b

38. The human body has approximately _____ functionally distinct kinds of cells.
    a. 12
    b. 24
    c. 100
    d. 200
    e. 1000

39. The fundamental question of importance to developmental biology that was being addressed using nuclear transplantation was
    a. why a frog embryo develops into a frog and not some other organism.
    b. if cell differentiation is due to loss of DNA from cells as they divide.
    c. if mankind can create many identical people.
    d. the effect of mitochondria on growth rates.
    e. All the above

40. The process of cells organizing to create the form of the multicellular organism is called
    a. morphogenesis.
    b. differentiation.
    c. determination.
    d. restriction.
    e. metamorphosis.

41. Pattern formation is necessary for
    a. morphogenesis.
    b. differentiation.
    c. determination.
    d. restriction.
    e. metamorphosis.

42. As cells become specialized, they must first
    a. be differentiated.
    b. be determined.
    c. lose broad developmental potential.
    d. gain a fate.
    e. become functionally distinct.

43. Once cells becomes fixed in a final functional and physical state, they are
    a. determined.
    b. committed.
    c. differentiated.
    d. totipotent.
    e. morphogized.

44. Differentiation is caused by
    a. loss of DNA.
    b. determination.
    c. morphogenesis.
    d. differential gene expression.
    e. nuclear transplantation.

45. The current theory on why the nucleus of an udder cell "dedifferentiated" to generate the famous lamb Dolly is
    a. the cells were cultured.
    b. the cells were starved and stalled at G1.
    c. the cells were starved and stalled at G2.
    d. the cells were fed and dividing.
    e. the egg failed to have its genetic material properly removed.

46. The fundamental question of importance to developmental biology that was addressed with the Dolly experiment was
    a. if nuclei of mammals irreversibly differentiate.
    b. the effect of cytozymes on cellular differentiation.
    c. if mankind can create many identical people.
    d. the effect of mitochondria on growth rates.
    e. All the above

47. An imaginal disc is removed from a fly larva and transplanted to an adult fly. Later the disc is removed and transplanted back into a larva. The disc no longer generates the appropriate body parts that it would have had it never been removed. This is an example of
    a. morphogenesis.
    b. determination.
    c. leftosis.
    d. incomplete metamorphosis.
    e. transdetermination.

48. Transdetermination provides more evidence that
    a. cells lose DNA during differentiation.
    b. determination depends on how and what genes are expressed.
    c. scientists must be controlled by politics.
    d. None of the above
    e. All the above

49. When the protein from the *MyoD1* gene is injected into a fat cell, the cell
    a. becomes a muscle cell.
    b. becomes a muscle cell only until the protein breaks down.
    c. becomes a hybrid fat–muscle cell called a factual cell.

    d. stays a fat cell because differentiation has already occurred.

    e. becomes a selector cell.

50. The anchor cell influences the differentiation and morphogenesis of several surrounding cells using the mechanism of
    a. cytoplasm determinants.
    b. A causes B.
    c. random events generator.
    ***d.* induction.**
    e. P granules.

51. Programmed cell death is crucial to normal development. The scientific term for this kind of cell death is
    ***a.* apoptosis.**
    b. program X.
    c. terminal differentiation.
    d. death by default.
    e. sonic hedgehog.

52. The genes *ced-3*, *ced-4*, and *ced-9* are all involved with regulating
    a. muscle cell differentiation.
    b. positional information.
    c. morphogenesis.
    d. egg polarization.
    ***e.* apoptosis.**

53. If the gene *ced-9* becomes nonfunctional, the result would be
    a. webbed feet and hands.
    ***b.* death.**
    c. a fetus with no muscle cells.
    d. loss of symmetry.
    e. loss of nerve function.

54. The morphogen produced by ZPA is called
    ***a.* sonic hedgehog.**
    b. ZPA effector protein.
    c. wingazyme.
    d. zone polarizing protein.
    e. None of the above

55. The reason retinoic acid can replace a removed ZPA and induce morphogenesis is
    a. retinoic acid is the morphogen.
    ***b.* retinoic acid stimulates cells to produce the protein coded for by the *sonic hedgehog* gene.**
    c. retinoic acid comes from vitamin A.
    d. the cytoskeleton has kept the cytoplasm determinants in their proper locations.
    e. None of the above

56. The mutant *Drosophila* called Antennapedia
    a. has legs growing out of its head.
    b. grows wings instead of eyes.
    c. is a homeotic mutant.
    ***d. *a* and *c* are correct**
    e. b and *c* are correct

16 Recombinant DNA and Biotechnology

Fill in the Blank

1. A **transgenic** animal has recombinant DNA integrated into its own genetic material.

2. Enzymes that cleave double-stranded DNA at specific sites are **restriction endonucleases**.

3. A short, single-stranded region at the end of a DNA fragment is a **sticky end**.

4. An enzyme that can covalently link two DNA fragments is **DNA ligase**.

5. A **cloning vector** is a virus or plasmid that can replicate its DNA and foreign DNA within a cell without being degraded.

6. A circular cloning vector that replicates autonomously within a cell is a **plasmid**.

7. A cloning vector that replicates and destroys a bacterial cell is a **bacteriophage**.

8. **Complementary DNA** is obtained by reverse transcription of RNA.

9. A collection of DNA molecules that represents a population of RNAs is a **cDNA library**.

10. DNA of the bacterial host is *not* cleaved by its own restriction endonucleases because of the activity of specific **methylases,** which add methyl groups to certain bases within the recognition sites.

11. Restriction endonucleases recognize sites by **a specific sequence of bases**.

12. An organism's DNA is isolated, cut by restriction endonucleases, and inserted into vectors where it is cloned. The result is a collection of clones called a **genomic library**.

13. In order to synthesize cDNA from mRNA, an important retroviral enzyme called **reverse transcriptase** is absolutely essential.

Multiple Choice

1. Restriction endonucleases are used naturally in a bacterial cell to
   a. **defend against foreign DNA.**
   b. cleave large sections of bacterial DNA.
   c. digest extra copies of plasmid DNA.
   d. replicate the bacterial chromosomal DNA.
   e. produce RNA from DNA.

2. Why is the DNA of the host cell *not* cleaved by the restriction endonucleases it produces?

   a. The restriction endonucleases are contained within lysosomes.
   b. The restriction endonucleases can only cleave RNA.
   c. The restriction endonucleases are made on the rough endoplasmic reticulum and exported out of the cell.
   d. **The bacterial DNA is altered by methylation and is not a substrate for restriction endonucleases.**
   e. None of the above

3. If the bacteriophage T7 DNA does *not* contain any *Eco*RI sites, how might an *E. coli* cell protect itself from T7 infection?
   a. The cell sequesters the phage DNA in a vesicle.
   b. **The cell makes a number of restriction endonucleases, each with a different recognition site.**
   c. The cell destroys its ribosomes and protein synthetic machinery.
   d. The cell destroys all of the cellular DNA replication enzymes.
   e. The cell cannot possibly protect itself.

4. A fragment of DNA with "sticky" ends
   a. **can form hydrogen bonds with another fragment with complementary "sticky" ends.**
   b. will be readily degraded in the test tube.
   c. is the starting point for RNA polymerase.
   d. is the recognition sequence for restriction endonucleases.
   e. None of the above

5. The enzyme that can join pieces of DNA together is
   a. RNA polymerase.
   b. DNA polymerase.
   c. **DNA ligase.**
   d. β-galactosidase.
   e. None of the above

6. Which of the following is *not* a usual characteristics of a cloning vector?
   a. It contains at least one restriction endonuclease recognition sequence.
   b. **It must integrate into the host chromosome.**
   c. It must be a relatively small piece of DNA.
   d. It must be able to replicate within the host cell.
   e. It carries a gene for antibiotic resistance.

7. Transfection of plant cells is more difficult than transfection of prokaryotic cells because
   a. plant cells are larger than prokaryotic cells.
   b. plant cells contain more DNA than prokaryotic cells.
   c. **DNA must get through plant cell walls.**
   d. *a* and *b*
   e. *a* and *c*

8. A second screening step is often necessary to find bacteria that have taken up plasmids with foreign DNA inserts because
   a. there are many cells without any plasmids at all.
   b. only the plasmids with foreign DNA are taken up into cells.
   c. **very often a plasmid without foreign DNA is taken up by a cell.**
   d. *a* and *b*
   e. *a* and *c*

9. At present, which of the following is *not* a way to insert DNA into cells?
   a. Microinjection
   b. Enclosure in an artificial membrane
   c. Projectiles coated with DNA
   d. Electroporation
   e. **None of the above**

10. From the list below, choose a reasonable sequence of steps for cloning a piece of foreign DNA into a plasmid vector, introducing the plasmid into bacteria, and verifying that the plasmid and insert are present.
    1. Transform competent cells
    2. Select for the lack of antibiotic resistance gene #1 function
    3. Select for the plasmid antibiotic resistance gene #2 function
    4. Digest vector and foreign DNA with *Eco*RI, which inactivates antibiotic resistance gene #1
    5. Ligate the digested DNA together
    a. **4, 5, 1, 3, 2**
    b. 4, 5, 1, 2, 3
    c. 1, 3, 4, 2, 5
    d. 3, 2, 1, 4, 5
    e. None of the above

11. A genomic library
    a. is a collection of foreign DNA molecules inserted into cloning vectors.
    b. ideally represents the entire genome of an organism.
    c. is a collection of RNA molecules inserted into cloning vectors.
    d. ***a* and *b***
    e. *b* and *c*

12. Sticky ends are "sticky" because they
    a. are single-stranded.
    b. are from one to four bases long.
    c. **are complementary to other sticky ends.**
    d. are poly A tails.
    e. are part RNA and part DNA.

13. The expression of a cloned gene in a bacterium is easily regulated if it is
    a. inserted anywhere in the bacterial chromosome.
    b. incorporated into a phage genome.
    c. spliced into a self-replicating plasmid.
    d. **spliced onto the *lac* operon promoter.**
    e. left free in the bacterial cell.

14. Which of the following is *not* usually a source of DNA for cloning?
    a. Pieces of genomic, chromosomal DNA
    b. DNA made by the reverse transcription of mRNA

    c. **DNA made by the reverse transcription of tRNA**
    d. DNA synthesized in the laboratory
    e. All of the above

15. The production of double-stranded cDNA utilizes
    a. hybridization between the poly A tails of mRNA and oligo dT.
    b. reverse transcriptase.
    c. DNA polymerase.
    d. *a* and *b*
    e. ***a*, *b*, and *c***

16. In the production of double-stranded cDNA, mRNA is used as
    a. **a template.**
    b. a tail.
    c. one strand of the completed cDNA.
    d. *a* and *b*
    e. *a* and *c*

17. DNA migrates in an electric field because
    a. it is positively charged.
    b. it is not charged.
    c. **it is negatively charged.**
    d. it combines with the gel molecules.
    e. it is hydrophobic.

18. Sanger's method of DNA sequencing is based on
    a. chemical modification and cleavage of DNA at specific bases.
    b. **termination of DNA polymerization by ddNTPs.**
    c. DNAs complementarity to RNA.
    d. restriction endonuclease sites.
    e. None of the above

19. The technique of PCR uses
    a. synthetic primers.
    b. reverse transcription.
    c. *Thermus aquaticus* DNA polymerase.
    d. *a* and *b*
    e. ***a* and *c***

20. The disease called crown gall is caused by
    a. **the insertion of a transposable element carried on the Ti plasmid.**
    b. the transcription of the Ti plasmid in the plant cells.
    c. the transfer of bacterial genomes into the plant cell genome.
    d. the rampant multiplication of *A. tumefaciens* bacteria within the plant.
    e. None of the above

21. In order to join a fragment of human DNA to bacterial or yeast DNA, both the human DNA and the bacterial (or yeast) DNA must be first treated with the same
    a. DNA ligase.
    b. DNA polymerase.
    c. **restriction endonuclease.**
    d. DNA gyrase.
    e. None of the above

22. *Eco*RI makes staggered cuts when it cleaves DNA, creating single-stranded tails called "sticky ends." These ends will form a complementary base pair. Which of the following conditions must exist for this to happen?
    a. Specific helicases
    b. High enough temperatures

  c. Methyl groups at each end
  ***d.* Low enough temperatures**
  e. None of the above

23. Genetic engineering is important to botanists, and the goal of this technology in plants is to
  a. alleviate the fears of the public regarding genetic engineering.
  ***b.* isolate beneficial genes from one species and introduce them into another plant species for a superior crop.**
  c. create new species of plants to replace the extinct ones.
  d. cause plant disease.
  e. None of the above

24. The specific purpose for cloning DNA fragments out of plants or animals is to
  a. splice other DNA fragments from dissimilar genomes together.
  ***b.* locate a specific gene and prepare DNA probes.**
  c. collect appropriate vectors for genetic engineering.
  d. prepare synthetic oligonucleotides.
  e. None of the above

25. If a cloned DNA molecule is inserted into either of the antibiotic sites in the pBR322 (amp^r or tet^r), the antibiotic gene then becomes
  a. transferred to a gene upstream of the site on the chromosome.
  b. immediately transcribed into mRNA.
  ***c.* inactivated.**
  d. complemented with a corresponding sequence.
  e. None of the above

26. Electrophoresis separates DNA fragments of different sizes, but this technique does not indicate which of the fragments contains the DNA piece of interest. This problem is solved by
  a. measuring the sizes of the bands on the gel.
  ***b.* removing the bands from the gel and hybridizing with a known strand of DNA complementary to the gene of interest.**
  c. knowing the isoelectric points of the piece in question.
  d. identifying the molecular weights of the fragments in question.
  e. None of the above

27. Which of the following could be used for detecting genetic disorders?
  a. cDNA probes that hybridize with DNA regions that are missing or have mutant nucleotide sequences near or within the DNA sequence responsible for the genetic disorder
  b. PCR amplification of the region responsible for the genetic disorder
  c. Restriction mapping of sequences that have different restriction sites and are associated with the disease
  d. *a* and *c*
  ***e.* All of the above**

28. Polymerase chain reaction is a technique that makes it possible to obtain large quantities of DNA pieces of interest. In order for this technique to be successful, scientists must

  a. have large quantities of DNA to start with.
  ***b.* know a sequence of about 15–20 bases at the 3′ end of the target sequence on each DNA strand.**
  c. grow the culture in large quantities of methionine.
  d. insure that the DNA is from a eukaryotic cell.
  e. None of the above

29. Complementary DNA (cDNA) is made using
  a. synthetic oligonucleotides.
  b. amino acid sequences from structural genes.
  c. DNA as the template.
  ***d.* mRNA as a template.**
  e. None of the above

30. Endonucleases were first identified in
  a. plant cells.
  b. eukaryotic cells.
  ***c.* prokaryotic cells.**
  d. blue-green algae.
  e. None of the above

31. Which of the following processes makes use of the nucleic acid base-pairing rules?
  a. DNA replication
  b. Transcription
  c. Translation
  d. Sequencing of genes
  ***e.* All of the above**

32. A cell or organism that contains foreign DNA inserted into its own genetic material is termed
  ***a.* transgenic.**
  b. polygenic.
  c. engineered.
  d. foreign.
  e. xenophobic.

33. Methods to insert foreign DNA into host cells include which of the following?
  a. Bacterial transformation
  b. Transfection
  c. Electroporation
  d. Microprojectiles
  ***e.* All of the above**

34. Principal sources of genes or DNA fragments used in recombinant DNA work include
  a. genomic libraries.
  b. cDNA samples.
  c. artificially prepared oligonucleotides.
  d.* *a, b,* and *c
  e. *a* and *b*

35. A critical element in the polymerase chain reaction is
  ***a.* a temperature-resistant DNA polymerase.**
  b. good quality RNA.
  c. a bomb calorimeter for the reaction.
  d. *a* and *c*
  e. None of the above

36. An important vector for manipulation of plant genes comes from the bacterium *A. tumefaciens* and is called
  a. an *Eco*RI plasmid.
  b. a raze bacteriophage.
  c. a pangene-site vector.
  ***d.* a Ti plasmid.**
  e. None of the above

37. The advantage of tissue plasminogen activator (TPA) over streptokinase is
    a. streptokinase is too effective.
    b. streptokinase might trigger an immune response.
    c. cost. Streptokinase costs much more than TPA.
    d. *a* and *b*
    e. *b* and *c*

38. The advantage of a viral vector over a plasmid vector is
    a. viral vectors can often carry much larger DNA inserts.
    b. plasmid vectors are unreliable.
    c. viral vectors lack the necessary origins of replication.
    d. viral vectors require less medium.
    e. All of the above

39. A plasmid is isolated, digested with a restriction enzyme, and run on a gel using electrophoresis. The fragments closest to the well where the DNA was loaded are _____ than the fragments farthest away.
    a. shorter
    b. longer
    c. duller
    d. less original
    e. We can only say that they are different.

40. The actual process of splicing DNA fragments into plasmids takes place
    a. in the bacteria.
    b. in the phage.
    c. in a test tube.
    d. All of the above
    e. None of the above

41. A YAC is
    a. a strange Siberian mammal, known to geneticists for its unusual mutations.
    b. a yeast artificial chromosome.
    c. a "chattering" sequence.
    d. prokaryotic vector.
    e. None of the above

42. Computational biology is a new branch of biology that
    a. attempts to calculate mutation frequencies in laboratory microorganisms.
    b. attempts to calculate mutation frequencies of organisms in natural populations.
    c. combines DNA computer databases with evidence collected at crime scenes.
    d. devises rules that will allow predictions of protein structures.
    e. All of the above

43. The most recent advance in DNA sequencing technology, which was a modification to Sanger's method, was
    a. using dideoxynucleotides that have fluorescent tags.
    b. using a laser beam to detect the terminal nucleotide as the DNA is electrophoresed.
    c. the development of a system called automated DNA sequencing.
    d. All of the above
    e. None of the above

44. When human DNA is amplified using PCR including primers for a gene sequence, the product DNA is run on a gel using electrophoresis, and stained specifically for DNA,
    a. a smear will probably be observed.
    b. just a background glow will be observed.
    c. nothing will be observed unless a probe is applied.
    d. bands will probably be observed.
    e. It is impossible to predict.

45. Antisense RNA is
    a. RNA that investigators find confusing.
    b. RNA that makes opposite sense.
    c. RNA that is complementary to a certain mRNA.
    d. DNA.
    e. the noncoding strand of the DNA molecule.

46. The first human drug made using recombinant DNA technology was
    a. glyphosatase.
    b. tissue plasminogen activator.
    c. insulin.
    d. human growth hormone.
    e. erythropoietin.

47. A single hair is found at the scene of a crime. What technology would you use to determine if the hair could have come from the suspect?
    a. PCR
    b. DNA sequencing
    c. Fragment cloning
    d. Probing
    e. Antisense RNA

17 Molecular Biology and Medicine

Fill in the Blank

1. In addition to defective alleles, a second major source of genetic ill health is **chromosomal aberrations**.

2. A genetic condition resulting from an additional chromosome 21 (trisomy 21) is known as **Down syndrome**.

3. The name of an X-linked recessive disorder resulting in muscular deterioration is **Duchenne's muscular dystrophy**.

4. **Hemophilia** is an X-linked recessive trait affecting the clotting of blood.

5. Not all hereditary diseases are products of recessive alleles. **Huntington's disease, hypercholesterolemia, or osteogenesis imperfecta** is an example of an autosomal dominant trait leading to premature death from nervous system degeneration.

6. **Gene therapy** is the process whereby an abnormal gene is replaced by a normal one.

7. Success was short-lived for a treatment of a patient with defective adenosine deaminase genes who had genes transferred to her white blood cells. It might have been more effective to transfer the genes to bone marrow **stem** cells.

8. In testing a fetus for harmful alleles, DNA obtained from amniocentesis can be amplified by an artificial cycling process called **the polymerase chain reaction, or PCR**.

9. The most thoroughly studied example of a disease that arises via somatic mutation is **cancer**.

10. A patient suffers the rupturing of a major blood vessel subsequent to exercise. This patient might have **Marfan** syndrome.

11. An estimated half of spontaneous abortions that occur during the first trimester of pregnancy are attributed to **chromosome abnormalities**.

12. The genomes of **two** species of bacteria and **one** species of yeast have been sequenced.

Multiple Choice

1. The probability of carrying a defective allele for a certain disease depends on
   a. environment.
   b. ancestry.
   c. genetic predisposition.
   d. unknown factors.
   e. None of the above

2. The frequency of cystic fibrosis in the human population is
   a. 1 in 2.5.
   b. 1 in 25.
   c. 1 in 250.
   d. 1 in 2,500.
   e. 1 in 25,000.

3. Tissues affected by cystic fibrosis include the
   a. lungs.
   b. liver.
   c. pancreas.
   d. gut.
   e. All of the above

4. Patients with cystic fibrosis often die in their
   a. 10s and 20s.
   b. 20s and 30s.
   c. 30s and 40s.
   d. 40s and 50s.
   e. None of the above

5. The gene that is affected by cystic fibrosis normally encodes a protein that
   a. stimulates mitochondria.
   b. regulates gene expression.
   c. synthesizes mucus.
   d. encodes a chloride channel in the cell membrane.
   e. None of the above

6. Victims of phenylketonuria cannot metabolize
   a. alanine.
   b. phenylalanine.
   c. glutamic acid.
   d. tryptophan.
   e. tyrosine.

7. The principle consequence of phenylketonuria is
   a. muscle atrophy.
   b. kidney failure.
   c. mental retardation.
   d. skeletal problems.
   e. None of the above

8. Following diagnosis of phenylketonuria, infants are restricted to a diet low in
   a. alanine.
   b. phenylalanine.
   c. glutamic acid.
   d. tryptophan.
   e. tyrosine.

9. Among the African-American population, the frequency of sickle-cell anemia is about
   a. 1 per 2,000.
   b. 1 per 600.
   c. 1 per 60.

d. 1 per 20.
e. None of the above

10. Individuals homozygous for the sickle-cell trait produce abnormal
    a. keratin.
    b. hemoglobin.
    c. myosin.
    d. tyrosinase.
    e. None of the above

11. The gene associated with Duchenne's muscular dystrophy
    a. codes for a protein called dystrophin.
    b. is recessive.
    c. encodes components of plasma membranes of skeletal muscle cells.
    d. is X-linked.
    e. All of the above

12. The most common form of inherited mental retardation is
    a. Tay-Sachs disease.
    b. cystic fibrosis.
    c. Down syndrome.
    d. fragile-X syndrome.
    e. None of the above

13. Which of the following genetic disorders is associated with triplet repeats?
    a. Fragile-X syndrome
    b. Myotonic dystrophy
    c. Huntington's disease
    d. All of the above
    e. None of the above

14. Triplet repeats
    a. occur in certain genetic disorders.
    b. are not limited to disease genes.
    c. expand readily in human genes but not in nonhuman genes.
    d. a and b
    e. a, b, and c

15. Triplet repeats are the subject of intense scientific research. Which of the following statements describes what we know about triplet repeats?
    a. Triplet repeats probably play important roles in normal alleles, but we have yet to identify those roles.
    b. Triplet repeats encode certain proteins.
    c. Triplet repeats are found on every human gene.
    d. a and c
    e. None of the above

16. Which of the following is a step that scientists take to deal with genetic disease?
    a. Characterize symptoms of the disease.
    b. Develop epidemiological data.
    c. Define the pattern of inheritance.
    d. Move from the Mendelian to the molecular level of analysis.
    e. All of the above

17. Differences in RFLP banding patterns indicate that
    a. the two different DNAs being tested possess different base pairs.
    b. mRNA is not transcribed.

c. the genes map to different chromosomes.
d. a and c
e. None of the above

18. RFLP is
    a. restriction fragment length polymorphism.
    b. inherited in a Mendelian fashion.
    c. can be used as a genetic marker.
    d. can be useful to help define a discrete gene.
    e. All of the above

19. When a patient with defective adenosine deaminase (ADA) was treated, which of the following steps was performed for gene therapy?
    a. Leukocytes were obtained from the patient.
    b. Leukocytes were transferred to culture dishes.
    c. Leukocytes were transfected with normal ADA genes.
    d. The transfected cells were returned to the patient.
    e. All of the above

20. Human gene therapy requires
    a. gene isolation.
    b. introduction of DNA into target cells.
    c. inclusion of a promoter sequence.
    d. a and b
    e. a, b, and c

21. Which of the following is an ethical issue that arises from screening fetuses for genetic diseases?
    a. The controversial option of abortion may be recommended when screening detects a defective allele in the fetus.
    b. The question of privacy of screening results, and the debate over who has rightful access to those results
    c. Should we humans be "playing God" with our genetic makeup in the first place?
    d. How will we determine which diseases "merit" gene therapy, and which genetic conditions warrant screening?
    e. All of the above

22. In addition to inherited genetic diseases, _____ can also undergo mutations that can result in genetic diseases that are *not* inherited.
    a. somatic cells
    b. germ cells
    c. white blood cells
    d. connective tissue cells
    e. All of the above

23. How do all forms of cancer differ from other diseases?
    a. Control of cell division is lost in cancerous cells: they divide rapidly and continuously.
    b. Cancers arise in several different tissues.
    c. Cancer cells metastasize.
    d. a, b, and c
    e. a and c

24. Noncancerous tumors that remain in place are termed
    a. malignant.
    b. benign.
    c. harmless.
    d. innocuous.
    e. None of the above

25. The spread of tumor cells through the body is termed
    a. cell focusing.
    b. metastasis.
    c. benign neglect.
    d. orientation.
    e. None of the above

26. Metastasis proceeds via
    a. the cancer cells' extension into surrounding tissues.
    b. the cancer cells' differentiation into normal cells.
    c. the cancer cells' entrance into the bloodstream or the lymphatic system.
    d. a, b, and c
    e. a and c

27. Carcinomas are defined as cancers that arise in
    a. muscle.
    b. bone.
    c. the liver.
    d. the lung, breast, colon, or liver.
    e. None of the above

28. Sarcomas are cancers of
    a. the brain.
    b. the skin.
    c. bone, blood vessels, or muscle tissue.
    d. the lung, breast, colon, or liver.
    e. None of the above

29. Lymphomas and leukemias affect the cells
    a. that give rise to the white and red blood cells.
    b. of the brain.
    c. of blood vessels.
    d. of connective tissue.
    e. of muscle and connective tissue.

30. The first cancer-causing virus to be identified was
    a. Epstein-Barr virus.
    b. T cell leukemia virus.
    c. Rous sarcoma virus in chickens.
    d. porcine sarc virus.
    e. benzene 1 virus.

31. Which of the following factors is thought to be involved in the development of cancer cells?
    a. Viruses
    b. Chemicals
    c. Ultraviolet radiation and X rays
    d. Excessive exposure to sunlight
    e. All of the above

32. The development of cancer cells results from changes in genes required for
    a. collagen synthesis.
    b. normal growth.
    c. the production of keratin sulfate.
    d. the production of myosin and actin.
    e. None of the above

33. Eighty-five percent of all human cancers fall into which of the following categories?
    a. Carcinomas
    b. Lymphomas
    c. Leukemias
    d. Sarcomas
    e. b and c

34. Agents that can cause mutations in the DNA of host cells include
    a. chemical carcinogens.
    b. radiation.
    c. tumor-induced viruses.
    d. enzyme kinetics.
    e. a, b, and c

35. Cell division is regulated in part by a group of proteins that circulate in the blood and trigger the normal division of cells. These proteins are called
    a. follicle-stimulating hormones.
    b. erythropoietins.
    c. anabolic steroids.
    d. growth factors.
    e. None of the above

36. Cancers originate in the activities of normal but potentially cancer-producing alleles called
    a. homeoboxes.
    b. hormones.
    c. retroalleles.
    d. a and c
    e. proto-oncogenes.

37. An oncogene arises by a mutation from a
    a. normal gene that controls some aspect of the normal development of the cell.
    b. proto-oncogene.
    c. class of cancer genes.
    d. a and b
    e. None of the above

38. Which of the following is a treatment for cancer?
    a. Allow the body's immune system to destroy the tumor.
    b. Radiotherapy
    c. Surgery
    d. Chemotherapy
    e. b, c, and d

39. A mutation that causes an amino acid change in one of the hemoglobin subunits
    a. always causes a disease when an individual is homozygous for the mutant allele.
    b. always causes a disease even when an individual is heterozygous.
    c. sometimes causes a disease when the individual is homozygous.
    d. is always a dominant mutation.
    e. None of the above

40. The human genome is expected to code for _____ different coding regions.
    a. 1,000
    b. 10,000
    c. 100,000
    d. 1,000,000
    e. 10,000,000

41. The reason most serious genetic diseases are rare is
    a. each person is unlikely to carry any genetic disease.
    b. each person is unlikely to carry the same mutant allele as the person with whom he or she mates.
    c. genetic diseases are usually dominant.
    d. genetic diseases are usually not serious.
    e. None of the above

42. Metabolic diseases such as PKU and alkaptonuria are caused by
    a. **a nonfunctional or missing enzyme.**
    b. an abnormal number of chromosomes.
    c. a mutation in the mitochondrial DNA.
    d. a virus.
    e. an abnormal structural protein.

43. Although the exact cause of the mental retardation associated with PKU is not known, what is known is that
    a. it occurs when the X chromosome breaks into pieces.
    b. **high levels of phenylalanine are involved.**
    c. it is currently untreatable.
    d. All of the above
    e. None of the above

44. The first genetic disease for which an amino acid abnormality was tracked down was
    a. PKU.
    b. alkaptonuria.
    c. fragile-X syndrome.
    d. **sickle-cell anemia.**
    e. hemophilia.

45. Sickle-cell anemia is actually the result of
    a. a deletion.
    b. a nonsense mutation.
    c. a frameshift mutation.
    d. **a base pair substitution.**
    e. a chromosomal deletion.

46. If a variant for a protein can be found at least 2% of the time, the protein is
    a. mutant.
    b. **polymorphic.**
    c. an RFLP.
    d. a marker.
    e. All of the above

47. Familial hypercholesterolemia is a genetic disease
    a. that causes elevated cholesterol levels in the blood.
    b. in which a liver cell membrane receptor is defective.
    c. that leads to higher likelihood of strokes.
    d. **All of the above**
    e. None of the above

48. An example of a genetic disease that causes a defect not in an enzyme, but in a structural protein is
    a. sickle cell disease.
    b. hemophilia.

    c. **Duchenne's muscular dystrophy.**
    d. PKU.
    e. All of the above

49. "Hot spots" are locations where
    a. there are cytosine residues.
    b. **there are 5-methylcytosine residues.**
    c. there are adenine residues.
    d. there is guanidine.
    e. All bases are equally prone to mutation.

50. Fragile-X associated mental retardation occurs when the number of CGG repeats
    a. declines to 6–54.
    b. increases to 52–200 copies.
    c. **increases to 200–1,300 copies.**
    d. All of the above
    e. b and c

51. The basic concept of the "two hit" hypothesis is that to get cancer
    a. **both copies of a tumor suppressor gene must mutate.**
    b. one tumor suppressor and one proto-oncogene must mutate.
    c. both copies of a proto-oncogene must mutate.
    d. All the above
    e. None of the above

52. The tumor suppressor gene, p53, codes for a product that
    a. kills cancerous cells.
    b. causes apoptosis.
    c. **stops cell division during G1.**
    d. triggers an immune response.
    e. All the above

53. Today, sufferers of hemophilia A are treated with
    a. screened human-derived blood products.
    b. porcine blood products.
    c. a yeast clotting factor.
    d. **a product generated from recombinant DNA technology.**
    e. None of the above

54. Of the approximately 30,000 human genes that have been isolated and partially sequenced, about _____ are genes for basic metabolism.
    a. **40%**
    b. 22%
    c. 12%
    d. 60%
    e. 5%

18 Natural Defenses against Disease

Fill in the Blank

1. The **humoral immune response** is a specific defense system against pathogens that is carried out by antibodies in the blood.

2. The blood fluid that carries leukocytes but not red blood cells is **lymph**.

3. **T cell receptors** are specific molecules on the surface of T cells that react to antigenic determinants.

4. **Immunological tolerance** is the term that describes the acceptance of foreign tissue.

5. The fusion of lymphocytes and myeloma cells in culture produces **hybridomas**, which make monoclonal antibodies.

6. When the immune recognition of self fails, an **autoimmune disease** results.

7. HIV-I, the retrovirus that causes AIDS, uses the enzyme **reverse transcriptase** to make a DNA copy of the viral genome.

8. The **constant** regions of the heavy chains of the antibody molecule determine whether the antibody remains part of the cell's plasma membrane or is secreted into the bloodstream.

9. The concept that antigenic determinants stimulate clones of B cells that were already making specific antibodies against those antigens is called the **clonal selection theory**.

10. The ability of the human body to remember a specific antigen explains why **immunization** has eliminated diseases such as smallpox, diphtheria, and polio.

11. Highly specified protein molecules called **antibodies (or immunoglobulins)** carry out the humoral immune response against invaders in the fluids.

12. Two broad groups of cells, the **B cells**, which sometime differentiate to produce antibodies, and the **T cells**, of which some types are involved with elimination of virus-infected cells, are the important cells of the immune system.

13. Tears, nasal drips, and saliva possess an enzyme called **lysozyme** that degrades the cell walls of many bacteria.

14. Types of defenses that provide general protection against a wide variety of pathogens are classified as **nonspecific** defenses.

Multiple Choice

1. The bacteria *E. coli* live in our large intestines and do not normally cause disease. These microorganisms are called
   a. pathogens.
   b. antibodies.
   c. normal flora.
   d. phagocytes.
   e. the complement system.

2. Which of the following is not one of the first lines of defense against invading pathogens?
   a. Skin
   b. Mucus secretion
   c. Lysozyme in tears
   d. T cell receptors
   e. Low pH of the stomach

3. Phagocytes kill pathogenic bacteria by
   a. endocytosis.
   b. production of antibodies.
   c. complement fixation.
   d. T cell stimulation.
   e. inflammation.

4. Which of the following is a nonspecific defense mechanism used to protect animals against pathogenic microorganisms?
   a. Sealing off the damaged tissue
   b. Production of phytoalexins
   c. Production of antibodies
   d. Humoral immune response
   e. Inflammation

5. An individual with influenza is unlikely to develop a second viral infection. This is because the infected cells are producing a glycoprotein called
   a. phytoalexin.
   b. interferon.
   c. immunoglobulin.
   d. antigen.
   e. IgG.

6. The humoral immune system primarily acts against
   a. intracellular viruses.
   b. circulating bacteria.
   c. tissue transplants.
   d. a and b
   e. a and c

7. Blood is a fluid tissue with a noncellular fluid called
   a. lymph.
   b. leukocyte.
   c. plasma.
   d. lymphocyte.
   e. immunoglobulin.

8. B cells will react
   a. nonspecifically with any foreign matter they encounter.
   b. only with tissue transplants.
   c. with all of the antigenic determinants on a specific antigen.
   d. with only one specific antigenic determinant on an antigen.
   e. only with specific antibody molecules.

9. When an individual is first exposed to the smallpox virus, there is a delay of several days before significant numbers of specific antibody molecules and T cells are produced. However, a second exposure to the virus causes a large and rapid production of antibodies and T cells. This is an example of
   a. antigenic determinants.
   b. phytoalexins.
   c. phagocytosis.
   d. interferon production.
   e. immunological memory.

10. When an animal encounters an antigen for the second time, it is capable of producing a massive and rapid immune response to the antigen. The cells responsible for this rapid response are called
    a. memory cells.
    b. effector cells.
    c. humoral cells.
    d. immunization cells.
    e. antigenic cells.

11. According to the clonal selection theory,
    a. antibodies are not produced until the animal encounters a specific antigen.
    b. antigens determine the three-dimensional structure of antibodies.
    c. all B cells have identical genotypes.
    d. an antigen stimulates the proliferation of a specific group of B cells.
    e. B cells give rise to specific T cells.

12. One explanation for the absence of antiself lymphocytes in the bloodstream is
    a. the presence of memory cells.
    b. immunological memory.
    c. clonal deletion.
    d. interferon production.
    e. destruction by natural killer cells.

13. Immunological tolerance occurs
    a. after an exposure to an antigen early in development.
    b. when an antigen has no antigenic determinants.
    c. during the clonal growth of B cells.
    d. as a result of class switching.
    e. when interleukins activate T cells.

14. Failure to distinguish "self" from "nonself" can result in
    a. clonal deletion.
    b. the production of suppressor T cells.
    c. the development of AIDS.
    d. an autoimmune disease.
    e. a deficiency in complement proteins.

15. When a T cell is activated by an antigen, it
    a. secretes antibodies.
    b. proliferates.
    c. dies.
    d. becomes a hybridoma.
    e. becomes a plasma cell.

16. Which of the following is not a characteristic of plasma cells?
    a. They arise from B cells.
    b. They are effector cells.
    c. They secrete antibodies.
    d. They survive in the animal for many years.
    e. They have large amounts of endoplasmic reticulum.

17. Hay fever is an allergic response to pollen. What type of antibody molecule is being produced?
    a. IgG
    b. IgM
    c. IgD
    d. IgA
    e. IgE

18. Monoclonal antibodies
    a. recognize a single antigenic determinant.
    b. are produced in the spleen.
    c. are produced by animals injected with a single antigen.
    d. are memory cells.
    e. are a complex mixture of different antibody classes.

19. The region of the antibody that binds to the antigen is
    a. the constant region of the heavy chain.
    b. the constant region of the light chain.
    c. the variable region.
    d. a and b
    e. b and c

20. Since the joining and random deletion mechanisms only account for a part of antibody diversity, what else is involved in producing vast numbers of unique immunoglobulins?
    a. Mutation
    b. Inversion
    c. Cell fusion
    d. Cell surface proteins
    e. None of the above

21. The genes for immunoglobulin heavy chains are _____ the genes for the light chains.
    a. located on the same chromosomes as
    b. spliced together with
    c. expressed when the cells are exposed to antigens, as are
    d. located on different chromosomes from
    e. rearranged during development as are

22. A plasma cell is producing IgM molecules that recognize an antigenic determinant on an influenza virus. After several days, the cell begins to produce IgG molecules that recognize the same antigenic determinant. This is called
    a. activation.
    b. RNA splicing.
    c. gene mutation.
    d. class switching.
    e. an autoimmune disease.

23. T cell receptors recognize and bind
    a. T$_C$ cells.
    b. B cells.
    c. processed antigens.
    d. T$_H$ cells.
    e. IgM antibodies.

24. The class I MHC proteins are _____ in mice.
    a. secreted by B cells
    b. only found on T cells
    c. on the surface of all nucleated cell types
    d. only produced early in development
    e. the same in all individuals of the same strain

25. The retrovirus HIV specifically destroys TH cells and thus disrupts
    a. the humoral immune response.
    b. the cellular immune response.
    c. both the humoral and cellular immune responses.
    d. the inflammatory response.
    e. the complement cascade.

26. Which of the following cells is not normally involved in the functioning of the immune system?
    a. Phagocytes
    b. Red blood cells
    c. Lymphocytes
    d. B cells
    e. T cells

27. Which of the following is not associated with a non-specific defense mechanism?
    a. Inflammation
    b. Mucous membranes
    c. Interleukins
    d. Phagocytes
    e. Interferons

28. Which of the following is not an adaptation for preventing a pathogen from penetrating the body surface?
    a. Presence of a normal flora
    b. Sneeze reflex
    c. Mucus-covered body surfaces
    d. Low pH
    e. Immunological tolerance

29. Which of the following activities is not normally involved in controlling pathogens from infecting the mucous membranes of vertebrate animals?
    a. Beating cilia
    b. Lysozyme production
    c. Secretion of HCl
    d. Production of bile salts in the small intestine
    e. Secretion of interferon

30. One of the following is not a characteristic of interferons. Select the exception.
    a. Bind to receptors on cell surfaces
    b. All are glycoproteins
    c. Prevent viral replication
    d. Found only in mammals
    e. Confer a generalized resistance to viral diseases

31. Select the following feature that is not shared by B cells and T cells.
    a. A kind of lymphocyte
    b. Found in the lymph
    c. Antibody secreting
    d. Give rise to both effector and memory cells
    e. Originate in bone marrow

32. Select the following feature that is characteristic of both the humoral and cellular immune responses.
    a. Leads to elimination of infected cells
    b. Cell–cell communication via interleukin
    c. Involved in immunological memory
    d. Activation of the complement system
    e. Activation of phagocytosis by macrophages

33. Which of the following cellular immune components is the functional equivalent of an immunoglobulin?
    a. T cell receptor
    b. An antigenic determinant
    c. An antibody
    d. Processed antigen bound to class II MHC proteins
    e. A complement cascade

34. IgG antibodies can bind to
    a. specific antigens.
    b. other IgG antibodies.
    c. B cells.
    d. macrophages.
    e. More than one of the above

35. A foreign cell can
    a. act as just one antigen.
    b. have only one antigenic determinant.
    c. elicit and bind only one antibody.
    d. lead to the production of more than one immunoglobulin.
    e. lead to the selection of only a single B cell.

36. A fundamental postulate of the clonal selection theory is that
    a. the antigen specifies the structure of the antibody directed against it.
    b. an antigen can only lead to the selection of a single line of B cells.
    c. a B cell makes only one specific immunoglobulin.
    d. the production of diverse antibodies was not genetically based.
    e. a mechanism must exist to prevent the production of antiself lymphocytes.

37. Which of the following statements about immunological memory is true?
    a. The time interval between the first and second exposure to the antigen cannot exceed three years.
    b. Antibody production after the second exposure is less because the antibodies are now specific to the antigen.
    c. The class of antibody produced after the second exposure differs from that produced after the first exposure.
    d. If an organism is exposed to a different antigen at the same time as its second exposure to the original antigen, antibody production against the new antigen will be enhanced.
    e. Immunological memory can be passed from parents to offspring.

38. Which of the following statements about the clonal selection theory is *not* true?
    a. An enormous variety of B cells exists in an animal.
    b. **All B cells have the same genotype, although the expression of that genotype always differs.**
    c. Differences in B cell specificities exist prior to an encounter with antigen.
    d. Exposure to an antigen leads to the production of both memory and effector cells.
    e. An antigen activates a preexisting lymphocyte with receptors for it.

39. Which of the following is *not* specifically involved with attempts to explain the body's recognition of self?
    a. **Monoclonal antibodies**
    b. The clonal deletion theory
    c. Presence of self-identifying cell surface proteins
    d. Immunological tolerance
    e. Nonidentical twin cattle with blood of mixed types

40. What would happen if a newborn mouse of strain A was injected with lymphoid cells of strain B, and later as an adult received a skin graft from closely related strain C?
    a. The mouse would die.
    b. The skin graft would be accepted.
    c. **The skin graft would be rejected.**
    d. The skin graft would be accepted if strain C is closely related to strain A.
    e. The skin graft would be accepted if strain C is unrelated to strain A.

41. Which of the following features is *not* characteristic of the structure of an immunoglobulin such as IgG?
    a. A protein molecule with a quaternary structure
    b. **Subunits held together by hydrogen bonding**
    c. A tetramer with two heavy chains and two light chains
    d. Variable and constant regions in each subunit
    e. Two antigen binding sites

42. Which of the following features is *not* characteristic of an immunoglobulin?
    a. The variable regions determine the type of antigen that will bind.
    b. The constant regions determine the class of the immunoglobulin.
    c. Disulfide bonds occur within and between the polypeptide chains.
    d. **The antigen binding sites are formed by the variable portions of the light chains only.**
    e. The two halves of an immunoglobulin are identical.

43. Which of the following features is characteristic of the immunoglobulin class IgG?
    a. Found in blood immediately after first exposure to the antigen
    b. Involved in inflammation and allergic reactions
    c. Form multi-immunoglobulin complexes
    d. **Produced in greatest amount after the second exposure to the antigen**
    e. Found in saliva, tears, milk, and gastric secretions

44. Which of the following features is characteristic of the immunoglobulin class IgM?
    a. Able to bind to receptors on macrophages
    b. Involved in inflammation and allergic reactions
    c. **Form multi-immunoglobulin complexes**
    d. Produced in greatest amount after the second exposure to the antigen
    e. Found in saliva, tears, milk, and gastric secretions

45. Which of the following features is *not* characteristic of the complement system?
    a. Consists of 20 different proteins
    b. Involved in a cascade of reactions
    c. Forms a lytic complex
    d. **Able to lyse foreign cells in the absence of antibody**
    e. Interacts with phagocytes to promote endocytosis

46. If a plasma cell that started out producing antibodies of class IgM undergoes class switching, it would next produce
    a. **IgG antibody.**
    b. IgA antibody.
    c. IgE antibody.
    d. IgD antibody.
    e. Any of the above

47. Which of the following features is *not* characteristic of T cell receptors?
    a. Consist of two polypeptide chains
    b. Are glycoproteins
    c. **Are able to bind to free antigen**
    d. Are able to bind MHC proteins
    e. Have constant and variable regions

48. Which of the following is *not* a normal activity of helper T cells?
    a. **Release of lytic signals when bound to processed antigen on the surface of a virus-infected cell**
    b. Binding to class II MHC and processed antigen on the surface of macrophage
    c. Binding to class II MHC and processed antigen on the surface of B cell
    d. Release of helping signals
    e. Proliferation and differentiation into memory and effector cells

49. With respect to AIDS, which of the following statements is true?
    a. Chances of getting the disease is far greater if one of the partners already has a sexually transmitted disease.
    b. AIDS stands for "autoimmune deficiency syndrome."
    c. The HIV virus of AIDS does not kill people directly.
    d. *a, b,* and *c*
    e. **a and c**

50. To ensure survival, pathogenic organisms must
    a. enter a host.
    b. multiply in the host.
    c. prepare to infect the next host.
    d. **a, b, and c**
    e. *a* and *b*

51. _____ is the generalized bodily response to infections and is accompanied by redness, swelling, heat (increased temperature), and pain.
    a. Shock
    b. **Inflammation**
    c. DNA repair
    d. AIDS
    e. None of the above

52. Interferons
    a. bind to cell receptors.
    b. are glycoproteins.
    c. inhibit the ability of viruses to replicate.
    d. are the subject of intense research because of their potential medical applications.
    e. **All of the above**

53. The immune system has two general types of responses against invaders,
    a. **the humoral immune response and the cellular immune response.**
    b. the humoral immune response and the antihumoral immune response.
    c. complementation and clonal deletion.
    d. the cellular immune response and the antihumoral immune response.
    e. None of the above

54. Antibodies are produced by
    a. fibroblasts.
    b. mesenchyme cells.
    c. adrenal cortex cells.
    d. cells of the anterior pituitary called antibodycytes.
    e. **plasma cells.**

55. A person can produce _____ of distinct antibodies directed against antigenic determinants that it has never encountered.
    a. dozens
    b. hundreds
    c. thousands
    d. **millions**
    e. billions

56. There appear to be two mechanisms of self-tolerance in the immune system. They are
    a. clonal selection and clonal deletion.
    b. **clonal deletion and clonal anergy.**
    c. clonal proliferation and suppressor T cell action.
    d. suppressor T cell action and clonal selection.
    e. None of the above

57. In order to synthesize antibodies, a plasma cell must
    a. be activated by the binding of specific antigens to the antibodies carried on the B-cell surface.
    b. interact with a helper T cell.
    c. develop an extensive endoplasmic reticulum and Golgi complex.
    d. a and b
    e. **a, b, and c**

58. The immunoglobulin class IgA is found
    a. circulating in the blood.
    b. **in mucus secretions.**
    c. near the surface of the skin associated with mask cells.

d. shortly after first exposure to an antigen.
e. All of the above

59. Nonspecific responses include all of the following components except
    a. macrophages.
    b. **antibodies.**
    c. eosinophils.
    d. neutrophils.
    e. interferons.

60. All but which of the following are necessary for a humoral response?
    a. T_C cells
    b. T_H cells
    c. B cells
    d. macrophages
    e. antigenic determinant

61. Which of the following is *not* a characteristic of an inflammatory reaction?
    a. Release of histamine
    b. Invasion of the region by phagocytes
    c. **Binding of IgG**
    d. Dilation of capillaries
    e. Escape of blood plasma

62. To make a hybridoma, a plasma cell is fused to a _____ cell.
    a. carcinoma
    b. B
    c. macrophage
    d. **myeloma**
    e. liver

63. To produce monoclonal antibodies, hybridoma cells are
    a. transferred to animals.
    b. transferred to culture.
    c. placed in deep freeze.
    d. **a and b**
    e. b and c

64. An important organ that helps to protect against autoimmune disease is
    a. the kidney.
    b. the spleen.
    c. **the thymus.**
    d. the adrenal gland.
    e. the tonsils.

65. Allergies involve
    a. **IgE**
    b. mast cells.
    c. basophils.
    d. histamine release.
    e. All of the above

19 *The History of Life on Earth*

Fill in the Blank

1. A massive extinction occurred at the end of the **Permian** period. It may have been caused over a long period of time (ten million years) by the coalescing of the continents into the supercontinent, Pangaea.

2. The period within the Mesozoic era in which frogs, salamanders, and lizards first appeared, and in which one lineage of dinosaurs gave rise to birds, is termed the **Jurassic** period.

3. The Earth's crust consists of solid **plates** approximately 40 kilometers thick that float on a liquid mantle. Often, when these structures come together, one slides under the other, creating **mountain ranges**. The movement of these structures and the continents they contain has had enormous effects on **climates** and **sea levels** and thus on the distributions of organisms.

4. Many patterns in the fossil record suggest that there are frequently long periods during which rates of morphological evolutionary change are extremely slow. These periods are called **stasis**.

5. There have been **six** mass extinctions in the history of life, and **three** periods of rapid diversification of organisms.

6. There is a general tendency for organisms to **increase** in size through the fossil record.

7. **Fauna** is the term for all the species of animals in a certain area; **flora** is the corresponding term for all the species of plants.

8. During the Jurassic period, the supercontinent Pangaea broke up into two continents called **Laurasia** and **Gondwanaland.**

9. Some ancient insects have been perfectly preserved in **amber** formed by tree sap.

Multiple Choice

1. Which of the following is *not* true concerning the history of the evolution of life?
   a. Organisms have increased in size.
   b. Organisms have increased in complexity.
   c. The number of species has increased.
   d. Life evolved early in the history of Earth.
   e. **There are more varieties of animal body plans now than there were 600 million years ago.**

2. The movements of continents were important causes of extinctions during the history of life on Earth. Which of the following results of continental movements was important in these extinctions?
   a. Changes in sea levels
   b. Separation of biotas
   c. Mixing of biotas
   d. **Changes in climates**
   e. All of the above

3. Over much of its history, the climate of Earth was
   a. about the same as it is today.
   b. considerably cooler than it is today.
   c. **considerably warmer than it is today.**
   d. unknown; we have no information about climates in the past.

4. Conditions for the preservation of organisms (fossilization) are best in environments _____ , and ecological communities in _____ environments are richest in species.
   a. **lacking oxygen; well oxygenated**
   b. with high levels of oxygen; well oxygenated
   c. lacking oxygen; poorly oxygenated
   d. with high levels of oxygen; poorly oxygenated
   e. moderate levels of oxygen; poorly oxygenated

5. The fossil record of horses depicts gradual changes over time in North America; however, new forms of horses appear suddenly in Asia. Which of the following hypotheses about horse evolution is accepted by most scientists?
   a. Horses evolved quickly in North America and periodically migrated into Asia.
   b. Horses evolved slowly in North America and quickly in Asia.
   c. Horses evolved quickly in both North America and Asia.
   d. **Horses evolved slowly in North America and periodically migrated into Asia.**
   e. Horses evolved quickly in Asia and migrated to North America, where they evolved slowly in the absence of predators.

6. If a new form of organism appears in the fossil record in a particular area, we could conclude that
   a. it evolved there rapidly.
   b. it migrated there from another location.
   c. there is a gap in the fossil record and it evolved there slowly.
   d. **All of the above**
   e. None of the above

7. Radioisotopes are often used to determine the time of death of fossilized remains. Tritium has a half-life of 12.3 years. That means that 24.6 years after an organism dies it will have what fraction of the original radioactive tritium?
   a. All
   b. One-half
   c. None
   d. One-quarter
   e. One-eighth

8. The evolution of hard skeletons occurred in many animal phyla during the Cambrian period. What major ecological factor is thought to have been responsible for this evolution?
   a. Competition with other species
   b. Predation pressure
   c. Protection from desiccation
   d. Hard skeletons permitted movement
   e. Hard skeletons provided protection from parasitism

9. The coal beds we now mine for energy are the remains of trees of the
   a. Precambrian period (600 million years ago).
   b. Cambrian period (600–500 million years ago).
   c. Ordovician period (500–440 million years ago).
   d. Silurian period (440–400 million years ago).
   e. Carboniferous period (about 300 million years ago).

10–13. The Mesozoic era (about 245–66 million years ago[mya]) had three periods: the Triassic, the Jurassic, and the Cretaceous. Match the following events with the correct time frame from the list below. Each choice may be used once, more than once, or not at all.
   a. Mesozoic era (about 245–66 mya)
   b. Triassic period (245–195 mya)
   c. Jurassic period (195–138 mya)
   d. Cretaceous period (138–66 mya)

10. The first mammals evolved from reptiles. **(b)**

11. Individual continents acquired distinctive terrestrial floras and faunas. **(a)**

12. Frogs, salamanders, and lizards first appeared. **(c)**

13. The first birds evolved from reptiles. **(c)**

14. The Cenozoic era (66 mya–present) is often called
   a. the age of bacteria, because they cause so many diseases.
   b. the age of fishes, because they are an important food resource.
   c. the age of mammals, because of the extensive radiation of this group.
   d. the age of plants, because they convert radiant energy from the sun into chemical energy.
   e. the age of viruses, because viruses such as AIDS have recently evolved.

15. The last glaciers retreated from temperate latitudes about _____ years ago. As a result, many temperate ecological communities have occupied their current locations for no more than _____ years.
   a. 10,000,000; a million

   b. 1,000,000; a hundred thousand
   c. 100,000; tens of thousands of
   d. 10,000; a few thousand
   e. 1,000; a few hundred

16. Evolutionists have long puzzled over why no new phyla have evolved since the Cambrian, about 500 million years ago. The most commonly accepted theory is
   a. that the Cambrian radiation took place in a world that contained few species, and the ecological setting was favorable for the evolution of many new body plans and different ways of life.
   b. that competition resulted in many different species, with different ecological requirements, that were able to coexist.
   c. that heavy predation pressures selected for differences among prey; therefore predators could not be as efficient at capturing many diverse species as they could one numerous species.
   d. that there are only a certain number of body plans that are functional and all of those types evolved during the Cambrian.
   e. that there has been relatively little genetic variation present since the Cambrian, so natural selection has been restricted to creating new species, genera, and families instead of phyla.

17. It appears from the fossil record that there has been a general tendency of increase in size during the course of evolution. Which of the following groups runs counter to that trend, that is, has *not* increased in size over time?
   a. Echinoderms
   b. Foraminiferans
   c. Insects
   d. Vertebrates
   e. Plants

18. One of the most controversial theories about the cause of the mass extinction at the end of the Cretaceous period is that a large meteorite or asteroid collided with Earth. The proposed ecological effect of this collision is
   a. that the impact changed the distance between Earth and the sun enough so that Earth became significantly cooler.
   b. that the collision threw enough dust into the atmosphere to darken the skies and lower temperatures worldwide.
   c. that the impact was like a giant earthquake and caused instantaneous death to most inhabitants of Earth.
   d. that the impact, which occurred in what is now the Caribbean Sea, caused giant tidal waves in the oceans. When the waves impacted the land, many organisms died.

19. Which of the following is *not* considered to be a plausible hypothesis concerning the causes of mass extinctions?
   a. Extraterrestrial causes, such as meteorite or asteroid collisions
   b. Glaciations
   c. Massive volcanic activity
   d. Competition among organisms
   e. All of the above

20. Data on sexual size dimorphism and mating systems among American blackbirds suggest that species that are polygynous (one male mates with several females) have larger size differences than monogamous (one male mates with one female) species. This suggests that male body size may be under stronger selection when there is more competition for mates. If this is a general trend, we might interpret an increase in size in the fossil record to indicate
    a. increasing monogamy.
    b. increasing territoriality.
    c. decreasing competition.
    d. increasing competition.
    e. no change in competition.

21. There are thought to be _____ mass extinctions in the history of life on Earth and _____ times when many new evolutionary lineages originated.
    a. 6; 0
    b. 3; 3
    c. 6; 6
    d. 3; 6
    e. 6; 3

22. The major difference between the earliest explosion in evolutionary lineages, the Cambrian, about 500 million years ago, and the Paleozoic and Triassic, which were more recent, is that
    a. during the Cambrian, new phyla originated.
    b. during the Paleozoic and Triassic, new phyla originated.
    c. the Cambrian explosions were probably caused by an asteroid colliding with Earth.
    d. there were more species, and thus more competition, when the Cambrian explosion occurred.

23. During the Mesozoic era (245–66 million years ago), Pangaea separated into individual continents. As a result
    a. many species became extinct.
    b. many new phyla evolved.
    c. individual continents acquired distinctive terrestrial floras and faunas.
    d. flight evolved to allow organisms to migrate among continents.
    e. All of the above

24. Which of the following continents is oldest?
    a. Laurasia
    b. Pangaea
    c. Antarctica
    d. Australia
    e. None of the above

25. During the majority of Earth's history, the climate has been
    a. about the same as it is now.
    b. uniform and considerably colder than it is now.
    c. considerably colder than it is now, but with numerous, short warm periods.
    d. considerably warmer than it is now.
    e. considerably warmer than it is now, but with numerous, short cold periods.

26. What can be said about the nature of a typical environment that will become the site for a rich fossil bed and the origin of organisms that will become fossils there?
    a. The environment is anaerobic and most organisms originate elsewhere.
    b. The environment is anaerobic and most organisms originate locally.
    c. The environment is aerobic and most organisms originate elsewhere.
    d. The environment is aerobic and most organisms originate locally.
    e. The environment is anaerobic and about equal numbers of organisms originate elsewhere or locally.

27. The half-life of a particular radioactive isotope of element X is 100 years. If you know that 400 years ago a fossil contained 10 milligrams of this isotope, how much would you expect to find in that fossil today?
    a. About 5 milligrams
    b. About 2.5 milligrams
    c. About 0.6 milligrams
    d. About 0.3 milligrams
    e. About 0.15 milligrams

28. Select the era of the Burgess Shale fauna.
    a. Proterozoic
    b. Paleozoic
    c. Mesozoic
    d. Cenozoic
    e. a and b

29. Select the time division during which the world biota became provincialized due to continental drift.
    a. Precambrian
    b. Paleozoic
    c. Mesozoic
    d. Cenozoic
    e. b and c

30. Select the time division during which the major diversification of angiosperms, birds, and mammals occurred.
    a. Precambrian
    b. Paleozoic
    c. Mesozoic
    d. Cenozoic
    e. c and d

31. Select the period during which the first insects and amphibians appeared on land.
    a. Cambrian
    b. Devonian
    c. Permian
    d. Triassic
    e. Tertiary

32. What types of organisms would you *not* expect to see during a walk in a Permian forest?
    a. Dragonflies
    b. Amphibians
    c. Club mosses and horsetails
    d. Flowering plants
    e. Gymnosperms

33. During this period, the "age of reptiles" began.
    a. Permian
    b. Triassic
    c. Jurassic

    d. Cretaceous
    e. Tertiary

34. During the Cretaceous period of the Mesozoic era
    a. the continents Laurasia and Gondwanaland first formed.
    b. horses first appeared.
    c. birds and mammals diverged from a common reptilian stock.
    ***d.* a second major extinction eliminated most land vertebrates.**
    e. bony fishes dominated the seas.

35. During the Tertiary period of the Cenozoic era
    a. the dinosaurs become extinct.
    b. the "age of fishes" began.
    c. many major "ice ages" occurred.
    d. the genus *Homo* first appeared.
    ***e.* the main radiation of birds took place.**

36. Which of the following statements about the Cambrian, Paleozoic, or modern faunas is true?
    a. The diversity of body plans is greatest in the modern fauna.
    b. Most of the phyla that were present in the Paleozoic fauna are now extinct.
    ***c.* The number of families has steadily increased in the modern fauna.**
    d. Most of the species that were part of the Cambrian fauna are alive today.
    e. None of the above

37. Which of the following statements about the diversity of the modern fauna is true?
    a. As the diversity of insect pests declined, the diversity of flowering plants increased.
    ***b.* The evolution of more complex ecological communities was important in creating the changes in diversity in the modern fauna.**
    c. Provinciality due to continental drift was an unimportant factor after the Cambrian era.
    d. As diversity increased, organisms became more specialized and less interdependent.
    e. The numbers of species during the Cenozoic has been rather stable.

38. Over a wide variety of lineages
    ***a.* size tends to increase with time.**
    b. size tends to decrease with time.
    c. size tends to remain constant with time.
    d. size tends to increase with time, but only in the insects.
    e. a decrease in size is usually followed by an adaptive radiation of many smaller forms.

39. Of the six mass extinctions on Earth, the extinction at the end of the _____ eliminated the most phyla.
    ***a.* Cambrian**
    b. Devonian
    c. Permian
    d. Triassic
    e. Cretaceous

40. Spine reduction in sticklebacks is usually related to
    a. genetic drift.
    b. a reduction in predation in fresh water.
    c. a dominant allele.

    d. movement into a more stable environment.
    ***e.* differences in the type of predators encountered in freshwater and marine habitats.**

41. Which of the following statements about evolution would be supported by most biologists?
    ***a.* Evolution is understandable, but not predictable.**
    b. Evolution results in greater complexity.
    c. Given specific initial conditions, evolution is predetermined.
    d. As evolution progresses, extinction rates decrease.
    e. Mutation is more important than environment in directing the course of evolution.

42. Of the four standard eons recognized by geologists (Hadean, Archean, Proterozoic, and Phanerozoic), which one is subdivided into eras (Paleozoic, Mesozoic, and Cenozoic)?
    a. Hadean
    b. Archean
    c. Proterozoic
    ***d.* Phanerozoic**
    e. None of the above

43. Earth's crust consists of solid tectonic plates approximately _____ km thick that "float" on a liquid mantle.
    a. 4
    ***b.* 40**
    c. 400
    d. 4,000
    e. None of the above

44. The 1979 Alvarez hypothesis of a collision of a large meteorite with Earth attempts to explain
    a. the theory of Gondwanaland.
    b. Alfred Wegner's concept of continental drift.
    ***c.* the mass extinction that occurred at the end of the Cretaceous period.**
    d. contemporary atmospheric conditions on Earth.
    e. None of the above

45. More than _____ of the species that have ever lived on Earth since the beginning of time are now extinct.
    a. 10%
    b. 50%
    c. 75%
    d. 90%
    ***e.* 99%**

46. Three times during the history of life, many new evolutionary lineages originated. Select the correct choice below that represents these periods.
    ***a.* Cambrian explosion, Paleozoic explosion, Triassic explosion**
    b. Jurassic revolution, Cambrian explosion, Hadean period
    c. Cambrian explosion, Archean extinction, Jurassic explosion
    d. Paleozoic explosion, Mesozoic explosion, Cenozoic explosion
    e. Archean extinction, Jurassic explosion, Triassic explosion

47. The most probable cause of the mass extinction at the end of the Permian period was

a. a volcano.

b. a giant meteorite hitting Earth.

c. major predation among species.

***d.* aggregation of the continents into Pangaea.**

e. a large dust cloud that cooled the earth's temperature.

48. Luis Alvarez and colleagues proposed that a meteorite was the cause of the mass extinction 66 million years ago. Their hypothesis was based on

    ***a.* finding high concentrations of iridium in the rock layer separating the Cretaceous and Tertiary periods.**

    b. fossils found in the rock layer separating the Cretaceous and Tertiary periods.

    c. the finding of a crater site.

    d. finding high concentrations of argon 39 in the rock layer separating the Cretaceous and Tertiary periods.

    e. finding evidence of unusual volcanic activity during that period.

49. Most kingdoms evolved prior to the Cambrian period (about 600 million years ago). Which kingdom evolved after the Cambrian, i.e., which was the last kingdom to evolve?

    a. Prokaryotes

    b. Protists

    c. Fungi

    ***d.* Plants**

    e. Animals

50. Since the mid-Cambrian period, about 530 million years ago, _____ new animal phyla are thought to have evolved, and _____ animal phyla are thought to have become extinct.

    a. 100; 100

    b. 10; 10–70

    c. 0; 0

    d. 10; 0

    ***e.* 0; 10–70**

51. The Ediacaran fauna consisted of mostly

    a. prokaryotic organisms.

    b. soft-bodied invertebrates similar to present-day forms.

    ***c.* soft-bodied invertebrates unlike present-day forms.**

    d. trilobites and other hard-shelled forms.

    e. primitive horses.

52. Which of the following is believed to be a cause of the mass extinction at the end of the Devonian period?

    a. Two continents colliding

    b. An asteroid colliding with Earth

    c. A large volcano

    d. The breaking apart of Pangaea and subsequent rise in the ocean

    e.* *a* and *b

53. Which of the following did *not* occur during the Cenozoic era?

    a. Australia and Antarctica were still attached.

    b. Flowering plants dominated world forests.

    ***c.* The climate got hotter and wetter.**

    d. There was an extensive radiation of mammals.

    e. *Homo sapiens* arrived in North and South America.

20 *The Mechanisms of Evolution*

Fill in the Blank

1. Evolution is the accumulation of **heritable** changes within populations over time.

2. The frequencies of the variants at each locus in a population are called **allele frequencies**.

3. The genetic expression of traits, for example, "homozygous recessive," is called the organism's **genotype**.

4. The physical expression of a trait, for example, height or eye color, describes an organism's **phenotype.**

5. Short-term changes in a population's gene pool are often called **microevolution**.

6. The **gene pool** is the sum total of genetic information present in a population at any given moment. It includes every allele at every locus in every organism in a population.

7. The relative reproductive contribution of an individual to subsequent generations is termed the individual's **fitness**.

8. When individuals with intermediate values of traits have the highest fitness, **stabilizing selection** is said to be operating.

9. A population that is not changing, that has constant genotype and allele frequencies from generation to generation, is said to be in **equilibrium**.

10. **Gene flow** involves movement of individuals to a new location, followed by breeding.

11. The differential contribution of offspring resulting from different heritable traits is called **natural selection.**

12. The idea of natural selection is most closely associated with **Charles Darwin**, who proposed it in his book, *The Origin of Species*, in 1859.

13. A change in a population's allele frequencies that results from chance as a result of small population size is called **genetic drift**.

14. Genetic drift can be brought about by either a severe reduction in population size, known as a population **bottleneck**, or a small number of individuals establishing a new population, which results in a **founder effect**.

15. Populations of a European clover, *Trifolium repens*, produce cyanide, which increases their resistance to herbivores such as mice and slugs; however, it also increases their susceptibility to frost. Populations from northeast Europe produce relatively little cyanide compared with populations in southwest Europe. This geographic change in phenotype is known as a **cline**.

16. A **mutation** is a change in DNA sequence.

Multiple Choice

1. Assume that a population is in Hardy–Weinberg equilibrium for a trait controlled by one locus and two alleles. If the frequency of the recessive allele is 0.90, what is the frequency of the dominant allele?
   a. **0.10**
   b. 0.19
   c. 0.81
   d. None of the above
   e. Not enough information to tell

2. There is a gene that causes people to have crumbly earwax. This gene is expressed as a complete dominant: individuals that are homozygous dominants (*CC*) or heterozygous (*Cc*) have crumbly earwax. Homozygous recessives (*cc*) have gooey earwax. On Paradise Island there are 100 people, of which 75 have crumbly earwax. Assuming Hardy–Weinberg conditions, what is the frequency of the *c* allele on Paradise Island?
   a. 0.25
   b. **0.50**
   c. 0.87
   d. None of the above
   e. Not enough information to tell

3. In a population at Hardy–Weinberg equilibrium, the frequency of heterozygotes is 0.64. What is the frequency of the homozygous dominants?
   a. 0.08
   b. 0.64
   c. 0.80
   d. None of the above
   e. **Not enough information to tell**

4. In a population at Hardy–Weinberg equilibrium, the frequency of the *a* allele is 0.60. What is the frequency of individuals heterozygous for the *A* gene?
   a. 0.16
   b. 0.24
   c. **0.48**
   d. None of the above
   e. Not enough information to tell

5. One in 10,000 babies in the U.S. is born with phenylketonuria (PKU), a metabolic disorder caused by a recessive allele. What proportion of that human population is likely to be a carrier of the PKU allele?
   a. **Approximately 0.02**
   b. Approximately 0.20
   c. Approximately 0.98
   d. None of the above
   e. Not enough information to tell

6. There are five conditions that must be met for a population to be in Hardy–Weinberg equilibrium. Which of the following is *not* one of those conditions?
   a. **Nonrandom mating**
   b. Large population size
   c. No migration
   d. No natural selection
   e. No mutations

7. Which of the following situations would demonstrate a population bottleneck?
   a. The population of El Paso, Texas, moves to Patagonia.
   b. **Eight male and eight female elephant seals survive the wreck of the Exxon Valdez.**
   c. A million male orangutans
   d. Six male orangutans collected from a natural population in Sumatra and moved to the San Diego Zoo
   e. A head of lettuce

8. _____ is the effect produced when a bee carries pollen from one population to another.
   a. **Gene flow**
   b. Population bottleneck
   c. Founder event
   d. Genetic equilibrium
   e. Assortative mating

9. In a large population, mutation pressure in the absence of selection
   a. **probably has little effect on the gene pool of a population.**
   b. produces major evolutionary changes.
   c. is what makes us different from the dinosaurs.
   d. is usually beneficial.
   e. never occurs.

10. Which of the following is *not* true of mutation?
    a. It creates the raw material that makes evolution possible.
    b. Most mutations are harmful or neutral.
    c. Mutation rates are very low for most loci.
    d. **Mutations are a likely cause of deviations from Hardy–Weinberg proportions in a population.**
    e. Mutations probably have little effect on the gene pool of a large population.

11. Over the long run, mutations are very important to evolution because
    a. **they are the original source of genetic variation.**
    b. once an allele is lost through mutation, another mutation to that same allele cannot occur.
    c. most mutation rates are one in a thousand.
    d. whether good or bad, mutations increase the fitness of an individual.
    e. mutations are usually beneficial to the progeny.

12–13. Choose the correct term from the following list to match the example given.
    a. Hardy–Weinberg equilibrium
    b. Directional selection
    c. Disruptive selection

12. Selection for variation in bill sizes in African finches. *(c)*

13. Selection for large dark peppered moths on dark trees. *(b)*

14. Which of the following agents of evolution adapts populations to their environment?
    a. Nonrandom mating
    b. **Natural selection**
    c. Migration
    d. Genetic drift
    e. Mutation

15. The raw material for evolutionary change is
    a. phenotypic variation.
    b. **genetic variation.**
    c. geographical variation.
    d. environmentally induced variation.
    e. behavioral variation.

16. A population evolves when
    a. environmentally induced variation is constant between generations.
    b. **individuals having different genotypes survive or reproduce at different rates.**
    c. the environment changes on a seasonal basis.
    d. members reproduce by cloning.
    e. juvenile and adult stages require different environments.

17. Selection acts on _____ ; however, evolution depends on _____.
    a. **phenotypic variation; genetic variation**
    b. genetic variation; phenotypic variation
    c. genetic variation; environmentally induced variation
    d. environmentally induced variation; phenotypic variation
    e. environmentally induced variation; genetic variation

18. Limpets growing high in the intertidal zone, where they experience heavy wave action, are more conical than individuals of the same species growing in the subtidal zone, where they are protected from waves. Individuals transplanted from the high intertidal zone to the subtidal zone add new growth, which produces a flatter, subtidal shape. This experiment suggests that
    a. **the difference in phenotypes is environmentally induced.**
    b. the difference in genotypes is environmentally induced.
    c. the difference in phenotypes is genetically based.
    d. the difference in genotypes is due to natural selection.
    e. Not enough information to tell

19. Which of the following is *not* an example of environmentally induced variation?
    a. Water fleas grown in cool or calm water develop round heads. If they are moved to warm or turbulent water, they develop pointed "helmets" on their heads.
    b. Plants collected along an altitudinal cline vary in size when grown in their native habitat, yet when grown in a greenhouse they are all the same size.
    c. Limpets growing high in the intertidal zone, where they experience heavy wave action, are more conical than individuals of the same species growing in the subtidal zone, where they are protected from waves. Individuals transplanted from the high intertidal zone to the subtidal zone add new growth, which produces a flatter, subtidal shape.

d. Leaves on the same tree or shrub often differ in shape and size. In oaks, leaves closer to the top receive more wind and sunlight and are more deeply lobed than leaves lower down.

e. **Beetles that feed on various host plants are different colors, yet when their offspring are grown on different plants than the parents, they are the same color as the parents.**

20–24. Suppose you have a population of flour beetles with 1,000 individuals. Normally the beetles are a red color; however, this population is polymorphic for a mutant autosomal body color, black, designated by *b/b*. Red is dominant to black, so *B/B* and *B/b* genotypes are red. Assume the population is in Hardy–Weinberg equilibrium, with f(*B*) = *p* = .5 and f(*b*) = *q* = .5.

20. What are the expected frequencies of the red and black phenotypes?
  a. .5 red; .5 black
  b. **.75 red; .25 black**
  c. .25 red, .75 black
  d. None of the above
  e. Not enough information to tell

21. What would be the expected frequencies of the homozygous dominant, heterozygous, and homozygous recessive after 100 generations if the population is under the conditions of Hardy–Weinberg?
  a. .75, .20, .05
  b. **.25, .5, .25**
  c. All red because it is the natural color
  d. All black because all red alleles would mutate to black
  e. .5, .2, .3

22. What would be the expected red and black allele frequencies if Hardy–Weinberg conditions were met except that 1,000 black individuals migrated into the population?
  a. .75, .25
  b. **.25, .75**
  c. .5, .5
  d. They would not change because the population would still be in Hardy–Weinberg equilibrium.
  e. None of the above

23. What would be the allele frequencies if a population bottleneck occurred and only four individuals survived, one female red heterozygote and three black males?
  a. .875, .125
  b. **.125, .875**
  c. .25, .75
  d. .75, .25
  e. .5, .5

24. If the population in Question 23 randomly mated, what would be the allele frequencies of their offspring?
  a. .875, .125
  b. **.125, .875**
  c. .25, .75
  d. .75, .25
  e. .5, .5

25. A gene pool is all the alleles
  a. of an individual's genotype.
  b. **present in a specific population at a given moment.**
  c. that occur in a species throughout its evolutionary existence.
  d. that contribute to the next generation of a population.

26. An important feature of sexual reproduction is that it
  a. is a reproductive process unique to animals.
  b. **produces variation through genetic recombination.**
  c. uses mitosis in producing gametes.
  d. is more efficient than asexual reproduction.
  e. is always associated with selective mating.

27. In a West African finch species, birds with large or small bills survive better than birds with intermediate-sized bills. The type of natural selection operating on these bird populations is
  a. directional selection.
  b. **disruptive selection.**
  c. stabilizing selection.
  d. nonrandom selection.
  e. deme selection.

28–29. Hardy–Weinberg frequencies for the three genotypes *AA*, *Aa*, and *aa*, at all values of *p* and *q*, are shown below.

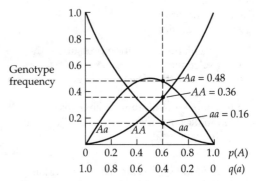

28. As the frequency of the homozygous recessive genotype increases, the frequency of the
  a. heterozygous genotype decreases continuously.
  b. homozygous dominant increases.
  c. heterozygous genotype increases continuously.
  d. dominant allele approaches 1.
  e. **recessive allele approaches 1.**

29. Which of the following is supported by the graph above?
  a. **When *aa* = 0.45, *Aa* = 0.45.**
  b. When *aa* = 0.16, *q* = 0.60.
  c. When *p* = 0.33, *q* = 0.77.
  d. When *Aa* = 0.49, *p* = 0.70.
  e. When *AA* = 0.50, *aa* = 0.

30. In a population of organisms, the frequency of the heterozygous genotype is always
  a. greater than the frequency of the homozygous recessive genotype.
  b. **equal to one minus the sum of the homozygous genotypes.**
  c. equal to two times the frequencies of the homozygous genotypes.
  d. greater than either homozygous genotype.
  e. equal to the frequency of the homozygous dominant genotype minus the frequency of the homozygous recessive genotype.

31. Cheetahs are a very homogeneous species. The lack of genetic variability among cheetahs may be attributed to
    a. gene flow.
    b. sexual selection.
    c. **population bottleneck.**
    d. high mutation rate.
    e. high immigration rates.

32. Genetic equilibrium in a population refers to
    a. equal numbers of dominant and recessive alleles.
    b. equal numbers of females and males.
    c. **unchanging allele frequencies in successive generations.**
    d. lack of mutations that affect the observed phenotypes.
    e. proportional numbers of each genotype.

33. An example of evolutionary change is
    a. a change from more conical to less conical shape as a limpet moves from turbulent water in the upper intertidal zone to less turbulent water in the subtidal zone.
    b. an increase in edge per unit surface area in sun-grown leaves versus leaves grown in the shade.
    c. the development of a pointed "helmet" on the head of *Daphnia* when it moves from cooler, calm water to warmer, turbulent water.
    d. **the development of "brussels sprouts" when *Brassica oleracea* was selected for lateral buds.**
    e. a tadpole metamorphosing into an adult frog.

34. Genetic drift as an evolutionary factor is
    a. **a greater force in a population with small numbers than in a population with large numbers.**
    b. a greater force in a population with much genetic variation than in a population with little genetic variation.
    c. a force because it is responsible for the selection of mutations.
    d. a force because it involves the movements of alleles between populations of a single species.
    e. a greater force the larger the size of the population.

35. In the Hardy–Weinberg equation, the homozygous dominant individuals in a population are represented by
    a. **p^2.**
    b. $2pq$.
    c. q^2.
    d. p.
    e. q.

36–39. The two graphs below represent phenotypic distribution of a character in a population of organisms over time. The first curve represents the distribution at an initial sampling time, while the second represents a sampling of a later generation.

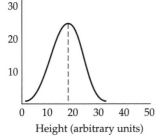

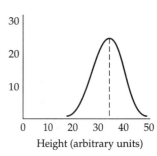

36. The average size of an individual at the initial sampling is
    a. 5 units.
    b. 10 units.
    c. 15 units.
    d. **20 units.**
    e. 25 units.

37. At the initial sampling, there are as many individuals 15 units tall as there are _____ units tall.
    a. 5 units
    b. 10 units
    c. 15 units
    d. 20 units
    e. **25 units**

38. The process illustrated by these graphs is called
    a. stabilizing selection.
    b. **directional selection.**
    c. disruptive selection.
    d. convergent evolution.
    e. divergent evolution.

39. The graphs suggest that
    a. taller individuals migrated away from the original population.
    b. shorter individuals migrated into the original population.
    c. **taller individuals were more successful at reproducing than short individuals.**
    d. in the initial population, there were more tall individuals than short ones.
    e. the individuals in the population continued to grow between sample times.

40. A mutation occurs in one of your lung cells. Which of the following is true?
    a. You have evolved to be better adapted to your environment.
    b. You will soon die because most mutations are lethal.
    c. You will be sterile and no longer be able to have children.
    d. The human species will have evolved because this mutation will be passed on to your children.
    e. **This mutation does not affect human evolution because it will not be passed on to your offspring.**

41. Natural selection acts directly on
    a. the genotype to produce new mutations.
    b. the phenotype to produce new mutations.
    c. the genotype to favor existing mutations.
    d. **the phenotype to favor traits due to existing mutations.**
    e. the genotype to inhibit new mutations.

42–45. Biologists use the term "fitness" when speaking of evolution. Below are descriptions of four male cats. Answer the following questions based on this information.

Name	Tabby	Chessy	Tony	Tiger
Size	12 lbs	10 lbs	6 lbs	8 lbs
Number of kittens fathered	19	25	20	20
Kittens surviving to adulthood	15	14	14	19
Age at death	13 yrs	16 yrs	12 yrs	9 yrs

Comments: Tabby is the largest and strongest cat

Chessy has mated with the most females

Tony was lost on a family vacation but adapted to street life and lived two more years

Tiger died from an infection after a cat fight

42. Who contributed the most genes to the next generation?
   a. Tabby
   b. Chessy
   c. Tony
   d. Tiger

43. Who contributed the most genes to the gene pool?
   a. Tabby
   b. Chessy
   c. Tony
   d. Tiger

44. Which cat contributed the most to the ongoing evolution of the species?
   a. Tabby
   b. Chessy
   c. Tony
   d. Tiger

45. The "fittest" cat is
   a. Tabby.
   b. Chessy.
   c. Tony.
   d. Tiger.

46–48. In a population of 200 individuals, 72 are homozygous recessive for the character of eye color (cc). One hundred individuals from this population die due to a fatal disease. Thirty-six of the survivors are homozygous recessive. Answer the following questions.

46. In the original population, the frequency of the dominant allele is
   a. 0.16.
   b. 0.36.
   c. 0.40.
   d. 0.48.
   e. 0.60.

47. In the new population, the frequency of the dominant allele is
   a. 0.16.
   b. 0.36.
   c. 0.40.
   d. 0.48.
   e. 0.60.

48. How many heterozygous individuals are expected in the new population?
   a. 16
   b. 36
   c. 40
   d. 48
   e. 60

49. Which of the following is *not* an effect sexual recombination has on alleles?
   a. Greater evolutionary potential
   b. Change in the frequency of specific alleles
   c. Greater genotypic variety
   d. Creates new combinations of genetic material
   e. Greater phenotypic variety

50. When male barn swallows had their tails artificially elongated, the females preferred them over males that had shorter tails. Why is this?
   a. The male barn swallows with long tails were more aggressive in their mating behaviors.
   b. The males with longer tails could fly farther and attracted more females.
   c. There is a correlation between long tails and mite resistance.
   d. The males with longer tails were more nurturing to their young.
   e. There is a correlation between long tails and virility.

51. In his book *The Blind Watchmaker*, Richard Dawkins describes the cumulative process of evolutionary change. The correct sequence is
   a. reproduction, development, mutation, evolution.
   b. development, reproduction, mutation, evolution.
   c. reproduction, mutation, development, evolution.
   d. evolution, development, reproduction, mutation.
   e. evolution, mutation, development, reproduction.

21 Species and Their Formation

Fill in the Blank

1. The process by which one evolutionary unit splits into two such units that thereafter evolve as distinct lineages is known as **speciation**.

2. **Geographic speciation** occurs when one species evolves into two daughter species after a physical barrier to movement develops within its range.

3. Two populations that are **reproductively isolated** are said to belong to different species.

4. The offspring of genetically dissimilar parents, such as parents from two geographically isolated populations, are called **hybrids**.

5. **Isolating mechanisms** are factors that reduce the possibility of interbreeding between geographically isolated populations that somehow come into contact.

6. **Prezygotic** isolating mechanisms reduce the probability that hybrids will be formed by preventing the formation of a zygote.

7. Once fertilization between isolated populations has occurred, **postzygotic** isolating mechanisms may prevent the hybrids from surviving.

8. A common means of sympatric speciation is **polyploidy**, in which the number of chromosomes is multiplied.

9. The fossil record and current distributions of organisms reveal that some species or groups have given rise to a large number of daughter species, a phenomenon called **evolutionary radiation**.

10. A change in a single lineage through time is called **vertical evolution, or anagenesis**.

11. A factor promoting cohesion of species is **gene flow**, the movement of individuals so that they reproduce in a different population from their original population.

12. Two hybrid species of sunflowers, *Tragopogon mirus* and *T. miscellus*, have four sets of chromosomes; this is called **tetraploidy**.

13. Long periods of little evolutionary change followed by periods of rapid change are called **punctuated equilibrium**.

14. Some blackbirds have evolved powerful bill-opening muscles for moving objects to expose food. This activity is called **gaping**.

Multiple Choice

1. Evolutionary biologists believe that all the species present today, along with all the species that lived in the past, are
   a. **descended from a single ancestral species.**
   b. descended from thousands of different origins.
   c. descended from one ancestral species per kingdom.
   d. descended from one ancestral species per phylum.
   e. descended from two or three ancestral species per kingdom.

2. Ernst Mayr's definition of *species* states that species are groups of
   a. actually interbreeding natural populations that are reproductively isolated from other such groups.
   b. potentially interbreeding natural populations that are reproductively isolated from other such groups.
   c. **actually or potentially interbreeding natural populations that are reproductively isolated from other such groups.**
   d. actually or potentially interbreeding natural populations that are geographically isolated from other such groups.
   e. potentially interbreeding natural populations that are geographically isolated from other such groups.

3. Ernst Mayr's definition of *species* does *not* include
   a. groups that actually or potentially interbreed.
   b. evolutionary units evolving separately from other units.
   c. gene exchange.
   d. **members in a single geographic location.**
   e. groups that are reproductively separated from other groups.

4. The phrase "natural population" is important to the definition of species because
   a. if two populations interbreed only in captivity, they are members of the same species.
   b. **if two populations co-occur but do not interbreed in nature, they are separate species.**
   c. if two populations do not co-occur, they must be different species because they cannot interbreed.
   d. if two populations can potentially interbreed, they are different species.
   e. if two populations interbreed for the first time, their offspring form a new species.

5–9. Match the following descriptions of speciation with the appropriate terms from the list below. Terms may be used once, more than once, or not at all.
  a. Geographic speciation
  b. Parapatric speciation
  c. Sympatric speciation
  d. Polyploidy

5. Two species of Japanese ladybird beetles occur in the same area. *Epilachnea niponica* feeds on thistles, and *Epilachnea yasutomii* feeds on other plants that grow with thistles. Adults of both species feed and mate on their own host plant. The two species hybridize in the laboratory, but not in nature. Although we cannot be sure which process actually produced speciation in this case, it is likely the species are a result of _____. *(c)*

6. In Wales there are soils heavily contaminated with lead very close to normal rich pasture land. The pasture grass *Agrostis tenuis* is common to both sites; however, there is a sharp gradient in lead tolerance among individuals separated by less than 20 meters. Nearly complete reproductive isolation exists between plants on contaminated and normal soil. This may eventually lead to _____. *(b)*

7. If the number of chromosomes of a species of lizard were to double (perhaps through an error in meiosis), resulting in a female offspring that could produce young from unfertilized eggs, we would say speciation had occurred through _____. *(d)*

8. The red tubular-flowered gilias of western North America are a group of species living in the Mojave Desert. Originally considered to be a single species, the group now contains five species: three diploids and two tetraploids (having four sets of chromosomes). These five species are similar in appearance and are sterile in all interspecific mating combinations. *(d)*

9. Different populations of platyfish live in the rivers of eastern Mexico. The subpopulations living in different streams have been diverging since an ancestral platyfish colonized all the streams. Some subpopulations have diverged so much that they cannot interbreed and produce viable offspring with individuals from other subpopulations. Thus, these populations of platyfish exist at various stages in the process of _____. *(a)*

10. Which of the following factors was probably *not* important in the formation of the 14 species of Darwin's finches on the Galapagos Islands?
  a. Geographical isolation from the mainland
  b. Polyploidy
  c. Different habitats on the different islands
  d. Geographical isolation among the Galapagos Islands
  e. Different food supplies on the different islands

11. Studies of island populations suggest that the smallest area within which a single species has split into two daughter species varies among types of organisms. Which of the following types of organisms requires the largest area to speciate?

  a. Mammals
  b. Birds
  c. Reptiles
  d. Amphibians
  e. Snails

12. Which of the following would *not* result in reproductive isolation between two populations reunited following geographic isolation?
  a. The two populations produce successful hybrids.
  b. The two populations have different breeding seasons.
  c. There are physiological differences between the two populations so that they cannot produce viable offspring.
  d. Members of one population do not find members of the other population attractive as mating partners.
  e. The two populations have different courtship behaviors.

13. Which of the following is a postzygotic isolation mechanism?
  a. Temporal isolation
  b. Behavioral isolation
  c. Reduced viability of hybrids
  d. Geographic variation in mating pheromones
  e. Differences in courtship behavior

14. Which of the following is a prezygotic isolation mechanism?
  a. Abnormal meiosis following fertilization
  b. Infertile hybrids
  c. Reduced viability of hybrids
  d. Abnormal mitosis following fertilization
  e. Geographic variation in mating pheromones

15. The development of reproductive isolation among members of a continuous population in the absence of a geographic barrier is called
  a. polyploidy.
  b. hybrid zonation.
  c. geographic speciation.
  d. anagenesis.
  e. parapatric speciation.

16. _____ involves the subdivision of a gene pool even though members of the daughter species overlap in their range during the speciation process. Often this is accomplished by a multiplication of the number of chromosomes.
  a. Polyploidy
  b. Hybrid zonation
  c. Sympatric speciation
  d. Geographic speciation
  e. Parapatric speciation

17. Which of the following is *not* true about polyploidy?
  a. Polyploidy is more common in animals than in plants.
  b. Many species of flowering plants arose by polyploidy.
  c. Animals that speciate by polyploidy are often parthenogenetic.
  d. Polyploidy can create new species quickly if the polyploid individuals can self-fertilize.
  e. Polyploid siblings are capable of reproducing with one another.

18. Which mode of speciation is thought to be most important in speciation of large animals?
    a. Polyploidy
    b. Hybrid zonation
    c. Sympatric speciation
    d. Geographic speciation
    e. Parapatric speciation

19. Which of the following is *not* true of the genetics of speciation?
    a. Sympatric species need not have diverged from each other much genetically.
    b. Speciation in *Drosophila* has not involved major reorganization of the genome.
    c. Species that show a lot of morphological variation may be relatively similar genetically.
    d. Gene flow is important in the creation of new species.
    e. Differences between closely related species occur due to the same mechanisms that operate within species.

20. Rapid speciation is thought to occur among animals with complex behavior patterns because
    a. the timing of reproduction is behaviorally mediated.
    b. behavioral differences usually reflect physiological differences.
    c. those organisms have difficulty identifying members of their own species and often mate with members of other species.
    d. those organisms make sophisticated discriminations among potential mates.
    e. those organisms find mates more quickly and efficiently.

21. The deer mouse, *Peromyscus maniculatus*, is the most widely distributed small mammal in North America. It varies greatly geographically, especially in coat color, tail length, and foot length. Over which area would you expect deer mice to be relatively uniform?
    a. In mountainous areas, where environmental conditions change dramatically from place to place
    b. On islands, where populations are isolated
    c. Over large areas with little topographic and vegetational change
    d. Between forests and deserts, with significant vegetational change
    e. Between forests and prairies, with significant vegetational change

22. Which of the following correlations between ecological factors and rates of speciation have been observed among the fossil record of large hoofed mammals of Africa?
    a. Speciation is correlated with birth rate.
    b. Speciation is correlated with population size.
    c. Speciation is correlated with social organization.
    d. Speciation is correlated with diet.
    e. Speciation is correlated with body size.

23. Which of the following is a reason why evolutionary radiation is likely to occur on islands?
    a. Islands may have fewer plant and animal groups and thus present more ecological opportunities.
    b. There are many more species on islands.
    c. Many organisms disperse easily over oceans.

d. Islands have more diverse habitats than do continents.
e. Islands tend to have more favorable habitats than continents, due to the moderating effects of the ocean.

24. The Hawaiian Islands are useful for studying evolutionary radiation because they are very isolated. Which of the following is *not* true about the biota of these islands?
    a. More than 90 percent of the plant species are endemic.
    b. Several groups of flowering plants are more diverse on these islands than their mainland counterparts.
    c. There are no endemic amphibians on these islands.
    d. There are no endemic reptiles on these islands.
    e. There are no endemic mammals on these islands.

25. The hypothesized sequence of events in geographic speciation is
    a. geographic barrier, reproductive isolation, genetic divergence.
    b. geographic barrier, genetic divergence, reproductive isolation.
    c. genetic divergence, geographic barrier, reproductive isolation.
    d. genetic divergence, reproductive isolation, geographic barrier.
    e. reproductive isolation, genetic divergence, geographic barrier.

26. Mules are the offspring of parents of two different species, a horse and a donkey. These hybrids exhibit
    a. polyploidy.
    b. shortened lifespan.
    c. behavioral isolation.
    d. reduced viability.
    e. sterility.

27. The modern polar bear species evolved from ancestral bear populations in southern Alaska that became separated by glaciers from bear populations in the rest of North America. This type of event is called
    a. allopatric speciation.
    b. temporal isolation.
    c. parapatric speciation.
    d. mechanical isolation.
    e. sympatric speciation.

28. A situation where two very similar animal species have overlapping distributions is most likely due to
    a. parapatric speciation.
    b. allopatric speciation followed by range expansion.
    c. sympatric speciation due to polyploidy.
    d. convergent evolution of unrelated species.
    e. mechanical isolation between the two species.

29. Which of the following is *not* true of a hybrid zone?
    a. It occurs where two different populations come into contact.
    b. It may shift in location due to environmental changes.
    c. Its habitat may differ from those favored by the parent populations.
    d. It may disappear if isolating mechanisms develop.
    e. It may occur among animals, but does not occur among plants.

30. American and European sycamores have been isolated from one another for at least 20 million years, but are morphologically very similar and can form fertile hybrids. According to Ernst Mayr's definition of *species,* these sycamores should belong to
    a. different species because they are geographically isolated.
    b. different species because they lack the opportunity to interbreed in nature.
    c. different species because they are evolving separately.
    d. the same species because they are morphologically similar.
    e. **the same species because they are capable of forming fertile offspring.**

31. Ginkgo trees occur in Asia and North America. Despite the geographic separation by the Pacific Ocean, biologists consider them the same species. What aspect of Ernst Mayr's definition of *species* accounts for this?
    a. They are reproductively isolated.
    b. **They are potentially capable of exchanging genes.**
    c. They are exchanging genes across the ocean.
    d. They have different evolutionary ancestry.
    e. They formed a large hybrid zone.

32. Why are biological species not always equivalent to taxonomic species?
    a. **Taxonomic species are based on appearance, not reproductive behavior.**
    b. Taxonomic species are based on reproductive behavior, not appearance.
    c. Biological species are based on appearance, not reproductive behavior.
    d. The question has a false premise; biological species are always equivalent to taxonomic species.
    e. Biological species are based on genetic information; taxonomic species are based on ecological information.

33. Which mode of speciation involves a geographic barrier that prevents exchange of genes?
    a. **Allopatric**
    b. Parapatric
    c. Sympatric
    d. Polyploidy
    e. Behavioral isolation

34. Six platyfish with different tail spotting patterns live in eastern Mexico. Five of them can interbreed and produce fertile offspring. One of them *cannot* interbreed with the other five. What can you conclude about these platyfish?
    a. They all belong to the same biological species.
    b. They are each a different biological species.
    c. **There are two biological species in the example above.**
    d. There is not enough data given to make a conclusion.
    e. There are six biological species in the example above.

35. What are the progeny of genetically dissimilar parents called?
    a. Infertile
    b. Fertile
    c. **Hybrids**

    d. Mutants
    e. Hermaphrodites

36. If two genetically differentiated populations reestablish contact and the resulting hybrid progeny are successful and reproduce with other members of the two populations, what will probably happen?
    a. Speciation will occur.
    b. Genetic differences will increase between the populations.
    c. Reproductive isolation will occur.
    d. **The two populations will amalgamate; no new species will form.**
    e. Two new species will form.

37. If two genetically differentiated populations reestablish contact and the resulting hybrid progeny are not as successful in reproduction as the members of the two parental populations, what will probably happen?
    a. **Reproductive isolation will occur.**
    b. A stable hybrid zone will form.
    c. The two populations will amalgamate; no new species will form.
    d. The populations will become genetically identical.
    e. Two new species will form.

38. What form of reproductive isolation occurs between a horse and donkey when they mate and produce a mule?
    a. Prezygotic isolation
    b. Abnormal zygote formation
    c. **Hybrid infertility**
    d. Hybrid vigor
    e. Hybrid fertility

39. What form of speciation tends to occur when a population exists across an important environmental discontinuity?
    a. Hybridization
    b. **Parapatric**
    c. Sympatric
    d. Dyspatric
    e. Extinction

40. Which of the following modes of speciation could occur most quickly?
    a. Allopatric
    b. Parapatric
    c. **Sympatric**
    d. Dyspatric
    e. Adaptive radiation

41. Because there are no physical barriers separating the species, many biologists believe that speciation of host-specific insects is thought to occur by
    a. **sympatric speciation.**
    b. parapatric speciation.
    c. allopatric speciation.
    d. hybridization.
    e. prezygotic isolation.

42. What is the most important factor promoting the cohesion of a species?
    a. Mutation
    b. Natural selection
    c. **Gene flow**
    d. Genetic drift
    e. Reproduction

43. Behavioral complexity and short generation time will tend to _____ the process of speciation.
   a. **accelerate**
   b. decelerate
   c. have no effect on
   d. obscure
   e. threaten

44. Some groups or species have given rise to large numbers of daughter species. This phenomenon is known as
   a. sympatric speciation.
   b. hybridization.
   c. **evolutionary radiation.**
   d. gene flow.
   e. evolution.

45. Which of the following statements is *not* true of the genetics of speciation?
   a. Speciation does not necessarily involve major reorganization of the genome.
   b. Small genetic differences can yield substantial morphological and physiological differences between species.
   c. Genetic differences among species are similar in kind to genetic differences found within species.
   d. **The genetic processes of speciation are different from those of intraspecific genetic differentiation.**
   e. Genetic changes within a short time period may result in speciation.

46. What group of organisms has the highest prevalence of sympatric speciation?
   a. Mammals
   b. Insects
   c. **Plants**
   d. Fungi
   e. Bacteria

47. A study of speciation among African mammals has shown that
   a. ecology and speciation are not correlated.
   b. **ecological specialists (grazers and browsers) speciate more frequently than ecological generalists (omnivores).**
   c. ecological generalists (omnivores) speciate more frequently than ecological specialists (grazers and browsers).
   d. generalists and specialists speciate approximately equally.
   e. The study of ecology cannot contribute to the understanding of speciation.

48. A species that is found only in a certain area of the planet and nowhere else is called
   a. endangered.
   b. **endemic.**
   c. extinct.
   d. emetic.
   e. exotic.

49. Which of the following is *not* true of ecology and the rate of speciation?
   a. Speciation rates are correlated with the diets of mammals.
   b. Speciation rates are not related to birth rates.
   c. Speciation rates differ markedly in the fossil record.

   d. **Speciation rates are consistent across all organisms yet studied.**
   e. Speciation rates are higher among animal species with more complex behavior patterns.

50. Islands are often called "natural laboratories for evolutionary studies." Why is this so?
   a. **Islands are isolated from other land masses.**
   b. Islands are geologically very young.
   c. Islands have very low speciation rates.
   d. Islands are all ecologically similar.
   e. Islands always have small numbers of species on them.

51. The adaptive radiation of Hawaiian silverswords has demonstrated what fact about speciation?
   a. **Major morphological changes can be produced by small genetic changes.**
   b. Major morphological changes require large scale genetic changes.
   c. Many different silversword ancestors colonized Hawaii.
   d. The Hawaiian silverswords are not different species.
   e. Plant speciation always requires polyploidy.

52. Allele frequencies of malate dehydrogenase-1 show significant differences among city block populations of the snail *Helix aspersa*. What is the most likely factor maintaining these frequency differences?
   a. Natural selection—the city blocks are very different ecologically.
   b. Genetic drift—founder effects of city block colonization.
   c. **Lack of gene flow among city blocks—cars run over snails in the roads.**
   d. Different mutation rates among city blocks.
   e. Parapatric effects due to different amounts of pollutants on different city blocks.

53. Of the following reproductive isolating mechanisms, which one is the most efficient at preventing waste of reproductive effort?
   a. Hybrid inviability
   b. Hybrid sterility
   c. Gamete incompatibility
   d. **Prezygotic isolation**
   e. Hybrid cross-reproduction

54. Which mode of speciation is most prevalent among larger animals?
   a. **Geographic**
   b. Parapatric
   c. Sympatric
   d. Polyploidy
   e. Genetic drift

55. Which of the following is *not* true of speciation?
   a. A new species can be formed in one breeding season.
   b. Animals with complex behavior and mating patterns speciate more rapidly.
   c. Members of the same species can live on separate continents.
   d. Speciation is more rapid on island archipelagos such as Hawaii and the Galapagos.
   e. **Sympatric speciation is the most common type among animals.**

56. The family Asteraceae contains several genera found on the Hawaiian islands. There is very little genetic variation between them, and they are thought to have evolved from a single ancestral species. What is responsible for the large morphological differences?
    a. **Each island has a unique environment that the plants adapted to.**
    b. Predation has eliminated all but the remaining genus on each island.
    c. Birds traveling between islands carried different types of seeds that started the plant growth.
    d. Polyploidy during reproduction created the variation in features.
    e. c and d

57. Populations, such as blue and snow geese, that occasionally interbreed and form hybrid zones
    a. cannot be different species.
    b. often result in evolutionary radiation.
    c. **can maintain species differences.**
    d. have little geographic variation.
    e. will produce polyploid offspring.

58. Why is it so difficult to obtain evidence supporting sympatric speciation?
    a. Sympatric speciation always takes a very long time to occur.
    b. Sympatric speciation always involves polyploidy.
    c. **It is hard to distinguish true sympatric speciation from allopatric speciation that occurred in the recent past.**
    d. It is impossible to show genetic differences in sympatric speciation.
    e. Sympatric speciation is a very rare event.

59. Niles Eldridge and Stephen Jay Gould proposed that most evolutionary changes take place at the time of speciation. Which of these do they suggest is the most common event leading to these changes?
    a. A large inbreeding population
    b. A large migrating population
    c. A small migrating population
    d. **A small population isolated from the source population**
    e. A small population interbreeding with a large population

22 Constructing and Using Phylogenies

Fill in the Blank

1. **Taxonomy** is the theory and practice of classifying organisms.

2. **Systematics** is the scientific study of the diversity of organisms.

3. **Carolus Linnaeus**, a Swedish biologist in the 1700s, developed the classification system used today.

4. A two-name classification system, referred to as **binomial nomenclature**, is used today throughout biology.

5. A trait, such as the modern horse's single toe, that differs from the trait in an organism's ancestors in the lineage is called a **derived trait**.

6. Any two traits derived from a common ancestral form are **homologous traits**.

7. Traits such as fins in aquatic mammals and fins in fishes exhibit **homoplasy**; that is, they evolved in different lineages, although they are not found in their most recent common ancestor.

8. Traits that evolve by **convergent evolution** were formerly very different, yet now resemble one another because they have undergone selection to perform similar functions.

9. **Cladistic** classification shows evolutionary relationships and expresses them in treelike diagrams.

10. The entire portion of a phylogeny that is descended from a common ancestor is called a **clade**.

11. The term "Rosaceae" is an example of the taxonomic category of **family**.

12. In the Linnaean system, classes are divided into **orders,** which are then divided into families.

13. When designing a cladogram, the operating rule that it is wiser to postulate a minimal number of changes in traits is called **parsimony**.

Multiple Choice

1. In modern systematics, each family name is based on
   a. the name of the order to which it belongs.
   b. a characteristic common to all members.
   c. **the name of a member genus.**
   d. the name of the largest member species.
   e. the Latin name for the organisms.

2. In the Linnaean system, the term "Aceraceae" refers to a
   a. genus of plants.
   b. genus of animals.
   c. **family of plants.**
   d. family of animals.
   e. order of plants.

3. North America and Great Britain both have birds called robins. These birds have brown backs and red breasts, but are incapable of interbreeding, are different sizes, and have different diets and habitats. Based on this information, one would expect these birds to belong to
   a. the same species.
   b. the same genus, but different species.
   c. the same family, but different genera.
   d. **different species, but more information is needed for further classification.**
   e. different genera, but more information is needed for further classification.

4. Within a family, the number of species placed into each genus is determined by
   a. **the evolutionary uniqueness among organisms.**
   b. an even distribution of the total number of species.
   c. an even distribution of the species based on their sizes.
   d. grouping species by their geographic ranges.
   e. clustering species based on their habitats.

5. Based on current systematic theory, crocodilians should be classified with
   a. lizards, based on their morphological similarities.
   b. lizards, based on consistency with classical systematics.
   c. **birds, based on evolutionary relationships.**
   d. aquatic birds, based on habitat similarities.
   e. salamanders, based on morphological and habitat similarities.

6–9. Use the cladogram below to answer the following questions.

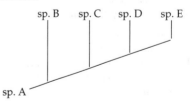

6. Assuming the cladogram includes modern species, which of these species is(are) alive today?
   a. A, B, C, D, E
   b. **B, C, D, E**
   c. C, D, E
   d. D, E
   e. E

7. Species D and E share _____ homologous and _____ homoplastic traits.
   a. many; many
   b. many; few
   c. few; many
   d. few; few
   e. no; no

8. Which species share(s) the most recent common ancestor with species E?
   a. Species A
   b. Species B
   c. Species C
   d. Species D
   e. All species share the same common ancestry.

9. The position of species B in the cladogram indicates that species B
   a. became extinct before species C, D, or E.
   b. has fewer derived traits than species C, D, or E.
   c. has a shorter evolutionary history than species C, D, or E.
   d. shares more common ancestors with species C than with species D or E.
   e. is less fit for its environment than species C, D, or E.

10. When compared to other taxa, a taxon that shares general homologous traits but lacks special homologous traits is called a(n)
    a. clade.
    b. genus.
    c. homoplasy.
    d. outgroup.
    e. population.

11. When comparing humans and chimps with dogs and cats, a general homologous trait would be
    a. hands specialized for grasping.
    b. presence of body hair.
    c. standing on two legs.
    d. lack of a tail.
    e. poor sense of smell.

12. When comparing humans and chimps with dogs and cats, a special homologous trait would be
    a. bony skeleton.
    b. presence of body hair.
    c. hands specialized for grasping.
    d. mouth containing teeth.
    e. presence of circulatory system.

13. Which two statements are often assumed to be true in the construction of cladograms?
    1. Derived traits appear only once in a lineage.
    2. Branching points are determined by the number of homologous traits.
    3. Derived traits are never lost.
    4. Descendant species often lose ancestral traits.
    a. 1 and 2
    b. 1 and 3
    c. 1 and 4
    d. 2 and 3
    e. 3 and 4

14–17. Use the table below to answer the following questions.

Ancestral and Derived Traits Among Five Species

Species	Trait				
	1	2	3	4	5
A	1	1	0	0	1
B	0	0	1	0	1
C	1	1	0	0	1
D	0	1	0	0	1

The ancestral form of each trait is coded 0, and the derived trait is coded 1.

14. In constructing a cladogram, the oldest divergence would separate
    a. A from B, C, and D.
    b. B from A, C, and D.
    c. C from A, B, and D.
    d. D from A, B, and C.
    e. A and D from B and C.

15. The two species that diverged most recently from one another are
    a. A and B.
    b. A and C.
    c. A and D.
    d. B and C.
    e. B and D.

16. Which trait is most recently derived?
    a. 1
    b. 2
    c. 3
    d. 4
    e. 5

17. Which trait is most ancestral?
    a. 1
    b. 2
    c. 3
    d. 4
    e. 5

18. What is the most appropriate use of knowledge about behavioral traits in the reconstruction of phylogenies?
    a. This knowledge is never relevant.
    b. This knowledge is more important than other traits among closely related species.
    c. This knowledge supports knowledge of other traits among closely related species.
    d. This knowledge supports knowledge of other traits among distantly related species.
    e. This knowledge is always relevant.

19. Systematists usually employ parsimony in reconstructing a phylogeny. Parsimony involves arranging the organisms such that
    a. the minimal number of changes in traits is postulated in determining the lineage.
    b. the maximum number of traits is used in establishing the lineage.
    c. molecular information is given priority over other traits in determining the lineage.
    d. larval traits are included in establishing the lineage.
    e. ancestral traits are given priority over derived traits in determining the lineage.

20. _____ is the study of biological diversity and its evolution.
    a. Phylogeny
    b. Systematics
    c. Taxonomy
    d. Classification
    e. Nomenclature

21. _____ is the science of biological classification.
    a. Phylogeny
    b. Systematics
    c. Taxonomy
    d. Classification
    e. Nomenclature

22. Classification systems have many uses. Which of the following is *not* a goal of biological classification?
    a. To depict convergent evolution
    b. To clarify relationships among organisms
    c. To help us remember organisms and their traits
    d. To clearly identify organisms being studied
    e. To provide predictive powers

23. Taxonomic systems used by biologists are hierarchical; that is,
    a. taxonomic groups reflect shared characters, not evolutionary relationships.
    b. each higher taxonomic group contains all the groups below it.
    c. taxonomic groups reflect common habitats.
    d. a hierarchy of traits is used to establish classifications.
    e. phylogenetic relationships do not help us understand evolution.

24. Classification systems serve four important roles. Which of the following is *not* one of those roles?
    a. To help us remember characteristics of a large number of different things
    b. To help us identify shared traits, such as hair, mammary glands, and constant high body temperature in mammals
    c. To reveal the harmony of nature
    d. To provide stable, unique, unequivocal names for organisms
    e. To help reconstruct evolutionary pathways

25. The biological classification system used today is based on the work of
    a. Charles Darwin.
    b. Barbara McClintock.
    c. Gregor Mendel.
    d. Lynn Margulis.
    e. Carolus Linnaeus.

26. The biological classification system used today is referred to as
    a. dichotomous taxonomy.
    b. dichotomous nomenclature.
    c. dichotomous keys.
    d. binomial taxonomy.
    e. binomial nomenclature.

27. The hierarchy of categories in the classification system used today is
    a. division or phylum, kingdom, order, family, class, genus, species.
    b. division or phylum, kingdom, order, class, family, genus, species.
    c. division or phylum, kingdom, class, order, family, genus, species.
    d. kingdom, division or phylum, order, class, family, genus, species.
    e. kingdom, division or phylum, class, order, family, genus, species.

28. In referring to an organism in writing, such as in a newspaper, textbook, or lab report, which of these rules should be followed?
    1. Underline or italicize genus
    2. Underline or italicize species
    3. First letter of species should be uppercase
    4. First letter of genus should be uppercase
    a. 1, 2, 4
    b. 1, 2, 3
    c. 2, 3, 4
    d. 1, 3, 4
    e. 1, 2, 3, 4

29. If you saw the name of a family of organisms and it ended with "-idae," such as Formicidae, you would know that it was a family of
    a. bacteria.
    b. fungi.
    c. plants.
    d. animals.
    e. protists.

30–33. Match the following descriptions of classification systems with their names from the list below.
    a. Linnaean systematics
    b. Cladistic systematics
    c. Early Hindu systematics

30. This system was the first to use binomial nomenclature. *(a)*

31. This classification system helps people to decide what plants and animals are desirable as food. *(c)*

32. The objective of this classification system is to determine the evolutionary histories of organisms and express those relationships in phylogenetic trees. *(b)*

33. This classification system groups organisms primarily on the basis of gross morphological similarities and differences. *(a)*

34. In a cladistic classification, each taxon
    a. includes several lineages from a single common ancestor.
    b. is a single lineage and includes all—and only—the descendants of a single ancestor.
    c. shares common morphologies that are homoplasic.
    d. shares common morphologies that are derived characters.
    e. includes several lineages that share common morphologies.

35. The excellent fossil record of horses shows that modern horses, which have one toe on each foot, evolved from ancestors that had multiple toes. A trait like the modern horse's single toe that differs from the ancestral trait in the lineage is called
    a. a derived trait.
    b. an ancestral trait.
    c. a morphological trait.
    d. a biochemical trait.
    e. a fundamental trait.

36. Any two structures derived from a common ancestral trait are said to be
    a. analogous.
    b. morphological traits.
    c. biochemical traits.
    d. **homologous.**
    e. homoplasic.

37. Homoplasy can result from convergent evolution. Under convergent evolution,
    a. **structures that were formerly very different come to resemble one another because they have undergone selection to perform similar functions.**
    b. the same character evolves in different lineages, often from a common basis.
    c. a structure evolves in a lineage that is not found in their common ancestor.
    d. the same character evolves in different lineages, from the same ancestral trait, due to similar selection pressures.
    e. the ancestral trait evolves to different characters in different lineages.

38. Genetic similarities among some vertebrates have been estimated by DNA hybridization. According to these data,
    a. humans are most closely related to gibbons.
    b. **humans are most closely related to chimpanzees.**
    c. humans are most closely related to baboons.
    d. humans are most closely related to galagos.
    e. humans are most closely related to rhesus monkeys.

39. The debate about whether the giant panda of China is more closely related to bears or raccoons was resolved by _____ data, which revealed that a single strand of panda DNA binds more tightly to a strand of bear DNA than to a strand of raccoon DNA.
    a. immunological distance
    b. electrophoresis
    c. **DNA hybridization**
    d. nucleic acid sequence
    e. amino acid sequence

40. In groups such as birds, _____ make(s) it difficult to resolve phylogenies using only morphological data because the use of different traits produces different phylogenies.
    a. reverse evolution
    b. parallel evolution
    c. **convergent evolution**
    d. genetic drift
    e. founder events

41. Which of the following statements about common names is *not* true?
    a. The same species can have two different common names in different areas.
    b. Two different species can have the same common name.
    c. **According to the Linnaean classification system, common names for two species within the same genus must be different.**
    d. Common names do not exist for some species.
    e. Common names can be both ambiguous and redundant.

42. A taxon is
    a. an archaic concept not used in modern classification systems.
    b. **any group of organisms treated as a unit.**
    c. a single species.
    d. a group of organisms that are reproductively isolated from other such groups.
    e. the smallest grouping in the Linnaean classification system.

43. Which of the following choices is the conventionally correct representation for the name of a sea star common to the rocky intertidal zone?
    a. Pisaster Ochraceous
    b. *Pisaster Ochraceous*
    c. **Pisaster ochraceous**
    d. *pisaster ochraceous*
    e. pisaster ochraceous

44. Which of the following choices is the conventionally correct way to refer to several species of fruit flies?
    a. **Drosophila spp.**
    b. *Drosophila* sp.
    c. *Drosophila's*
    d. Drosophila spp.
    e. Drosophila sp.

45. When writing a scientific paper, which of the following choices shows the conventionally correct way to refer to the binomial name of the English bluebell after the name has been cited earlier?
    a. *Edymion nonscriptus*
    b. Edymion n.
    c. **E. nonscriptus**
    d. E. n.
    e. E. nonscriptus

46. In the Linnaean classification system, which one of the following taxa usually ends in "-idae" when used with animals?
    a. Genus
    b. Order
    c. Division
    d. Class
    e. **Family**

47. Which of the following statements is *not* true?
    a. Members of a family are less similar than members of an included genus.
    b. An order has more members than the number of members in an included genus.
    c. **Families have more members than phyla.**
    d. Members of a family share a common ancestor in the more distant past than members of an included genus.
    e. The number of species in a taxon depends on their relative degree of similarity.

48. Homoplasy can result from all but one of the following. Select the exception.
    a. Convergent evolution
    b. Parallel evolution
    c. Reverse evolution
    d. **Descent from a common ancestor**
    e. Similar selection pressures

49. The classification unit in cladistic classification is the_____ , which is the portion of a phylogeny descended from _____ .
    a. clade; several ancestors
    b. clade; a common ancestor
    c. taxon; several ancestors
    d. taxon; a common ancestor
    e. cladogram; a common ancestor

50. In horses, presence of multiple toes is the _____ trait, while one toe is the _____ trait.
    a. ancestral; derived
    b. derived; ancestral
    c. ancestral; homologous
    d. homologous; ancestral
    e. ancestral; monophyletic

51. Which of the following statements about DNA hybridization is *not* true?
    a. This technique cannot disclose the exact sequence of the DNA.
    b. This technique is especially good for studying relatively unrelated taxa.
    c. DNA hybridization depends on complementary base pairing.
    d. The degree of mismatching in the interspecific double helices lowers the temperature at which dissociation occurs.
    e. Unrelated species would have greater mismatching in their DNA.

52. Which of the following techniques for studying the biochemical traits of organisms was used to understand the evolution of cytochrome *c*?
    a. Immunological distance determination
    b. Amino acid sequencing
    c. DNA hybridization
    d. Nucleic acid base sequencing
    e. cpDNA and mtDNA comparisons

53. DNA hybridization shows that the giant panda is a _____ , and the lesser panda is more closely related to a _____ , and that the two are _____ related.

    a. bear; bear; closely
    b. bear; raccoon; not closely
    c. raccoon; raccoon; closely
    d. raccoon; bear; not closely
    e. bear; raccoon; closely

54. In systematics and phylogeny, the fossil record is especially important
    a. because most groups are well represented.
    b. because it provides the absolute timing of evolutionary events.
    c. because random mutations make most biochemical methods unreliable.
    d. because DNA can be extracted from the fossils and analyzed.
    e. because it is the only type of data useful in reconstructing the past.

55. In reconstructing phylogenies, an outgroup is used to
    a. distinguish general homology from special homology.
    b. exclude a taxon from the phylogenic group.
    c. distinguish homoplasy from convergent traits.
    d. gain knowledge of reverse evolution.
    e. b and c

56. Larval stages can help determine relationships among organisms, but care must be taken in that
    a. not all organisms have a larval stage.
    b. the larval form is too morphologically different from the adult form to be useful.
    c. the larval form may closely resemble an organism that the adult stage does not.
    d. larval forms all have a notochord and all adult forms do not.
    e. All of the above

57. Reconstructed phylogenies can be useful for all of the following except
    a. evolution of human language.
    b. the migration of human populations.
    c. determining if divergent traits occurred from reverse evolution.
    d. predicting future trends in evolution.

23 Molecular Evolution

Fill in the Blank

1. The mechanisms of morphological change were not understood until discoveries were made in **biochemistry**.

2. Molecular evolution studies the patterns of **evolutionary change** in molecules and the process by which this is achieved.

3. The first protein that was sequenced was **insulin**.

4. Most of the variability in molecules being measured by evolutionists does not affect their functioning. This is the hypothesis of **neutral evolution**.

5. Genes that are responsible for development of specific body segments are called **homeobox** genes.

6. A molecule similar to hemoglobin, **myoglobin**, has the greater affinity for oxygen.

7. Lungfishes and lilies have more **total** DNA than humans but less of it is **coding** DNA.

8. If a protein has consistent changes over time, then its **molecular clock** ticks at a constant rate.

9. DNA that has changed little during evolution probably has the same **function** in all species.

10. A visual representation of the evolutionary relationship between species is called a **gene tree**.

11. Morphological traits are not useful for comparisons between bacteria and humans because they lack **comparable structures**.

12. Noncoding regions of DNA that are probably the remains of transposable elements are called **introns**.

13. In studying insulin, it was found that most amino acid substitutions do not affect its biological **activity**.

14. Evidence suggests that the three families of globin genes arose as gene **duplications**.

15. Langurs and cows share a very similar **lysozyme** molecule.

16. To infer times of the most ancient splits in lineages, scientists must use **molecules** that are found in all organisms and evolve slowly.

17. Scientists can only clone small segments of ancient DNA because it **deteriorates** with time.

18. Mitochondrial DNA is **maternally** inherited.

Multiple Choice

1. Molecular evolution
   a. studies the patterns of evolutionary change in molecules.
   b. studies the processes of molecular changes.
   c. compares the molecules of living organisms.
   d. can determine the linkages of various species based on molecular changes.
   e. All of the above

2. Why is the protein cytochrome *c* important in the study of molecular evolution?
   a. It is found in all eukaryotes.
   b. Its mutation rate is very low.
   c. Its mutation rate fluctuates to a great extent.
   d. It is found in all animals.
   e. All of its mutations are adaptive changes.

3. Using the polymerase chain reaction, DNA has been amplified from fossils up to
   a. 3,500 years old.
   b. 13 million years old.
   c. 135 million years old.
   d. 135 billion years old.
   e. DNA cannot be extracted from fossils.

4. Which of the following statements is *not* true of homeobox genes?
   a. All of these genes probably have a common evolutionary origin.
   b. They arose only once in animal development.
   c. They occur in the same order along the chromosome in various organisms.
   d. They are responsible for development.
   e. They vary greatly from species to species.

5. Which of the following is *not* an example of a gene that arose from an ancient gene duplication?
   a. rRNA
   b. tRNA
   c. Hemoglobin
   d. Pyruvate kinase
   e. Lactate dehydrogenase

6. Which of the following is *not* true of lysozyme?
   a. It is produced in tears and saliva.
   b. It can digest bacterial cell walls.
   c. It can digest plant matter.
   d. Langur lysozyme must function at a low pH.
   e. It is found in the whites of bird eggs.

7. What changes did scientists discover when they studied the amino acid sequence of lysozyme in langurs and cows?
   a. The changes were on the active site of the molecule.
   b. Arginine was changed to lysine, making it more resistant to trypsin.
   c. Lysine was changed to arginine, making it less resistant to trypsin.
   d. The changes are all neutral and do not affect the function of the molecule.
   e. That their lysozyme can break down plant matter.

8. Cows and langurs have nearly identical lysozyme. Why?
   a. Lysozyme is nearly identical in all animals, not just cows and langurs.
   b. They both ferment leafy food in their stomachs.
   c. They both share a common ancestor.
   d. The bacteria in their foregut caused mutations in their lysozyme.
   e. None of the above

9. Select the group that shares the most similar lysozyme molecule.
   a. Horse, cow, pig
   b. Langur, baboon, human
   c. Deer, antelope, horse
   d. Neotropical cuckoo, sparrow, cow
   e. Neotropical cuckoo, cow, langur

10. Select the correct order, from smallest to largest, of total DNA in these organisms.
    a. Bacterium, lungfish, human
    b. Fruit fly, bacterium, human
    c. Lungfish, human, fruit fly
    d. Bacterium, human, lungfish
    e. Yeast, bacterium, fruit fly

11. Which of the following is true of how the genome size differs between lungfish and humans?
    a. Lungfish have less total DNA than humans but more coding DNA.
    b. Lungfish have more total DNA than humans but less coding DNA.
    c. Lungfish have more total DNA than humans because they have more genes.
    d. Lungfish have less total DNA than humans because they are simpler organisms.
    e. Lungfish have the same amount of total DNA but less coding DNA.

12. Most changes in nucleic acids or amino acids
    a. change the function of the molecule.
    b. occur at constant locations.
    c. can be predicted.
    d. do not change the function of the molecule.
    e. are influenced by natural selection.

13. A molecular clock is useful for
    a. estimating lineage divergence during evolution.
    b. figuring the rates of change of a molecule.
    c. predicting where a molecule will mutate.
    d. timing when a new gene duplication will arise.
    e. timing the aging in an organism.

14. The consistency of molecular clocks is checked by
    a. finding the rate of mutation of a molecule.
    b. using well-dated fossil records to compare.

    c. artificially creating mutations of molecules in the lab and timing them.
    d. comparing similar molecules.
    e. comparing against cytochrome c.

15. Molecules that evolve slowly are used to study
    a. very similar species.
    b. relationships of all organisms.
    c. extinct organisms.
    d. duplicate genes.
    e. organisms that diverged long ago.

16. Molecules that have a rapid mutation rate are used to study
    a. very similar species.
    b. relationships of all organisms.
    c. extinct organisms.
    d. duplicate genes.
    e. organisms that diverged long ago.

17. Ribosomal RNA makes a good molecule to study evolution because it
    a. mutates frequently.
    b. mutates very slowly.
    c. is found only in a few organisms.
    d. greatly varies from organism to organism.
    e. does not have functional constraints.

18. The times of lineage splits between the major branches of life, Bacteria, Archaea, and Eukarya, have been estimated using
    a. lysozyme.
    b. cytochrome c.
    c. 16S rRNA.
    d. hemoglobin.
    e. genome size.

19. The DNA studies done on the flightless moas and kiwis of New Zealand showed that
    a. the two birds have a very recent divergence from each other.
    b. kiwis evolved on New Zealand and then migrated to Australia.
    c. Moas originated on Australia.
    d. Rheas and moas are more similar than kiwis and moas.
    e. Kiwis and moas are not each other's closest relative.

20. Which of the following statements about why ancient dinosaurs cannot be recreated (as in *Jurassic Park*) is *false*?
    a. Their fossilized DNA is degraded.
    b. It is unknown how to regulate expression of DNA during development.
    c. We don't know how to piece together the fragmented DNA.
    d. Their DNA is different from modern DNA; we couldn't make proteins from it.
    e. Only small amounts of DNA can be amplified.

21. The "out of Africa" hypothesis suggests that
    a. *Homo sapiens* diverged from chimpanzees.
    b. *Homo sapiens* in Europe and Asia arose from a single origin in Africa.
    c. *Homo sapiens* arose from Europe, Asia and Africa simultaneously.
    d. *Homo erectus* arose in Africa, but later *Homo sapiens* arose in Europe.
    e. *Homo sapiens* arose in Africa, died off, and then simultaneously arose in Europe, Asia, and Africa.

22. What finding led to the belief in the "out of Africa" hypothesis?
    a. Fossil records
    b. Y chromosomal DNA sequencing of modern humans
    c. Mitochondrial DNA sequencing of modern humans
    d. Cytochrome c sequencing of modern humans
    e. Total DNA comparisons between fossils and modern humans

23. DNA studies have shown that the ancient inhabitants of Easter Island are most closely related to
    a. Polynesians.
    b. South American Indians.
    c. Early Africans.
    d. Europeans.
    e. Australians.

24. If there are about 100 substitutions in the myoglobin molecule per 500 million years, then two organisms differing by 75 amino acids split about
    a. 500 million years ago.
    b. 125 million years ago.
    c. 380 million years ago.
    d. 485 million years ago.
    e. 65 million years ago.

25. How did many different bacteria acquire the Yops gene?
    a. They all have a recent common ancestor.
    b. All bacteria carry the Yops gene.
    c. Mutations in the DNA created the gene in many bacteria.
    d. It was transferred by phages and plasmids.
    e. It was due to convergent evolution.

26. Which of the following is *not* true of gene duplication?
    a. All duplications produce functional genes.
    b. It can result in evolution of novel functions in proteins.
    c. It created much of the diversity in our genome.
    d. rRNA and tRNA arose from duplications.
    e. It can result in increased complexity.

27. To achieve more accurate estimates of lineage divisions,
    a. molecular data is combined with morphological data.
    b. molecular data is combined with fossil data.
    c. the data from two or more molecules is combined.
    d. a and b
    e. a, b, and c

28–31. The next four questions are based on the hypothetical similarity matrix below:

Human	Horse	Cow	Langur	Baboon	
	9	11	1	0	**Human**
1		1	12	13	**Horse**
1	18		14	12	**Cow**
14	0	0		1	**Langur**
16	0	0	17		**Baboon**

28. What do the numbers on the upper right stand for?
    a. Similarities in amino acid sequence
    b. Differences in amino acid sequence
    c. Number of total amino acids
    d. Number of deletions in amino acids
    e. Number of additions in amino acids

29. What do the numbers on the bottom left stand for?
    a. Similarities in amino acid sequence
    b. Differences in amino acid sequence
    c. Number of total amino acids
    d. Number of deletions in amino acids
    e. Number of additions in amino acids

30. Which of the following are the most similar?
    a. Human and horse
    b. Horse and cow
    c. Human and baboon
    d. Langur and cow
    e. Human and langur

31. Which of the following are the most different?
    a. Human and horse
    b. Horse and cow
    c. Human and baboon
    d. Langur and cow
    e. Human and langur

32. The polymerase chain reaction allowed biologists to
    a. determine sequences of DNA from fossils.
    b. determine sequences of DNA from dried skins.
    c. recreate a genome from fragmented DNA.
    d. a and b
    e. a, b, and c

33. Humans have three families of globin genes. The three arose from
    a. spontaneous mutations.
    b. gene duplication.
    c. ancestral DNA.
    d. differential splicing of DNA.
    e. transposable elements.

34. Select the statement that is incorrect about mitochondrial DNA.
    a. It is maternally inherited.
    b. It was used to give validation to the "out of Africa" hypothesis.
    c. It mutates very slowly.
    d. It is useful for studying recent evolution of closely related species.
    e. None of the above

35. Transposable elements are
    a. vectors that transfer DNA between bacteria.
    b. noncoding regions of DNA.
    c. duplicate genes.
    d. segments of DNA that can be inserted in various points of the DNA molecule.
    e. segments of DNA that have mutated and acquired a new function.

36. Which of the following is *not* true of the genomes of organisms?
    a. Eukaryotes have more coding DNA than do prokaryotes.

b. Mutation rates of DNA can vary within a species.

c. Some genes are accidentally duplicated.

d. **We know how much genetic information is needed to program development of a structure.**

e. Not all DNA is transcribed into RNA.

37. The hypothesis of neutral evolution asserts that

    a. the rate of mutations in molecules is influenced by natural selection.

    b. **the variability in structures of molecules does not affect their functioning.**

    c. closely related species have more similar molecular structures than distantly related species.

    d. organisms evolved through neutral changes in their molecules.

    e. mutations neither add nor subtract amino acids from molecules.

38. The earliest organisms known to have both α- and β-globin genes lived about

    a. **500 million years ago.**

    b. 1 billion years ago.

    c. 100 million years ago.

    d. 50 million years ago.

    e. 1 million years ago.

39. Homeobox elements have been identified in

    a. flies only.

    b. animals.

    c. plants.

    d. bacteria.

    e. **b and c**

40. Sometimes molecular data can mislead scientists to conclude that two species are more closely related than other evidence suggests. A reason for this is

    a. **convergent evolution.**

    b. divergent evolution.

    c. neutral evolution.

    d. gene duplication.

    e. None of the above

41. Which of the following is *not* true of the parallel origins theory of human evolution?

    a. *Homo sapiens* arose simultaneously in Europe, Africa, and Asia.

    b. Ancient human fossils have been found in Europe, Indonesia, and Africa.

    c. **Mitochondrial DNA evidence suggests that this theory is correct.**

    d. It requires 1 million years of divergence from the last common ancestor.

    e. *Homo sapiens* arose from *Homo erectus*.

42. In 1952 the first complete sequence of _____ was determined.

    a. **insulin**

    b. cytochrome *c*

    c. hemoglobin

    d. myoglobin

    e. lysozyme

43. X ray crystallography allows investigators to determine

    a. the sequence of amino acids in proteins.

    b. the sequence of nucleic acids in DNA.

    c. **a molecule's three-dimensional shape.**

    d. the age of fossils.

    e. the composition of ancient bones.

44. The invariant positions 14, 17, 18, and 80 in cytochrome *c* interact with

    a. the positions on the molecule that change rapidly.

    b. the other invariant positions.

    c. **the heme group.**

    d. hemoglobin.

    e. myoglobin.

45. If there were mutations in the invariant positions of a cytochrome *c* molecule, what would be the result?

    a. The molecule would have a new function.

    b. **The molecule would no longer be functional.**

    c. The molecule would still bind the iron-containing group.

    d. The changes would not affect the function of the molecule.

    e. The molecule would bind to another metal-containing group.

46. Genes that are nonfunctional and probably resulted from gene duplication are called

    a. invariant genes.

    b. homeobox genes.

    c. homologous genes.

    d. **pseudogenes.**

    e. exons.

47. Which of the following is *not* true of hemoglobin?

    a. **It is a simpler molecule than myoglobin.**

    b. It binds four molecules of oxygen.

    c. It is a more refined molecule than myoglobin.

    d. It has a lower affinity for oxygen than myoglobin.

    e. It is a tetramer of two α and two β chains.

48. By studying the similarities of molecules between species, scientists can

    a. determine common ancestry.

    b. determine the position of branches in a cladogram.

    c. determine a similar function in widely divergent species.

    d. *a* and *b*

    e. **a, b, and c**

49. Which of the following must be true in order to use molecular clocks to date events?

    a. There must be relatively few adaptive changes in the molecule.

    b. Scientists must know what proportion of molecular clocks tick at a constant rate.

    c. The molecular clock of the molecule being studied must tick very slowly.

    d. **a and b**

    e. *a, b,* and *c*

50. Ancient lineage divisions can be estimated by

    a. molecular data.

    b. morphological data.

    c. fossil data.

    d. *a* and *c*

    e. **a, b, and c**

24 The Origin of Life on Earth

Fill in the Blank

1. **Louis Pasteur** performed an experiment in 1862 that finally convinced most people that spontaneous generation does not occur.

2. **Stanley Miller** was one of the first scientists to perform experiments simulating conditions similar to those believed to have prevailed on Earth when life first evolved.

3. The formation of living organisms from nonliving matter, often called **spontaneous generation**, was accepted by most people as an obvious fact of nature for more than 2,000 years.

4. **Coacervates** are drops that spontaneously form in solutions and have several properties relevant to the origin of life. They can contain enzyme molecules, absorb substrates, catalyze reactions, and let the products diffuse back out into the solution.

5. The universe is estimated to be between **10 and 20 billion** years old.

6. Life on Earth is thought to have arisen **1 billion** years after the planet was formed.

7. Oxygen-producing photosynthesis first evolved among prokaryotes called **cyanobacteria**.

8. Early organisms reproduced by simple **fusion**.

9. Before life existed on Earth, free oxygen was not available for chemical reactions in the atmosphere; this is known as a(n) **reducing** environment.

10. **Ribozymes** were believed to be the first catalysts for synthesizing proteins in early life forms.

11. The larger, eukaryotic organisms were only able to flourish after **oxygen** levels increased.

12. In order for eukaryotic organisms to replicate large amounts of DNA quickly, and for better organization, genes were formed into **chromosomes**.

13. **Mitochondria or Chloroplasts** evolved from a mutually beneficial relationship between large bacteria and the smaller bacteria that they engulfed.

14. The polymerization of large molecules is achieved by a class of reactions called condensations, or **dehydrations**.

15. Cells with a **lower** surface area-to-volume ratio need more oxygen than cells with a **higher** surface area-to-volume ratio.

Multiple Choice

1. According to the "big bang" and related theories, our part of the universe developed in this order:
   a. **galaxy, sun, Earth.**
   b. sun, Earth, galaxy.
   c. Earth, galaxy, sun.
   d. sun, galaxy, Earth.
   e. galaxy, Earth, sun.

2. Earth's crust is thinnest under the _____ and has a density that is _____ than Earth's mantle.
   a. continents; less
   b. continents; more
   c. continents; the same as
   d. **oceans; less**
   e. oceans; more

3. Earth's greatest concentration of iron and nickel is found in its
   a. soils.
   b. crust.
   c. mantle.
   d. outer liquid core.
   e. **central core.**

4. Earth's early atmosphere contained all of the following gases except
   a. ammonia.
   b. carbon dioxide.
   c. methane.
   d. nitrogen.
   e. **oxygen.**

5. The first seas formed from the
   a. upwelling of liquid water from vents in the Earth's crust.
   b. **condensation of water vapor from the atmosphere.**
   c. melting of polar ice caps.
   d. formation of water from hydrogen and oxygen gases.
   e. Earth's assimilation of asteroids composed of ice.

6. In the presence of oxygen, most anaerobic bacteria
   a. are unaffected.
   b. switch to aerobic respiration.
   c. **are poisoned.**
   d. convert the oxygen to carbon dioxide.
   e. are stimulated.

7. The presence of which feature of Earth is due most directly to the activities of organisms?
   a. Oceans
   b. Continents
   c. Atmospheric nitrogen
   d. Ozone layer
   e. Volcanoes

8. Of the organelles in the eukaryotic cell, which two most clearly originated as endosymbiotic bacteria?
   a. Nucleus and ribosomes
   b. Chloroplast and mitochondrion
   c. Vacuole and flagellum
   d. Nucleus and vacuole
   e. Chloroplast and ribosome

9. Stanley Miller showed that, given conditions similar to those of early Earth,
   a. inorganic molecules would react to form organic molecules.
   b. RNA would self-replicate.
   c. organic molecules would form primitive cells.
   d. an oxygen atmosphere would develop.
   e. DNA would be synthesized.

10. Polymerization of molecules is *not* associated with
    a. solutions of hydrogen cyanide.
    b. drying bodies of water containing organic molecules.
    c. solutions of formaldehyde.
    d. monomers attached to phosphates.
    e. solutions of glucose.

11. A ribozyme is
    a. a small organelle found in all cells.
    b. an enzyme that synthesizes ribose sugar.
    c. an RNA molecule that catalyzes reactions.
    d. a type of proto-cell containing both sugars and nucleic acids.
    e. a phosphorylated ribose solution capable of polymerization.

12. In which order did the following evolutionary events most likely occur?
    1. RNA catalyzes the synthesis of lipids and proteins.
    2. Proteins catalyze chemical reactions.
    3. RNA catalyzes its own replication.
    a. 1, 2, 3
    b. 2, 3, 1
    c. 3, 2, 1
    d. 2, 1, 3
    e. 3, 1, 2

13. Which of the following is *not* a characteristic of coacervates?
    a. They have lipid coats.
    b. Their interiors are high in protein.
    c. They can catalyze reactions.
    d. They contain DNA.
    e. They can absorb substrate molecules.

14. The theory that DNA evolved after RNA is supported by the fact that

a. **DNA is stable only in hydrophobic environments.**
b. not all modern cells have DNA, but all have RNA.
c. DNA is synthesized by RNA in modern cells.
d. DNA is more similar among different groups of organisms than RNA is.
e. DNA is found in cells' nuclei, and early cells lacked nuclei.

15. Purple sulfur bacteria are capable of growing in sediments that _____ oxygen and _____ exposed to light.
    a. lack; may or may not be
    b. have; are
    c. have; are not
    d. either have or lack; are not
    e. lack; are

16. Cyanobacteria liberate oxygen molecules by splitting molecules of
    a. carbon dioxide.
    b. glucose.
    c. sulfate.
    d. nitrate.
    e. water.

17. The "Milky Way" refers to our
    a. solar system.
    b. galaxy.
    c. universe.
    d. nebula.
    e. constellation.

18. Our solar system, including the sun, Earth, and our sister planets, took form
    a. about 10–20 billion years ago.
    b. about 5 billion years ago.
    c. about 3.5 billion years ago.
    d. about 5 million years ago.
    e. about 3.5 million years ago.

19. Lightning and other energy sources converted early atmospheric gases into
    a. proteins.
    b. DNA.
    c. simple organic molecules.
    d. RNA.
    e. chlorophyll *a*.

20. Which of the following types of information is *not* used as a basis of plausible theories of the origins of life?
    a. The preserved remains of early organisms
    b. The study of a single common ancestor of all present-day organisms
    c. Laboratory experiments investigating chemical reactions under conditions similar to those thought to have existed on early Earth
    d. The observation that all organisms share the same genetic machinery
    e. The observation that all organisms have similar basic cellular metabolic pathways

21. Oparin's theory about the order of events in the origin of life was
    a. **cells first, metabolism second, and replication third.**
    b. metabolism first, cells second, and replication third.
    c. cells first, replication second, and metabolism third.
    d. metabolism first, replication second, and cells third.
    e. replication first, cells second, and metabolism third.

22. Coacervate drops that contain enzymes
    a. do not preferentially concentrate substances.
    b. have lipids in their interiors.
    c. have polysaccharide boundaries with membrane-like structures.
    d. **absorb substrates, catalyze reactions, and release products back into the solution.**
    e. are usually unstable.

23. Oparin believed
    a. **that coacervate drops were possible precursors to cells.**
    b. that the first coacervates were complex and they changed into simpler structures.
    c. that coacervates could not evolve.
    d. that complex metabolic reactions could not occur in coacervates.
    e. that genes appeared before coacervates.

24. Eigen's work with RNA supports the theory that life originated according to the sequence
    a. **replication first, metabolism second, and cells third.**
    b. metabolism first, cells second, and replication third.
    c. cells first, metabolism second, and replication third.
    d. replication first, cells second, and metabolism third.
    e. cells first, replication second, and metabolism third.

25. Molecular biologists were attracted to the idea that genes evolved first because
    a. without replication there is no life.
    b. RNA is the heredity molecule.
    c. DNA is the heredity molecule.
    d. **RNA is simpler than proteins and might evolve more readily.**
    e. RNA can replicate without templates or enzymes.

26. A problem central to understanding the evolution of life is to determine how the process of transcription of DNA into RNA and the subsequent translation of RNA into proteins evolved. That process provides
    a. stability of proteins.
    b. **a hereditary basis for metabolic pathways.**
    c. versatility of nucleic acids.
    d. a mechanism of nutrient cycling.
    e. the ability to use RNA as the hereditary molecule.

27. The oldest remains of living things discovered so far are
    a. **3.5 billion years old.**
    b. 350 million years old.
    c. 35 million years old.
    d. 3.5 million years old.
    e. 5 million years old.

28. Of the following, the first bacteria to increase the oxygen content of the early atmosphere are the
    a. archaebacteria.
    b. **cyanobacteria.**
    c. green sulfur bacteria.
    d. purple sulfur bacteria.
    e. purple nonsulfur bacteria.

29. Some purple nonsulfur bacteria obtain the hydrogen atoms that they need from
    a. decaying organisms.
    b. plant material.
    c. **hydrogen gas.**
    d. water.
    e. ammonia.

30. During the first billion years of life on Earth, living things included all of the following, except
    a. anaerobic photosynthesizers.
    b. heterotrophic bacteria.
    c. sulfate reducers.
    d. **eukaryotes.**
    e. archaebacteria.

31. Louis Pasteur designed an experiment to prove that
    a. bacterial organisms cannot be killed by heat.
    b. **life does not arise spontaneously from nonliving matter.**
    c. Earth was really much older than people of the time thought.
    d. the half-life of uranium-238 is 10 billion years.
    e. maggots grow in meat.

32. Coacervates are
    a. the first catalysts for synthesizing proteins.
    b. **a mixture of protein and polysaccharides in an aqueous drop believed to be precursors to cells.**
    c. the fossils of what is believed to be the earliest life form.
    d. molecules that are precursors to DNA.
    e. early organisms that obtain energy without using oxygen.

33. Two organisms responsible for creating the atmospheric environment conducive to life today are
    a. **cyanobacteria and anaerobic bacteria.**
    b. anaerobic bacteria and archaebacteria.
    c. filamentous bacteria and archaebacteria.
    d. cyanobacteria and coacervates.
    e. coacervates and blue-green algae.

34. Diploid cells arose from
    a. fusion of two haploid cells.
    b. failure of the cell to divide following duplication of DNA.
    c. double duplication of DNA.
    d. environmental stresses to the cell.
    e. **a and b.**

THE ORIGIN OF LIFE ON EARTH **123**

35. Which of the following is *not* a benefit of reproduction by sexual recombination?
    a. Better repair of damaged chromosomes
    b. Better adaptability to different environments
    c. Faster reproduction
    d. Greater variation in offspring
    e. Longer evolutionary life

36. Life forming from nonlife, polymerization of biological molecules, does not happen today because
    a. there is too much oxygen in the atmosphere.
    b. cyanide and formaldehyde are present in lower concentrations today.
    c. those biological molecules are being consumed by existing life.
    d. the temperature of Earth is much lower today, those reactions cannot take place.
    e. All of the above

37. The key change necessary for aerobic life to evolve was
    a. the ability of organisms to use water as a source of hydrogen.
    b. Earth's reducing environment became an oxidizing environment.
    c. the ability of organisms to break down carbon dioxide.
    d. the ability of organisms to break down methane.
    e. the extinction of anaerobic bacteria.

38. Green and purple sulfur photosynthetic bacteria obtain hydrogen from
    a. water.
    b. hydrogen sulfide.
    c. sulfur dioxide.
    d. ethanol.
    e. hydrogen gas.

39. The purpose of red and yellow carotenoids in anaerobic photosynthetic bacteria is to
    a. convert light into mechanical energy.
    b. make oxygen.
    c. absorb wavelengths of light not absorbed by chlorophyll.
    d. make chlorophyll.
    e. convert chlorophyll into bacteriochlorophyll.

40. Fossils of cyanobacteria are called _____ and are still forming in salty areas of the earth today.
    a. stromatolites
    b. stalactites
    c. carotenoids
    d. coal
    e. carbon deposits

41. Which of the following is *not* true of haploid organisms?
    a. They can produce diploid cells when under stress.
    b. They can form a diploid cell by fusion of two haploid cells.
    c. Most are unicellular organisms.
    d. They can repair chromosome damage.
    e. They contain a single DNA molecule.

42. _____ appear(s) to be the main cause of ozone destruction.
    a. Chlorine compounds
    b. Phosphate compounds
    c. Carbon dioxide
    d. Methane
    e. Ultraviolet radiation

43. Ultraviolet radiation increases the incidence of mutations and cancer because it
    a. breaks down proteins.
    b. damages DNA.
    c. causes sunburns.
    d. interrupts translation of RNA into proteins.
    e. causes cells to rupture.

44. Which is the correct order, from surface inward, of the layers of the earth?
    a. Core, mantle, crust
    b. Mantle, core, crust
    c. Mantle, crust, core
    d. Crust, core, mantle
    e. Crust, mantle, core

45. Before life evolved, a new atmosphere on Earth was created by the release of _____ by the crust and mantle.
    a. carbon dioxide
    b. nitrogen
    c. heavy gasses
    d. a and b
    e. a, b, and c

46. Polymers that came to predominate in the prebiotic soup
    a. formed faster than others.
    b. were more stable than others.
    c. were larger molecules than others.
    d. a and b
    e. a, b, and c

47. The synthesis of sugars was important to early life because they
    a. are components of nucleotides.
    b. form cell walls.
    c. form alcohols and carboxylic acids.
    d. are components of lipids.
    e. catalyzed the first chemical reactions.

48. Eigen, in his RNA experiments, found that when he added _____ to the reaction, larger sequences of RNA could be formed.
    a. ribonuclease P
    b. DNA
    c. zinc
    d. nickel
    e. tRNA

49. In the textbook, which of the following molecules was shown to cut a pre-tRNA molecule at the correct spot?
    a. Ribonuclease P
    b. RNA
    c. Zinc
    d. Endonuclease
    e. None of the above

50. Which of the following is *not* true of proteins as enzymes?
    a. They are better catalysts than RNA.
    b. They are capable of more diverse specifications.
    c. They took over most enzymatic functions from RNA.
    d. They catalyze the actions of ribozyme.
    e. They are translated by RNA.

51. Large coacervate drops
    a. have a smaller capacity for replication than small drops.
    b. are less stable than smaller drops and can be broken.
    c. are more efficient and need less energy than smaller drops.
    d. b and c
    e. a, b, and c

52. Aerobic metabolism _____ than anaerobic metabolism.
    a. is faster
    b. is more efficient
    c. uses more energy
    d. a and b
    e. a and c

25 Bacteria and Archaea: The Prokaryotic Domains

Fill in the Blank

1. The most abundant organisms on Earth are bacteria that fall into the kingdoms Archaebacteria and **Eubacteria.**

2. Bacteria unable to survive for extended periods of time in the absence of oxygen are termed **obligate aerobes**.

3. Photoautotrophs use light as their source of energy and **carbon dioxide** as their source of carbon.

4. With respect to the four nutritional categories of bacteria, most bacteria are **chemoheterotrophs**, as are all animals, fungi, and most protists.

5. **Hans Christian Gram** developed a staining process for bacteria in 1884 that is still the single most common tool in the study of bacteria.

6. **Invasiveness** is the ability of a bacterial pathogen to enter into and multiply within the body of the host.

7. **Toxigenicity** is the ability of a bacterial pathogen to produce chemical substances injurious to the tissues of the host.

8. Archaea are closer, evolutionarily, to **eukarya** than to bacteria.

9. *Treponema pallidum,* the causative agent of syphilis, belongs to the group of bacteria known as the **spirochetes.**

10. *Agrobacterium tumefaciens* is an important bacteria, useful for inserting genes into **plant hosts.**

11. One of the smallest bacteria, *Chlamydia*, has a complex intracellular reproductive cycle.

12. The important bacteria *Escherichia coli* has a **rod** shape and a **negative** Gram stain.

13. Actinomycetes have a branched system of filaments and were once thought to be **fungi**.

14. Many gram-negative bacteria have **endotoxins** comprised of lipopolysaccharides that can cause fever and vomiting.

15. Some gram-positive bacteria have potent **exotoxins** made of proteins that can be fatal.

Multiple Choice

1. Which of the following is *not* one of Koch's postulates?
   a. The microorganism is always found in the diseased individual.
   b. The microorganism taken from the diseased host can be grown in pure culture.
   c. A sample of the pure culture of the microorganism produces the disease when injected into an uninfected host.
   d. A host infected by injection from the cultured microorganism yields a new culture of the microorganism identical to the original culture.
   e. The microorganism can be transmitted by insect vectors.

2. Pathogenic bacteria often face difficulty establishing themselves in a host because of the host's immune system. The bacterium that causes diphtheria, *Corynebacterium diphtheriae,* is able to multiply only within the throat of the host. We say such a bacterium has
   a. low invasiveness.
   b. high invasiveness.
   c. low toxigenicity.
   d. high toxigenicity.
   e. moderate toxigenicity.

3. One factor that determines the consequences of a bacterial infection for the host is the ability of the bacterium to produce chemical substances injurious to the tissues of the host. The anthrax-causing bacterium, *Bacillus anthracis,* produces few toxins; however, it can multiply readily and ultimately invades the entire bloodstream. We say such a bacterium has
   a. low invasiveness and high toxigenicity.
   b. high invasiveness and high toxigenicity.
   c. low invasiveness and low toxigenicity.
   d. high invasiveness and low toxigenicity.
   e. moderate invasiveness and moderate toxigenicity.

4. Bacteria are known for the many roles they play in biological communities. Which of the following roles is the rarest for this group of organisms?
   a. Pathogens
   b. Digestive aids
   c. Nitrogen and sulfur processing in soils
   d. Decomposers
   e. Uses in industry and agriculture

5. Which one of the following is *not* a motivation for classification of biological organisms?
   a. Convenience
   b. Showing evolutionary affinity
   c. Facilitating identification
   d. Displaying diversity
   e. None of the above

6. Which of the following is *not* a basic unit of a prokaryotic cell?
   a. DNA
   b. RNA
   c. Enzymes for transcription and translation
   d. A system for generating ATP
   e. An immune system

7. Which of the following would be found in both eukaryotes and prokaryotes?
   a. Organelles
   b. A system for generating ATP
   c. A nucleus
   d. Chromatin
   e. Histones

8. Which of the following is *not* a characteristic of prokaryotic cells?
   a. Mesosome
   b. Photosynthetic membrane system
   c. Peptidoglycan
   d. Circular DNA
   e. Organelles

9. An ecosystem based on chemoautotrophs exists 2,500 meters below the ocean surface near the Galapagos Islands. These bacteria
   a. use light as energy and use carbon dioxide for carbon.
   b. use light as energy and get organic compounds from other organisms.
   c. oxidize inorganic substances for energy and use carbon dioxide for carbon.
   d. get both energy and carbon from organic compounds.
   e. oxidize organic compounds for energy and use carbon dioxide for carbon.

10. Bacteria reproduce
    a. only asexually.
    b. only sexually.
    c. asexually and sexually by transformation, conjugation, and/or transduction.
    d. following mitosis.
    e. *a* and *d*

11. Which of the following is *not* a property of the Archaebacteria?
    a. They include some species that live in environments with extreme salinity and low oxygen.
    b. They have peptidoglycan in their cell walls.
    c. Their rRNA is as different from that of other prokaryotes as from that of eukaryotes.
    d. They include some species that are obligate anaerobes, which produce all the methane in the atmosphere, including that associated with human flatulence.
    e. They include some species that love heat and acid and may die of "cold" at 55°C (131°F).

12. Which of the following statements is true?
    a. Gram-positive bacteria have a lot of peptidoglycan in their cell walls and stain blue to purple.
    b. Gram-positive bacteria have relatively little peptidoglycan in their cell walls and stain pink to red.
    c. Gram-positive bacteria weigh more than a gram.
    d. Gram-positive bacteria weigh more than a milligram.
    e. Gram-negative bacteria have no peptidoglycan in their cell walls.

13–18. Match the following descriptions of groups of bacteria with their names from the list below.
    a. Cyanobacteria
    b. Spirochetes
    c. Chlamydia
    d. Gram-negative rods
    e. Actinomycetes
    f. Mycoplasmas

13. Gram-positive bacteria; form mycelium; once classified as fungi; include the bacteria that cause tuberculosis and produce streptomycin; most antibiotics come from bacteria in this group *(e)*

14. Photosynthesize using chlorophyll *a;* a homogeneous grouping with similar rRNA sequences; contain photosynthetic lamellae or thylakoids *(a)*

15. Use axial filaments to move; include the bacterium that causes syphilis *(b)*

16. Not a natural group with close evolutionary ties; lots of diversity among the members *(d)*

17. Small intracellular parasites; have a unique, complex reproductive cycle; include strains that cause eye infections and venereal disease *(c)*

18. Lack cell walls; smallest cellular creatures; mostly plant and animal parasites *(f)*

19. A disease-causing bacterium kills many cells within the host, but spreads poorly from one individual to another. It can be said to have
    a. low invasiveness.
    b. high invasiveness.
    c. high toxigenicity.
    d. low invasiveness and high toxigenicity.
    e. high invasiveness and high toxigenicity.

20. Which of the following is *not* a characteristic of prokaryotic cells?
    a. Lack any internal membrane systems
    b. Some have cell walls composed of peptidoglycan.
    c. DNA not organized into chromatin
    d. A circular chromosome
    e. Cell division by binary fission

21. The purple sulfur bacteria use H_2S as an electron donor and release pure sulfur as a waste product. They are examples of
    a. photoautotrophs.
    b. photoheterotrophs.
    c. chemoautotrophs.
    d. chemoheterotrophs.
    e. deep sea, volcanic vent bacteria.

22. Which of the following nutritional categories of bacteria can exist independently of other organisms?
    a. **Photoautotrophs**
    b. Photoheterotrophs
    c. Photochemotrophs
    d. Chemoheterotrophs
    e. More than one of the above

23. The majority of bacteria are
    a. photoautotrophs.
    b. photoheterotrophs.
    c. chemoautotrophs.
    d. **chemoheterotrophs.**
    e. disease-causing.

24. The Gram method is useful for classifying all bacteria that
    a. have two plasma membranes.
    b. have cell walls.
    c. **have cell walls with at least some peptidoglycan.**
    d. are prokaryotic.
    e. form endospores.

25. Which of the following is *not* characteristic of endospores?
    a. **Parent cells can produce more than one**
    b. Can survive harsh environmental conditions
    c. Contain some cytoplasm and replicated nucleic acid
    d. Enclosed within a tough cell wall
    e. A resting structure, not a reproductive structure

26. Which of the following is *not* characteristic of Archaebacteria?
    a. Live in harsh environments
    b. Cell walls without peptidoglycan
    c. Unlike most other bacteria
    d. Similarities in base sequences of ribosomal RNAs
    e. **A recently evolved group**

27. Which of the following areas or conditions would be favored by thermoacidophiles?
    a. The stomachs of many herbivores
    b. Hot, alkaline springs
    c. Anaerobic conditions
    d. **Hot, sulfur springs**
    e. Deep sea volcanic vents

28. Which of the following is *not* characteristic of the methanogens?
    a. **Methane is their preferred carbon source.**
    b. They are associated with mammalian flatulence.
    c. They prefer anaerobic conditions.
    d. Some are thermophilic.
    e. Some live in volcanic vents on the ocean floor.

29. If you were looking for a new heat-tolerant DNA polymerase enzyme, you should investigate members of the
    a. Thermoacidophiles
    b. Methanogens
    c. Strict halophiles
    d. ***a* and *b***
    e. All of the above

30. Which one of the following bacterial groups includes the greatest number of species?
    a. Archaebacteria
    b. **Proteobacteria**
    c. Gram-positive
    d. Mycoplasmas
    e. Cyanobacteria

31. Gliding bacteria
    a. were once classified as fungi.
    b. **form erect fruiting structures.**
    c. include spirochetes.
    d. are known to glide using cilia.
    e. contain less DNA than any other organism.

32. Spirochetes
    a. are all free-living.
    b. are the only spiral-shaped bacteria.
    c. can form chains of cells.
    d. **all possess structures called axial filaments.**
    e. are all parasitic on other bacteria.

33. Gram-negative rods
    a. are all closely related.
    b. **include many human pathogens, but also *Escherichia coli.***
    c. can only reproduce within the cells of other organisms.
    d. all possess structures called axial filaments.
    e. include the important genera *Bacillus* and *Clostridium.*

34. Chlamydias
    a. can form heterocysts.
    b. form a branched, filamentous mycelium.
    c. contain less DNA than any other organism.
    d. **can only reproduce within the cells of other organisms.**
    e. were once classified as fungi.

35. Cyanobacteria
    a. are a diverse, unrelated group of bacteria.
    b. **use chlorophyll *a* and liberate oxygen during photosynthesis.**
    c. are the only group of photosynthetic bacteria.
    d. can reproduce sexually.
    e. are obligate aerobes.

36. Gram-positive cocci
    a. include nitrogen-fixing bacteria, such as *Rhizobium.*
    b. include the important genus *Staphylococcus.*
    c. have members that produce extremely potent toxins.
    d. form chains of cells by mitosis.
    e. ***b* and *c***

37. Actinomycetes
    a. are photoheterotrophs.
    b. **are the source of many important antibiotics.**
    c. are gram-negative bacteria.
    d. were once classified as protists.
    e. are the only bacteria to divide by mitosis.

38. Mycoplasmas
    a. **contain less DNA than any other organism.**
    b. have peptidoglycan cell walls.
    c. form a branched, filamentous mycelium.
    d. include elaborate internal membrane systems.
    e. can be controlled by penicillin.

39. The earliest fossils of bacteria date back at least
    _____ years.
    a. 35,000
    b. 350,000
    c. 3.5 million
    d. 3.5 billion
    e. 3.5 trillion

40. Which of the following is a broad nutritional category
    of bacteria that is recognized by biologists?
    a. Physioautotrophs
    b. Heteroautotrophs
    c. Chemoautotrophs
    d. a and b
    e. a and c

41. Bacteria differ from one another
    a. structurally.
    b. metabolically.
    c. reproductively.
    d. a and c
    e. a, b, and c

42. An extremely important set of rules for the determi-
    nation of bacterial disease transmission is
    a. Ehrlich's optimal law.
    b. Koch's postulates.
    c. Occam's razor.
    d. Zeno's paradox.
    e. Darwin's law of evolution based on natural selection.

43. Bacteria participate in
    a. digestion in animals.
    b. processing nitrogen and sulfur in soils.
    c. decomposition in all ecosystems.
    d. many industrial and commercial processes.
    e. All of the above

44. The pathogenic bacteria *Clostridium* and *Bacillus*
    a. are gram-positive cocci.
    b. are gram-negative cocci.
    c. have endospores.
    d. have endotoxins.
    e. a and c

45. Among the modes of locomotion for bacteria are
    a. flagella, gas vesicles, and rolling.
    b. flagella, cilia, and axial filaments.
    c. axial filaments, rolling, and pseudopods.
    d. cilia, pseudopods, and axial filaments.
    e. pseudopods, flagella, and cilia.

46. Some cyanobacteria show heterocysts, which are
    a. the method of locomotion for the bacteria.
    b. a cell type specialized for nitrogen fixation.
    c. a reproductive state.
    d. an endospore.
    e. None of the above

26 Protists and the Dawn of the Eukarya

Fill in the Blank

1. All eukaryotes that are not plants, fungi, or animals are defined as **protists**.

2. Animal-like protists are also called **protozoans**.

3. The concept of different organisms living together one inside the other is called **endosymbiosis**.

4. The process whereby a diploid generation that produces spores alternates with a haploid generation that produces gametes is called **alternation of generations**.

5. Members of the protozoan phylum **Foraminifera** secrete shells of calcium carbonate. These protozoans are valuable in the geological dating of sedimentary rocks and in oil prospecting.

6. **Contractile vacuoles** keep some freshwater protists from exploding by taking on too much water.

7. Radiolarians contain photosynthetic algae as **endosymbionts**. The algae provide food for the radiolarians, and the radiolarians provide protection for the algae. Both the algae and the radiolarians are protists.

8. Ciliates, such as paramecia, often contain two kinds of nuclei, a single **macronucleus** with DNA that is translated and transcribed, and several **micronuclei**, which are typical eukaryotic nuclei.

9. Paramecia have an elaborate sexual behavior called **conjugation**, in which sexual recombination but not sexual reproduction occurs.

10. The form taken by acellular slime molds under adverse environmental conditions is called a **sclerotium**. This structure rapidly becomes a **plasmodium** again upon restoration of favorable conditions.

11. Many dinoflagellates are **bioluminescent;** that is, cultures of these organisms in complete darkness emit a faint glow.

12. **Chlorophyll *b*** is found only in plants and green algae; thus it is thought that plants evolved from green algae.

13. Many **diatoms** deposit silicon in their cell walls; architectural magnificence on a microscopic scale is a hallmark of this group, and their taxonomy is based entirely on their cell wall patterns.

14. Giant kelp belongs to the phylum Phaeophyta, whose members are commonly called **brown algae**.

15. **Foraminiferans** secrete shells made of calcium carbonate.

16. *Trypanosoma*, the causative agent of African sleeping sickness, is transmitted by an insect, the **tsetse fly**.

17. *Paramecium* uses a precise form of locomotion; by beating its **cilia**, it can move forward or backward.

Multiple Choice

1. Plants, fungi, and animals all evolved from
   a. eubacteria.
   b. archaebacteria.
   c. protists.
   d. None of the above
   e. *a* and *b*

2. Organisms thought to have evolved from protists include which of the following?
   a. Protozoans
   b. Algae
   c. Slime molds
   d. Fungi
   e. All of the above

3. The overall size that unicellular protists can achieve is limited by their
   a. energy-producing potential.
   b. metabolism.
   c. mitochondria.
   d. surface area-to-volume ratio.
   e. None of the above

4. Which of the following is a sexual reproductive process common to organisms in the kingdom Protista?
   a. Simple splitting of the cell
   b. Multiple fission
   c. Budding and spore formation
   d. Union of male and female gametes
   e. All of the above

5. Perhaps the best-known _____ are the malarial parasites of the genus *Plasmodium*.
   a. turbellarians
   b. mastigophorans
   c. dinoflagellates
   d. apicomplexans
   e. poriferans

6. Which of the following are classified as funguslike protists?
   a. Slime molds
   b. *Paramecium*
   c. *Giardia*
   d. Algae
   e. All of the above

7. Which of the following modes of nutrition is used by protists to fuel their metabolism?
   a. Autotrophic
   b. Absorptive heterotrophic
   c. Ingestive heterotrophic
   d. Both autotrophic and heterotrophic
   e. All of the above

8. Protists are found in which of the following habitats?
   a. Marine
   b. Freshwater aquatic
   c. Within the body fluids of other organisms
   d. Damp soil
   e. All of the above

9. Contractile vacuoles help some protists cope with a(n) _____ environment. These protists have a more negative osmotic potential than their freshwater environments and constantly take on water by osmosis.
   a. hypoosmotic
   b. hyperosmotic
   c. isoosmotic
   d. acidic
   e. basic

10. The _____ are beautiful marine protists that harbor photosynthetic protists as endosymbionts.
    a. algae
    b. protozoa
    c. radiolarians
    d. flagellates
    e. dinoflagellates

11. Some algae and funguslike protists demonstrate the phenomenon of alternation of generations, in which a multicellular diploid, spore-producing organism gives rise to a multicellular haploid, gamete-producing organism. Which of the following statements about alternation of generations is *not* true?
    a. The haploid and diploid organisms may or may not differ morphologically.
    b. The haploid and diploid organisms differ genetically.
    c. Only the haploid organism may also reproduce asexually.
    d. Haploid gametes can only produce new organisms by fusing with other gametes.
    e. Diploid sporophytes may undergo meiosis to produce haploid spores.

12. Paramecia contain two types of nuclei: a large macronucleus and as many as 80 micronuclei. The micronuclei are typical eukaryotic nuclei, essential for genetic recombination. The macronucleus
    a. is important in sexual recombination (conjugation).
    b. contains several micronuclei.
    c. contains many copies of the genetic information.
    d. contains DNA that is not transcribed.
    e. contains DNA that is transcribed but not translated.

13. Paramecia have an elaborate sexual behavior in which they line up against each other and fuse. This is followed by an extensive reorganization and exchange of nuclear material. This process is called
    a. isogamous reproduction.

b. alternation of generations.
c. sexual reproduction.
d. conjugation.
e. None of the above

14. Many ciliates have an exceptional degree of cytoplasmic organization, considering that they are unicellular. Which of the following types of organization is *not* found in this group?
    a. Fused cilia, cirri, that move in an independent, but coordinated fashion
    b. Nervelike neurofibrils that coordinate locomotion
    c. Musclelike fibers, myonemes, that cause retraction of stalks in some forms
    d. A "gut" with "mouth," "esophagus," and "anus"
    e. All of the above

15–18. Match the phyla in the list below with the correct description of their members.
    a. flagellates
    b. Myxomycota
    c. Dictyostelida
    d. Acrasida
    e. Oomycota

15. Cellular slime molds. An amoeboid cell is the vegetative unit. Myxamoebas, which have a single haploid nuclei, engulf food particles. *(c)*

16. Acellular slime molds. Their bodies, called plasmodia, contain many nuclei enclosed in a single plasma membrane. When conditions become unfavorable, they can form either a sclerotium, a hardened irregular mass of cell-like components, or sporangiophores, also known as fruiting structures. *(b)*

17. Another group of cellular slime molds. They do not appear to use cAMP as an aggregation signal. *(d)*

18. This phylum includes the water molds and their funguslike terrestrial relatives, such as the downy mildews. The mold that brought about the great Irish potato famine of 1845–1847 is included in this group. *(e)*

19. The feature that distinguishes acellular from cellular slime molds is
    a. that cellular slime molds are motile; acellular ones are not.
    b. the number of nuclei contained within one plasma membrane.
    c. that acellular slime molds ingest food by endocytosis; cellular ones do not.
    d. that cellular slime molds ingest food by endocytosis; acellular ones do not.
    e. that acellular slime molds prefer cool, moist habitats; cellular slime molds prefer dry, hot conditions.

20–24. Match the phyla in the list below with the correct description of their members.
    a. Pyrrophyta
    b. Chrysophyta
    c. Phaeophyta
    d. Rhodophyta
    e. Chlorophyta

20. Diatoms and their relatives make up this phylum. Some species are single-celled; others are filamentous. They may be radially or bilaterally symmetrical and reproduce both sexually and asexually. *(b)*

21. This phylum consists of the red algae, almost all of which are multicellular. The accessory pigment phycoerythrin gives this group its characteristic reddish color; however, many species also contain phycocyanin. The color of individuals in this group can vary depending on environmentally variable light conditions. *(d)*

22. This phylum includes unicellular algae. The dinoflagellates, a major group of marine photosynthetic producers, are in this group. Dinoflagellates may be endosymbionts; they are the cause of red tide, and may be luminescent. *(a)*

23. Multicellular brown algae make up this group. The brown algae are composed of either branched filaments or leaflike growths called thalli. Giant kelps, which may be over 60 meters long, and *Sargassum,* which forms dense mats of vegetation in the Sargasso Sea in the mid-Atlantic, are famous members of this group. *(c)*

24. This phylum is commonly known as the green algae. In this group are the only protists that contain the full complement of photosynthetic pigments characteristic of plants. *(e)*

25. Algae, the plantlike protists, probably carry out _____ of all the photosynthesis on the planet.
    a. **50–60%**
    b. 25–30%
    c. 10–25%
    d. 5%
    e. 1%

26. The plant kingdom is thought to have evolved from the Chlorophyta, the green algae. Which of the following statements is *not* a basis for this belief?
    a. Plants and green algae both contain chlorophyll *a.*
    b. **The photosynthetic storage products of both plants and green algae include oils.**
    c. Chlorophyll *b*, found in plants, is found in no other group of algae.
    d. The principal photosynthetic storage product of both groups is starch.
    e. The carotenoids found in green algae are characteristic of those found in plants.

27. The animal kingdom is thought to have evolved from
    a. **the protozoa.**
    b. the Eubacteria.
    c. the algae.
    d. the funguslike protists.
    e. None of the above

28. Algal life cycles show extreme variation. Only one phylum, the _____, does *not* have flagellated motile cells in at least one stage of the life cycle.
    a. Chlorophyta
    b. **Rhodophyta**
    c. Phaeophyta
    d. Chrysophyta
    e. Pyrrophyta

29. Chromatic adaptation, found in red algae and several other groups of algae, is
    a. the ability to become heterotrophic when light levels are low.
    b. the ability to bioluminesce when disturbed.
    c. the ability to change the wavelength of light to that useful to their photosynthetic pigments.
    d. **the capacity to change the relative amounts of their various photosynthetic pigments depending on the light conditions where they are growing.**
    e. None of the above

30. Radiolarians contain photosynthetic algae as endosymbionts. The algae provide _____ for the radiolarians, and the radiolarians provide _____ for the algae. Both the algae and the radiolarians are protists.
    a. **food, protection**
    b. protection, food
    c. food, essential nutrients
    d. essential nutrients, protection
    e. essential nutrients, food

31. Protists are believed to be ecologically and evolutionary important for many reasons. Which of the following is *not* true of this group?
    a. Multicellular kingdoms evolved from protists.
    b. Photosynthetic protists play a major role in the energy balance of the living world.
    c. **None of the protists are parasites.**
    d. Saprobic protists are among the important decomposers and thus play a major role in the nutrient cycles of the living world.
    e. Many protists have highly differentiated bodies even though they consist of but a single cell.

32. Which of the following terms does *not* apply to notable members of the kingdom Protista?
    a. Autotroph
    b. Absorptive heterotroph
    c. Ingestive heterotroph
    d. Aquatic
    e. **Dikaryotic**

33. Select the group of organisms *not* assigned to the kingdom Protista.
    a. Protozoa
    b. Algae
    c. Slime molds
    d. **Sponges**
    e. Giant kelp

34. You place two different species of protist into a solution with an unknown osmotic potential. Protist A has a contractile vacuole firing rate of 5 per minute; protist B has a contractile vacuole firing rate of 12 per minute. Select a reasonable conclusion based on these observations.
    a. Protist A has a more negative osmotic potential than protist B.
    b. **Protist B has a more negative osmotic potential than protist A.**
    c. The solution has a more negative osmotic potential than protist A.
    d. The solution has a more negative osmotic potential than protist B.
    e. No conclusions can be made unless we know the osmotic potential of the solution.

35. In the alternation of generations, the gametophyte generation is
    a. haploid and produces spores.
    b. haploid and produces gametes.
    c. diploid and produces spores.
    d. diploid and produces gametes.
    e. can be either haploid or diploid, but always produces spores.

36. What do the protists that are responsible for sleeping sickness and malaria both have in common?
    a. They are from the same phylum.
    b. They both have insect vectors for transmission to humans.
    c. They cause the same symptoms.
    d. They both have gametocyte life stages.
    e. All of the above

37. *Plasmodium,* the organism that causes malaria, is in the group
    a. Zoomastigophora.
    b. Sarcodina.
    c. Apicomplexa.
    d. Choanoflagellida.
    e. None of the above

38. Which characteristic does *not* apply to some species of amoebas?
    a. Autotrophic
    b. Free-living
    c. Predators
    d. Parasitic
    e. Shelled

39. In the disease malaria caused by the protist *Plasmodium,* the gametocytes
    a. develop into merozoites.
    b. inhabit the salivary glands of *Anopheles* mosquitoes.
    c. are the infective stage obtained from the insect vector.
    d. are found inside red blood cells.
    e. give rise to zygotes within the mammalian circulatory system.

40. Which of the following statements concerned with the micronucleus and macronucleus is false?
    a. More than one macronucleus may be present.
    b. The micronucleus is involved in genetic recombination.
    c. Multiple copies of macronuclear genes are common.
    d. Transcription and translation mostly involves genes found in the macronucleus.
    e. The micronucleus and macronucleus are unique to members of the phylum Ciliophora.

41. Of the following structures select the one that is *not* associated with movement or feeding in some members of the phylum Ciliophora.
    a. Cirri
    b. Myonemes
    c. Cilia
    d. Pseudopodia
    e. None of the above

42. Which of the following processes is *not* part of conjugation in *Paramecium?*
    a. Meiosis
    b. Mitosis
    c. Cytokinesis
    d. Fusion of haploid nuclei
    e. Breakdown of some micronuclei

43. Select the following event that is *not* normally associated with the life cycle of an acellular slime mold.
    a. Spores develop into myxamoebas
    b. Formation of a sclerotium when conditions are adverse
    c. Active feeding by the plasmodium as it engulfs food particles
    d. Meiosis to form sporangiophores
    e. Swarm cells may fuse to form a diploid zygote

44. Which of the following features is *not* characteristic of the plasmodium of an acellular slime mold?
    a. Coenocyte
    b. Composed of haploid nuclei
    c. Shows cytoplasmic streaming
    d. Formed by mitosis without cytokinesis
    e. Can become a sclerotium

45. In the cellular slime molds, cAMP causes
    a. aggregation of swarm cells to form a plasmodium.
    b. formation of sporangia.
    c. release of myxamoebas from fruiting bodies.
    d. the onset of cytoplasmic streaming.
    e. aggregation of myxamoebas to form a pseudoplasmodium.

46. The algal protist groups differ from each other in terms of the
    a. principal photosynthetic storage product.
    b. unicellular/multicellular body plan.
    c. possession of flagella.
    d. principal photosynthetic pigments.
    e. All of the above

47. Which phylum includes the dinoflagellates?
    a. Pyrrophyta
    b. Chrysophyta
    c. Rhodophyta
    d. Phaeophyta
    e. Chlorophyta

48. Which phylum includes the diatoms?
    a. Pyrrophyta
    b. Chrysophyta
    c. Rhodophyta
    d. Phaeophyta
    e. Chlorophyta

49. Select the following statement that does *not* apply to diatoms.
    a. Diatom cell walls may be impregnated with silicon.
    b. During mitosis, the top and bottom of the cell become the tops of the two new cells.
    c. Prior to meiosis, the cell sheds its cell wall and increases in size.
    d. Zygotes (auxospores) are formed by gametes that lack cell walls.
    e. Diatoms can show bilateral or radial symmetry.

50. The characteristic color of the red algae is due to the presence of
   a. **phycoerythrin.**
   b. fucoxanthin.
   c. β-carotene.
   d. chrysolaminarin.
   e. chlorophyll *b*.

51. Select the algal phylum that has some members showing chromatic adaptation and others that produce mucilaginous polysaccharides important in the production of agar.
   a. Pyrrophyta
   b. Chrysophyta
   c. **Rhodophyta**
   d. Phaeophyta
   e. Chlorophyta

52. Which of the following statements about the life cycle of the green alga *Ulva* is *not* true?
   a. *Ulva* has an isomorphic life cycle.
   b. The sporophyte and gametophyte can only be differentiated microscopically.
   c. **All species of *Ulva* are isogamous.**
   d. Gametophytes produce sperm only or eggs only, never both.
   e. The diploid sporophyte produces flagellated spores.

53. An ecologically and evolutionarily important algae phylum, which is a common endosymbiont of coral, is
   a. **Pyrrophyta.**
   b. Rhodophyta.
   c. Phaeophyta.
   d. Chrysophyta.
   e. Chlorophyta.

54. Which of the following criteria is *not* used in separating protists from animals?
   a. Unicellular versus multicellular
   b. Having an ingestive metabolism
   c. Having an embryonic stage
   d. Having an extracellular matrix of actin and collagen
   e. **Having membrane-enclosed organelles**

27 Plants: Pioneers of the Terrestrial Environment

Fill in the Blank

1. In plants and some algae like *Ulva*, the sporophyte and gametophyte exhibit different levels of ploidy. Sporophytes are **diploid**, while gametophytes are **haploid**.

2. Broadly defined, a **plant** is a multicellular, photosynthetic eukaryote that develops from an embryo protected by tissues of the parent.

3. The vascular system can be said to have been launched by a single evolutionary event. Sometime during the Paleozoic era, the sporophyte generation of a now long-extinct green algae produced a new cell type, the **tracheid**.

4. In the strictest sense, a **leaf** is a flattened photosynthetic structure emerging laterally from a main axis or stem and possessing true vascular tissue.

5. Cycadophyta, Ginkgophyta, Gnetophyta, and Coniferophyta are all phyla of the **gymnosperms**, of which there are only about 750 living species.

6. Most gymnosperms are wind-pollinated but most angiosperms are **animal**-pollinated.

7. A French botanist, E. A. O. Lignier, proposed a currently accepted hypothesis about the origin of **roots**. His hypothesis states that they were derived from dichotomous branches, some of which entered the soil, anchoring the plant and absorbing water and nutrients.

8. **Ferns** are plants with large leaves (fronds) and no seeds, which require water as a medium for the transfer of male gametes. The sporophytes and gametophytes grow independently of each other.

9. One group of seed plants is called the **gymnosperms**. These plants include pines and their relatives and have active secondary growth.

10. The phylum **Angiospermae** is also known as the flowering plants. These plants produce seeds, have double fertilization, and have vessel elements.

11. The tissues that conduct the products of photosynthesis from sites of production to sites of utilization or storage are called the **phloem**.

12. The **sporophyte** generation extends from the zygote through the adult, multicellular, diploid plant.

13. The reproductive organ of angiosperms is the **flower**.

14. The ovary of a flowering plant may develop into a **fruit**. This structure may consist only of the mature ovary and its seeds, or it may include other parts of the flower.

15. Also known as seed leaves, **cotyledons** can digest endosperm or become photosynthetic, or both.

16. In a flower, the male organs are contained in the **stamen,** and the female organs are contained in the **pistil**.

17. **Petals** are modified leaves on a flower that can be quite showy to attract animals.

18. Plants that have an internal transport system are called **tracheophytes**. The older term was vascular plants.

19. The vascular system in plants consists of specialized tissues. The **xylem** transports water and minerals from soil to aerial parts. The **phloem** transports products of photosynthesis from sites of production to storage sites.

Multiple Choice

1. It is widely agreed that the plant kingdom arose from
   a. Eumycota.
   b. Chrysophyta.
   c. Phaeophyta.
   d. Rhodophyta.
   e. Chlorophyta.

2. Characteristics of green algae make them attractive candidates for the origin of plants. Which of the following describes such a characteristic?
   a. Photosynthetic pigments in plastids
   b. Active stomata
   c. Starch as a major storage compound
   d. Cellulose in cell walls
   e. a, c, and d

3. Nonvascular plants have never evolved to the size of vascular plants. The most probable explanation is that they could not solve the problems posed by
   a. photosynthesis.
   b. respiration.
   c. nutrient and water transportation within the plants themselves.
   d. nutrient and water absorption.
   e. All of the above

4. In some of the liverworts there are special elongated cells called elaters that possess a helical thickening of the cell wall. As elaters lose water, they shrink longitudinally and compress the helical thickening like a spring. When the stress reaches a critical point, the compressed "spring" snaps back to its resting position, liberating hundreds of _____ in all directions.
   a. moisture particles
   b. spores
   c. sperm
   d. ova
   e. rhizoids

5. Two important evolutionary consequences of plants having tissues composed of tracheids are
   a. a plant vascular system and structural support.

b. structural support and increased growth.

c. enhanced photosynthesis and structural support.

d. enhanced photosynthesis and a plant vascular system.

e. None of the above

6. Which of the following is *not* consistent with Lignier's hypothesis on the origin of roots?

a. Ancestors of the first vascular plants branched dichotomously.

b. Roots evolved from a symbiotic relationship with simple, avascular plants.

c. Roots were originally stems that just went underground.

d. The different selective pressures between branches in the air and branches in the ground led to the differences in root and stem structure that we see today.

e. All of the above

7. Although Earth is estimated to be 5 billion years old, and although life first appeared about 4 billion years ago, plants did not appear until about _____ years ago.

a. 3 billion

b. 300 million

c. 30 million

d. 3 million

e. 300 thousand

8. Plants can be broadly defined as multicellular photosynthetic eukaryotes. There are several additional features of this group. Which of the following is *not* a characteristic of plants?

a. They develop from embryos protected by tissues of the parent plant.

b. Their cell walls contain cellulose.

c. Their chloroplasts contain chlorophylls *a* and *b*.

d. Their storage carbohydrate is starch.

e. Their respiration is anaerobic.

9. A universal feature of the life cycle of plants is

a. morphologically identical haploid and diploid stages.

b. genetically identical haploid and diploid stages.

c. alteration of generations between heteromorphic haploid gametophytes and diploid sporophytes.

d. All of the above

e. None of the above

10. It was on land that plants first appeared and evolved. Two challenges of occupying the land were

a. (1) development of photosynthetic pigments that are *not* dependent on an aqueous environment and (2) transport of water and minerals to aerial parts.

b. (1) development of starch for carbohydrate storage and (2) transport of water and minerals to aerial parts.

c. (1) physical support and (2) transport of water and minerals to aerial parts.

d. (1) development of photosynthetic pigments that are not dependent on an aqueous environment and (2) development of starch for carbohydrate storage.

e. (1) development of photosynthetic pigments that are not dependent on an aqueous environment and (2) physical support.

11–13. Match the names from the list below with the following descriptions:

a. Hepaticophyta (liverworts)

b. Anthocerophyta (hornworts)

c. Bryophyta (mosses)

d. Eumycota (fungi)

e. Lichens

11. Sporophytes are shorter than the other groups, contain elaters but do not contain stomata *(a)*

12. Archegonia are embedded in the gametophyte tissue instead of being borne on stalks; grows in a manner similar to indefinite growth and thus can be as tall as 20 centimeters; has a single large chloroplast in each cell; has internal cavities filled with mucilage, often populated by cyanobacteria that covert atmospheric nitrogen gas into a form usable by the host plant *(b)*

13. The gametophyte is a branched filamentous plant; the sporophytes of this group usually contain stomata; includes *Sphagnum*, a species found in northern bogs and tundra and which probably is the dominant species of this group in terms of biomass *(c)*

14. Vascular plants are thought to be the result of a single evolutionary event: the evolution of a wholly new cell type, the tracheid. This cell type

a. provides a mechanism for the storage of a new type of carbohydrate, starch.

b. is the first cell type to contain chloroplasts.

c. permits fertilization in the absence of water, thus permitting plants to invade dry habitats.

d. forms the seed.

e. is the principal water-conducting element of the xylem in all vascular plants except the angiosperms.

15–19. Match the names from the list below with the following descriptions:

a. Lycophyta

b. Sphenophyta

c. Pterophyta

d. Gymnosperms

e. Anthophyta

15. Club mosses. These plants have microphyllous leaves and true roots, and include heterosporous and homosporous species. The leaves are arranged spirally on the stems. *(a)*

16. Flowering plants. These plants have megaphyllous leaves; water is not required for the transfer of male gametes; all species are heterosporous. *(e)*

17. Ferns. These plants have megaphyllous leaves, true roots, and large sporophytes; gametophytes are independent of the sporophytes; there are no seeds; water is required for the transfer of male gametes. *(c)*

18. Pines and their relatives. These plants have megaphyllous leaves; only a few species require water for reproduction; all species are heterosporous. *(d)*

19. Horsetails or scouring rushes. These plants have microphyllous leaves and true roots, and include heterosporous and homosporous species. The leaves are arranged in whorls on the stems. *(b)*

20. The dominant vegetation type 200 million years ago, when dinosaurs roamed Earth, was

a. early whisk ferns.

b. horsetail–tree fern forests.

c. lycopod–fern forests.

d. gymnosperm forests.

e. angiosperm forests.

21. There has been a change in the dominant vegetation type since plants first invaded the terrestrial environment about 400–500 million years ago. What was the order in which the following vegetation types were dominant, from earliest to modern day?
    I. Gymnosperm forests
    II. Horsetail–lycopod–fern forests
    III. Whisk fern forests
    IV. Angiosperm forests
    a. I, II, III, IV
    b. II, III, I, IV
    c. **III, II, I, IV**
    d. IV, III, I, II
    e. I, III, II, IV

22. Several important ecological changes occurred that helped to make the invasion of land by plants permanent. Which of the following is *not* one of those changes?
    a. Evolution of a water-impermeable cuticle
    b. **Evolution of a carbohydrate energy storage molecule**
    c. Evolution of protective layers for the gamete-bearing structures
    d. Initial absence of herbivores
    e. All of the above

23. The major difference between a simple leaf and a complex leaf is that in a complex leaf,
    a. **the vascular system of the leaf creates a major alteration in the architecture of the stem vascular system.**
    b. the leaf can be over one centimeter long.
    c. the leaf possesses true vascular tissue.
    d. the chloroplasts are the only photosynthetic structures.
    e. the vascular strand departs from the vascular system of the stem so that there is scarcely any perturbation in the stem's vascular system.

24. A seed of a flowering plant or gymnosperm may contain tissues from _____ generation(s).
    a. 1
    b. 2
    c. **3**
    d. 4
    e. 5

25. One reason for the enormous evolutionary success of seed plants is their possession of
    a. complex leaves that can photosynthesize at a faster rate so that they can outcompete non-seed-producing plants.
    b. seeds with food reserves for the young sporophyte.
    c. seeds with a resting stage that can remain viable for many years, germinating when conditions are favorable for growth of the sporophyte.
    d. **b and c**
    e. All of the above

26. The most abundant gymnosperm order today contains the cone-bearing plants such as pines. These plants belong to the group called
    a. cycads.
    b. ginkgos.
    c. Gnetales.
    d. **conifers.**
    e. sporophylls.

27. A universal feature of plant life cycles is
    a. the seed.
    b. the archegonium.

    c. the elater.
    d. the tracheid.
    e. **alternation of generations.**

28. One generation of a plant life cycle is called the gametophyte generation. The gametophyte is a multicellular haploid plant that produces haploid gametes. In some plants, the gametophyte is free-living and photosynthetic. Which group does *not* have a free-living gametophyte generation?
    a. Ferns
    b. Gymnosperms
    c. Angiosperms
    d. **b and c**
    e. None of the above

29. All plant life cycles have alternation of generations, alternating between the gametophyte generation and the _____. This generation extends from the zygote through the adult, multicellular, diploid plant.
    a. heteromorphic generation
    b. **sporophyte generation**
    c. vascular generation
    d. archegonium generation
    e. antheridium generation

30. Coniferous gymnosperms such as pines depend primarily on _____ for pollination, thus the plants produce large quantities of pollen that disperse over large areas during the spring.
    a. insects
    b. birds
    c. water
    d. **wind**
    e. mammals

31. An evolutionary trend that runs throughout the plant kingdom is that the sporophyte generation _____ and is more independent of the gametophyte, and the gametophyte generation _____ and is more dependent upon the sporophyte.
    a. becomes smaller; becomes smaller
    b. **becomes larger; becomes smaller**
    c. becomes smaller; becomes larger
    d. becomes larger; becomes larger
    e. does not change in size; becomes larger

32. Angiosperms differ from other plants in that two male gametes, contained within a single male gametophyte, participate in fertilization events. One sperm nucleus combines with the egg to produce a diploid zygote. The other sperm nucleus combines with two other haploid nuclei of the female gametophyte. This process is called
    a. biparental inheritance.
    b. multiple paternity.
    c. **double fertilization.**
    d. biparental fertilization.
    e. multiple fertilization.

33. The reproductive organ of angiosperms is the
    a. sporangium.
    b. **flower.**
    c. cone.
    d. archegonium.
    e. sporophyte.

34. Plant species that produce truly male and female plants are said to be
    a. dicots.

b. heterozygous.
c. perfect.
d. monoecious.
e. dioecious.

35. The ovary of a flowering plant, together with its seeds, develops into a fruit after fertilization. This structure may consist only of the mature ovary and its seeds, or it may include other parts of the flower or structures closely related to it. A fruit that develops from several carpels of a single flower, such as a raspberry, is a(n)
a. aggregate fruit.
b. simple fruit.
c. multiple fruit.
d. accessory fruit.
e. perfect fruit.

36. The two major groups of angiosperms are called monocots and dicots. These plants differ in
a. the number of sperm involved in fertilization.
b. the number of sexes per plant; monocots have one sex per plant, dicots have both.
c. the number of sexes per plant; dicots have male and female plants, monocots have both sexes in one plant and thus have one type of plant.
d. the number of seed leaves.
e. None of the above

37. The bryophytes are dependent on water for reproduction because
a. sperm are passively transported to eggs by water.
b. gametogenesis only occurs when the plants are moist.
c. eggs and sperm are released into water and then unite.
d. sperm must swim through water to reach and fertilize eggs.
e. None of the above

38. Great forests were widespread across the face of Earth by the Carboniferous period, approximately _____ years ago.
a. 345–290
b. 3,450–2,900
c. 3,450,000–2,900,000
d. 345,000,000–290,000,000
e. 3,450,000,000–2,900,000,000

39. Unlike the broad, loose definition, the narrower definition states that plants
a. are multicellular.
b. have photosynthetic pigments.
c. are eukaryotic.
d. develop from embryos.
e. have a vascular system.

40. Several evolutionary adaptations to land are shared by all plants. These shared adaptations do *not* include
a. waxy protective coverings.
b. support against gravity.
c. means of taking up water from the soil.
d. protective structures for the new sporophyte.
e. water transport by xylem.

41. All plants produce _____ by mitosis and _____ by meiosis.
a. spores; gametes
b. gametes; gametes
c. gametes; spores

d. spores; spores
e. spores; gametes and spores

42. Which of the following might you find in an archegonium?
a. Sperm
b. An egg
c. An egg and a sperm
d. An embryo
e. An egg, a sperm, and an embryo

43. You find a green, "leafy" bryophyte growing on your neighbor's front lawn. It is most likely a
a. liverwort.
b. hornwort.
c. moss.
d. lichen.
e. fern.

44. Fossil vascular plants can be recognized by the presence of
a. tracheids.
b. antheridia.
c. archegonia.
d. protonemata.
e. vessels.

45. One important consequence associated with the evolution of xylem was
a. sugar transport.
b. sperm transport.
c. prevention of water evaporation.
d. rigid structural support.
e. None of the above

46. One can identify sporangia as part of the fossil *Rhynia* since meiosis produces
a. two cells.
b. four cells.
c. two cells, but only one is functional.
d. four cells, but only one is functional.
e. two cells, but they remain attached.

47. Which of the following plants have simple leaves?
a. Moss
b. Conifers
c. Club mosses
d. Ferns
e. Some angiosperms

48. In heterosporous plants
a. microgametophytes produce eggs.
b. microgametophytes produce sperm.
c. microgametophytes produce eggs and sperm.
d. megagametophytes produce sperm.
e. megagametophytes produce eggs and sperm.

49. Heterospory probably affords selective advantages since
a. it evolved a number of times.
b. it is only found in the most advanced plants.
c. it is simpler than homospory.
d. it is found in the most primitive plants.
e. it evolved along with swimming sperm.

50. In your textbook, ferns are placed in which phylum?
a. Lycophyta
b. Anthocerophyta
c. Hepaticophyta
d. Pterophyta
e. Bryophyta

51. From an evolutionary standpoint, pollen is
    a. a microsporophyll.
    b. a megasporophyll.
    c. a microgametophyte.
    d. a megagametophyte.
    e. a microspore.

52. The prominent elements of Earth's modern land flora in most areas are
    a. angiosperms.
    b. gymnosperms.
    c. ferns.
    d. bryophytes.
    e. club mosses.

53–55. Answer the following three questions based upon this diagram of a pine seed.

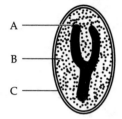

53. Which structure represents the embryo?
    a. A
    b. B
    c. C
    d. A and C
    e. B and C

54. Which structures are diploid?
    a. A
    b. B
    c. C
    d. A and C
    e. B and C

55. The integument portion of the ovule developed into part _____.
    a. A
    b. B
    c. C
    d. A and C
    e. B and C

56. In angiosperms, double fertilization results in the development of
    a. two embryos.
    b. two embryos, but only one survives.
    c. the embryo and the endosperm.
    d. the embryo and the seed coat.
    e. the embryo and the megagametophyte.

57. The seeds in angiosperms are
    a. on the upper surface of the sporophylls.
    b. on the lower surfaces of the sporophylls.
    c. buried within the sporophylls.
    d. enclosed in the ovule.
    e. None of the above

58. Flowers that produce both megasporangia and microsporangia are called
    a. perfect and monoecious.
    b. perfect and dioecious.
    c. imperfect and monoecious.
    d. imperfect and dioecious.
    e. imperfect and monoecious or dioecious.

59. Flowers with insect pollination have showy _____.
    a. petals
    b. sepals
    c. tepals
    d. petals or tepals
    e. sepals or tepals

60. In angiosperms, pollen is transferred from the _____ to the _____.
    a. anther; style
    b. filament; ovary
    c. anther; stigma
    d. filament; ovary
    e. anther; ovule

61. The diploid zygote in angiosperms develops into the
    a. embryonic axis.
    b. embryonic axis and cotyledons.
    c. embryonic axis and endosperm.
    d. embryonic axis, cotyledons, and endosperm.
    e. embryonic axis, cotyledons, endosperm, and seed coat.

62. Some herbaceous (nonwoody) plants predominate in such extreme environments as tundras because
    a. they grow slowly throughout the winter.
    b. they store food in underground rhizomes.
    c. photosynthesis in these plants uses carotenoids.
    d. the wood is only composed of tracheids.
    e. they are nonflowering plants.

63. The pistil consists of
    a. anthers, filaments, and stamen.
    b. ovary, archegonium, and embryo.
    c. stigma, style, and ovary.
    d. sepals, petals, and corolla.
    e. embryo, endosperm, and cotyledons.

64. Seed plants are all
    a. heterosporous.
    b. dioecious.
    c. monoecious.
    d. dicots.
    e. required to rely on animals for fertilization.

65. One difference between gymnosperms and angiosperms is
    a. gymnosperms do not form seeds.
    b. gymnosperms do not form flowers.
    c. gymnosperms do *not* have tracheid cells.
    d. gymnosperms rely on animals for fertilization, angiosperms do not.
    e. None of the above

66. The pollen tube is used to
    a. eject pollen from the microsporangium.
    b. direct pollen to the megasporangium.
    c. digest the sporophyte tissue toward the female gametocyte.
    d. produce pollen.
    e. attract animals to the plant to spread the pollen.

28 Fungi: A Kingdom of Recyclers

Fill in the Blank

1. Organisms living in mutually beneficial symbiosis with other organisms are called **mutualists**.

2. Sexual reproduction in the fungi is accomplished when two different mating types **fuse**.

3. The cell walls of all fungi consist of the polysaccharide **chitin**, which is also found in the cell walls of certain funguslike protists and in the skeletons of arthropods.

4. The body of a multicellular fungus is called a **mycelium**.

5. The kingdom **Fungi** is defined as heterotrophic organisms with absorptive nutrition.

6. The cells of a body of a multicellular fungus are organized into rapidly growing individual filaments called **hyphae**.

7. Individual filaments that anchor saprobic fungi to their substrate are called **rhizoids**.

8. There are two types of parasitic fungi, **facultative parasites**, which can grow parasitically but can also grow by themselves, and **obligate parasites**, those that can only grow on their specific hosts.

9. Sexual reproduction in fungi occurs between different **mating types**, which are genetically different, yet can function as either a male or a female.

10. Fungi often have a life stage that is **dikaryotic**, that is, one cell has two haploid nuclei, being neither haploid (n) nor diploid ($2n$). This stage is often described as ($n + n$).

11. One traditional means of combating black stem wheat rust (*P. graminis*) was to remove **barberry**, the intermediate host, from the life cycle.

12. Hyphae of conjugating fungi (zygomycetes) are attracted to one another due to the release of chemicals called **pheromones**.

13. Lichens can **reproduce** by producing a thallus or a soredia.

14. Biologists are exploring ways to use fungi to break down toxic substances in the environment. This process is known as **bioremediation**.

15. Mildew is a(n) **obligate** parasite.

16. The most ancient phylum of fungi, and one that was until recently classified as a protist, is the **Chytridiomycota**.

Multiple Choice

1. The cell walls of all fungi consist of the polysaccharide
   a. **chitin.**
   b. cellulose.
   c. starch.
   d. silica.
   e. pectin.

2. A major role of saprobic fungi in terrestrial ecosystems is to
   a. trap atmospheric carbon dioxide.
   b. **return carbon and other elements to the environment for further cycling.**
   c. parasitize animals.
   d. parasitize plants.
   e. parasitize protists.

3. Many fungi are _____, associating with photosynthetic organisms to form mycorrhizae or lichens.
   a. **mutualistic**
   b. parasitic
   c. saprobic
   d. photosynthetic
   e. predatory

4. Fungi can be parasitic on
   a. animals.
   b. plants.
   c. protists.
   d. other fungi.
   e. **All of the above**

5. The polysaccharide chitin is found in
   a. the cell walls of all fungi.
   b. the cell walls of some funguslike protists.
   c. the exoskeletons of arthropods.
   d. **All of the above**
   e. None of the above

6. Which of the following is *not* a basis for distinguishing the classes of the Eumycota (fungi)?
   a. Methods of sexual reproduction
   b. The structures used for sexual reproduction
   c. The presence or absence of cross-walls separating their cells
   d. The presence or absence of fleshy fruiting bodies
   e. **The presence or absence of flagella**

7–9. Fill in the blanks in the following descriptions with the correct terms from the list below.
   a. Dikaryon
   b. Hyphae
   c. Rhizoids
   d. Mycelium

7. The body of a multicellular fungus is called a _____. *(d)*

8. Individual filaments that anchor saprobic fungi to their substrate are called _____. *(c)*

9. The cells of the body of a multicellular fungus are organized into rapidly growing individual filaments called _____. *(b)*

10. One adaptation that fungi have for absorptive nutrition, in which nutrients are absorbed across the cell surfaces, is
    a. lack of a cell wall.
    b. a low surface area-to-volume ratio.
    c. **a high surface area-to-volume ratio.**
    d. tolerance of low temperatures.
    e. tolerance of high temperatures.

11. Fungi are more common than bacteria in a jar of jelly in your refrigerator because
    a. fungi have a lower tolerance for highly hypotonic environments.
    b. fungi have a lower tolerance for highly hypertonic environments.
    c. fungi have a higher tolerance for highly hypotonic environments.
    d. **fungi have a higher tolerance for highly hypertonic environments.**
    e. None of the above

12. Predatory fungi may trap prey with
    a. a constricting ring that contracts and traps the prey.
    b. sticky substances secreted by hyphae.
    c. mycorrhizae.
    d. **a and b**
    e. All of the above

13. Lichens are _____ associations of a fungus with _____.
    a. **mutualistic; an alga or bacterium**
    b. saprobic; an alga or bacterium
    c. parasitic; an alga or bacterium
    d. mutualistic; plant roots
    e. parasitic; plant roots

14. Mycorrhizae are _____ associations of a fungus with _____.
    a. mutualistic; an alga or bacterium
    b. **mutualistic; plant roots**
    c. parasitic; an alga or bacterium
    d. parasitic; plant roots
    e. saprobic; an alga or bacterium

15–20. Match the descriptive terms from the list below with the following descriptions of fungal interactions. Each term may be used once, more than once, or not at all.
    a. Saprobic
    b. Competitive
    c. Predation
    d. Parasitic
    e. Mutualistic

15. Fungus decaying a fallen tree. *(a)*

16. Black stem rust draws nutrition from wheat. The rust damages the wheat plant. *(d)*

17. Fungi grow in association with the roots of soybeans, providing the plants with nitrogen. *(e)*

18. A constricting ring formed by *Arthrobotrys* traps a nematode. Fungal hyphae invade and digest the nematode. *(c)*

19. Seed germination in most orchid species depends on the presence of a specific fungus species. The fungus derives nutrients from the seed and seedling. *(e)*

20. Some leaf-cutting ants farm fungi, feeding the fungi and later harvesting and eating them. The ants may even "weed" the fungal gardens by removing other fungal species. *(e)*

21. Fungal mating types are
    a. haploid, genetically identical, and can function as male or female.
    b. diploid, genetically identical, and can function as male or female.
    c. haploid and can function as either male or female.
    d. diploid and can function as either male or female.
    e. **haploid, genetically different, but any mating type can function as either male or female.**

22. The fusion of two different mating types forms a dikaryon or heterokaryon. The term "heterokaryon" emphasizes the fact that
    a. it is haploid.
    b. two nuclei fused in its formation.
    c. **the two mating types are different.**
    d. the two nuclei are different.
    e. None of the above

23. Which of the following is *not* true about fungal reproduction?
    a. **Gamete cells are produced.**
    b. Only gamete nuclei are produced.
    c. There is no true diploid tissue.
    d. Gametes are not motile and thus water is not required for reproduction.
    e. All of the above

24. The black stem rust of wheat, *Puccinia graminis*, causes a major agricultural disease. This fungus has a complicated life cycle including
    a. **utilization of wheat as a summer host.**
    b. utilization of corn as a winter host.
    c. dikaryotic hyphae as a minor segment of the life cycle.
    d. winter spores that are sensitive to freezing, thus extreme cold in winter diminishes the disease.
    e. All of the above

25. Wheat rust, *Puccinia graminis*, is a major agricultural pest. The wind-borne uredospores are the primary agents for
    a. **spreading the rust from wheat plant to wheat plant.**
    b. overwintering.
    c. spreading the rust from wheat to barberry plants.
    d. initiating the sexual cycle of the rust.
    e. spreading the rust from barberry to wheat.

26–30. Match the classes of Eumycota from the list below with the following descriptions.
 a. Zygomycetes
 b. Ascomycetes, subgroup Euascomycetes
 c. Ascomycetes, subgroup Hemiascomycetes
 d. Basidiomycetes
 e. Deuteromycetes

26. No fleshy fruiting body; no hyphal cross-walls; reproduce sexually by conjugation. *(a)*

27. Perforated cross-walls; no specialized fruiting structures; includes baker's and brewer's yeast. *(c)*

28. Common name is club fungi; complete cross-walls; includes puffballs, mushrooms, wheat rust, smut fungi, mycorrhizae. *(d)*

29. Perforated cross-walls produced in a specialized fruiting structure (called the perithecium); includes molds, parasites such as Dutch elm disease, and epicurean delights such as morels and truffles. *(b)*

30. No known sexual stages, it is likely the sexual stages were lost in evolution; includes the fungi that give Camembert and Roquefort their flavors, and *Penicillium. (e)*

31. Virtually all oak trees and pine trees depend on mycorrhizal fungi to absorb nutrients. The relationship between the tree and the fungus is an example of
 a. saprobism.
 b. parasitism.
 c. **mutualism.**
 d. heterotropism.
 e. commensalism.

32. Fungi have a larger surface area-to-volume ratio than most other multicellular organisms because
 a. **every cell along a hypha may be in contact with the environment.**
 b. an individual mycelium can grow very large.
 c. hyphae grow together to form a mycelium.
 d. must fungi are microscopic organisms.
 e. chitinous cell walls are more permeable than cellulose cell walls.

33. Which of the following is *not* an economically useful aspect of fungi? Some species
 a. are used commercially to flavor foods.
 b. are edible.
 c. produce alcohol via fermentation.
 d. **produce oxygen via fermentation.**
 e. produce antibiotics.

34. Rusts and smuts are pathogens of cereal grains classified in the
 a. zygomycetes.
 b. ascomycetes.
 c. **basidiomycetes.**
 d. deuteromycetes.
 e. lichens.

35. An unknown fungus was examined microscopically and found to lack cross-walls in its hyphae. It must belong to the
 a. **zygomycetes.**
 b. ascomycetes.

 c. basidiomycetes.
 d. deuteromycetes.
 e. lichens.

36. Cells of ascomycetes typically have
 a. cellulose cell walls.
 b. **perforated cross-walls.**
 c. no sexual stages.
 d. diploid nuclei.
 e. mycorrhizae.

37. Dikaryotic cells
 a. have two identical nuclei per cell.
 b. contain pairs of homologous chromosomes.
 c. divide only by meiosis.
 d. **contain two genetically different nuclei per cell.**
 e. contain diploid nuclei.

38. *Puccinia graminis,* wheat rust, is a major pathogen of wheat causing millions of dollars worth of damage per year. A means of control first employed in the 1930s was to remove all barberry plants from the wheat belt states. In theory this would control the disease by
 a. removing the source of overwintering rust spores.
 b. removing the food source for insects that carried the disease.
 c. preventing meiosis from occurring on the barberry plants.
 d. **preventing dikaryotic cells from forming during the life cycle.**
 e. This was a misguided strategy; barberry has nothing to do with wheat rust.

39. A sexually produced spore that buds from the surface of a basidium is a
 a. zygospore.
 b. ascospore.
 c. conidiospore.
 d. **basidiospore.**
 e. uredospore.

40. Ascomycetes produce conidia
 a. **asexually.**
 b. sexually.
 c. in response to harsh environmental conditions.
 d. Conidia are produced by basidiomycetes, not ascomycetes.
 e. Conidia are produced by zygomycetes, not ascomycetes.

41. The gills of a mushroom (basidiomycetes) are specialized for
 a. respiration.
 b. food production.
 c. defense.
 d. **reproduction.**
 e. water storage.

42. The algal partner in a lichen symbiosis is responsible primarily for
 a. respiration.
 b. **food production.**
 c. defense.
 d. reproduction.
 e. water storage.

43. Which of the following statements is sufficient by itself to identify an unknown organism as belonging to the kingdom Fungi?
    a. It is multicellular and nonphotosynthetic.
    b. It has cell walls and reproduces by spores.
    c. It has filamentous growth and obtains its food by absorption.
    d. It has prokaryotic cells, and cell walls made of chitin.
    e. It is unicellular and eukaryotic.

44. Lichens obtain nutrients by
    a. photosynthesis.
    b. engulfing other organisms.
    c. absorbing nutrients from the environment.
    d. decaying organic material.
    e. parasitizing flowering plants.

45. In a lichen, which part of the fungus is involved directly in the symbiosis?
    a. Fruiting body
    b. Mycelium
    c. Spores
    d. Spore cases
    e. Blue-green bacterium

46. What benefit goes to the photosynthetic partner in a lichen?
    a. Increased store of starch
    b. Increased store of glucose
    c. Increased availability of mineral nutrients
    d. Increased availability of lipids
    e. Increased exposure to light

47. Which chemical should one expect to find as part of the fungal partner in a lichen?
    a. Chitin
    b. Chlorophyll
    c. Reverse transcriptase
    d. Silica
    e. Cellulose

48. What is the function of a fruiting structure for the fungus of which it is a part?
    a. It distracts predators away from the essential underground parts.
    b. It is an important organ for gas exchange with the atmosphere.
    c. It is an organ of asexual reproduction.
    d. It provides hallucinogens for rodents and mammals.
    e. It serves as a landing pad for fungal pollinators.

49. Ecologically speaking, most soil fungi are to plant roots as
    a. ergot is to rye grains.
    b. water mold is to potato plants.
    c. viruses are to animals.
    d. *Escherichia coli* is to humans.
    e. mosquitoes are to mammals.

50. Which of the following plant diseases or parasitic fungi is a basidiomycete?
    a. Dutch elm disease
    b. Chestnut blight
    c. Powdery mildew
    d. Green fruit mold
    e. Bracket fungus

51. Sexual reproduction may be found in all of the following except
    a. zygomycetes.
    b. ascomycetes.
    c. basidiomycetes.
    d. deuteromycetes.
    e. lichens.

52. Soredia are characteristic of some
    a. zygomycetes.
    b. ascomycetes.
    c. basidiomycetes.
    d. deuteromycetes.
    e. lichens.

53. The names of fungal classes are based on important and characteristic structures associated with
    a. reproduction.
    b. nutrition.
    c. ecology.
    d. vegetative growth.
    e. cell division.

54. Which of the following is true of fungi?
    a. They have been with us for almost a billion years.
    b. They return elemental carbon and other elements to the environment.
    c. They are a rich source of food for livestock.
    d. *a* and *b*
    e. *a*, *b*, and *c*

55. _____ are organisms that live on dead matter.
    a. Parasites
    b. Saprobes
    c. Anaerobes
    d. Aerobes
    e. Autotrophs

56. Plants with active mycorrhizae
    a. benefit nutritionally from this arrangement.
    b. display enhanced absorption of water and minerals (especially phosphorus).
    c. are heavily parasitized and die.
    d. *a* and *b*
    e. None of the above

57. Forms of asexual reproduction in fungi include
    a. production of haploid spores within sporangia.
    b. conjugation.
    c. production of conidia.
    d. cell division.
    e. All of the above

58. What is *not* a way to distinguish slime molds from fungi?
    a. The presence of flagella
    b. Phagocytosis versus absorption for taking up nutrients
    c. Multicellular
    d. Cellulose versus chitin in cell walls
    e. None of the above

59. What distinguishes the phylum Chytridiomycota from all the other fungi?
    a. It reproduces only asexually.
    b. Its haploid gametes have flagella.
    c. It is the only one that is parasitic.
    d. It contains a fruiting body.
    e. It contains a thallus.

29 Protostomate Animals: The Evolution of Bilateral Symmetry and Body Cavities

Fill in the Blank

1. Because they lack rigid **cell walls**, animals cannot use high osmotic pressures to control the exchange of materials within their environments as plants do.

2. **Bilateral** symmetry is strongly correlated with the development of sense organs and central nervous tissues at the anterior end of the animal, a process known as cephalization.

3. Animals probably arose from protists whose cells remained together after division, forming a multicellular colony. Among those alive today, the animals that are most similar to the probable ancestral colonial protists are the **sponges**.

4. The phylum **Placozoa** consists of small animals less than 3 mm in diameter that have only four cell types, display no symmetry, and whose body plan consists of a flat plate made up of two layers of flagellated cells that enclose a fluid-filled area with fibrous cells. Only two species have been described in this group, and they are often observed in aquariums.

5. Cnidaria possess **radial** symmetry, having a cylindrical form with one main axis around which body parts are arranged.

6. Platyhelminthes, nematodes, rotifers, annelids, and arthropods possess a body plan that shows **bilateral** symmetry.

7. **Protostomes** are animals in which cleavage of the fertilized egg is determinate; that is, if the egg is allowed to undergo cell divisions, and the cells are then separated, each cell will develop into only a partial embryo.

8. Animals that lack an internal body cavity, such as Porifera, Cnidaria, and Platyhelminthes, are called **acoelomates**.

9. In Precambrian times, evolution resulted in animals with a chitinous external body covering. This is referred to as an **exoskeleton** and is a characteristic of arthropods.

10. Growth in arthropods is accomplished by **molting**, a periodic shedding of the exoskeleton followed by the rapid hardening of a new and larger exoskeleton that has formed under the old one.

11. A class of arthropods that has few members living in the oceans is the **insects**; however, in fresh water and on land they are a dominant group.

12. A distinguishing feature of mollusks is the **mantle**, a sheet of specialized tissue that covers the internal organs like a body wall. This structure secretes the shell.

13. When a body radiates around a central point, this is called **spherical** symmetry.

14. The **ventral** side of the bilaterally symmetrical animal is the surface with the mouth.

15. The oligochaetes are **hermaphrodites;** each organism has both male and female sex organs.

16. The thick, protective body covering acquired by Precambrian flatworm lineages is called **chitin.**

17. The immature stages of insects between molts are called **instars**.

18. Cephalopods and gastropods are lineages of the phylum **Mollusca**.

Multiple Choice

1. A phylum that includes animals with subdivided coeloms is
   a. Nematoda.
   b. Porifera.
   c. Annelida.
   d. Platyhelminthes.
   e. Cnidaria.

2. Animals can be divided into two major groups based on major evolutionary lineages that separated in the Cambrian. Those two lineages differ fundamentally in their
   a. mode of reproduction.
   b. early embryological development.
   c. mode of obtaining and storing energy.
   d. environmental conditions needed to live.
   e. metabolism.

3–6. Refer to the following diagram of the cross section of a sponge.

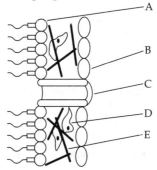

3. Which of the above structures is a choanocyte? *(a)*

4. Which of the above structures is an osculum? *(c)*

5. Which of the above structures is an epidermal cell? *(b)*

6. Which of the above structures is not a cell? *(e)*

7. Which of the following statements is true concerning nematocysts?
   a. **They are used in the capture of prey by cnidarians.**
   b. They are excretory organs in Platyhelminthes.
   c. They are reproductive cells in cnidarians.
   d. They are ciliated cells in Porifera.
   e. They are excretory cells in Porifera.

8. Which class within the Cnidaria is important geologically because the growth of its members can result in the formation of islands and atolls in tropical oceans?
   a. Class Hydrozoa
   b. Class Scyphozoa
   c. **Class Anthozoa**
   d. Class Turbellaria
   e. Class Ctenophora

9. Animals in the phylum Platyhelminthes are often parasitic. Which of the following is an adaptation found in Platyhelminthes to cope with a parasitic lifestyle?
   a. A flat shape
   b. A highly branched gastrovascular cavity
   c. An oxygen transport system
   d. **Absorption of nutrients through the body surface**
   e. *b* and *c*

10. The phylum Platyhelminthes differs from the phylum Cnidaria in that
    a. platyhelminthes are radially symmetrical; cnidarians are bilaterally symmetrical.
    b. **platyhelminthes have more complex internal organs than cnidarians do.**
    c. platyhelminthes are diploblastic, whereas cnidarians are triploblastic.
    d. platyhelminthes have two openings to the gastrovascular cavity, whereas cnidarians have only a single opening to the gastrovascular cavity.
    e. platyhelminthes have an excretory system, whereas cnidarians do not.

11. An interesting feature of many members of the class Trematoda is their complex life cycle. Understanding the life cycles of trematode parasites is of great importance to humans because
    a. **it is often possible to find a portion of the life cycle where the parasite is vulnerable and can be controlled.**
    b. although these parasites are of little importance in modern society, they must be studied before they are completely eliminated.
    c. it is always important to know the biology of the diseases of humans.
    d. it helps us to understand the evolutionary relationships between the different parasites.
    e. it emphasizes the interrelationship between humans and their environment.

12. In which phylum did bilateral symmetry first appear?
    a. Porifera
    b. Cnidaria
    c. **Platyhelminthes**
    d. Nematoda
    e. Rotifera

13. Which phyla contain acoelomate animals?
    a. Porifera and Rotifera
    b. Porifera and Cnidaria
    c. **Porifera, Cnidaria, and Platyhelminthes**
    d. Porifera, Cnidaria, Platyhelminthes, and Nematoda
    e. Porifera, Cnidaria, Platyhelminthes, Nematoda, and Rotifera

14. Which phyla contain coelomate animals?
    a. Arthropods and Rotifera
    b. **Arthropods and Annelida**
    c. Arthropods, Annelida, and Rotifera
    d. Arthropods, Annelida, Rotifera, and Nematoda
    e. Arthropods, Annelida, Rotifera, Nematoda, and Platyhelminthes

15. Which phylum(la) contain(s) animals with no digestive tract?
    a. **Porifera**
    b. Porifera and Cnidaria
    c. Porifera, Cnidaria, and Platyhelminthes
    d. Porifera, Cnidaria, Platyhelminthes, and Nematoda
    e. Porifera, Cnidaria, Platyhelminthes, Nematoda, and Rotifera

16. Which phyla contain animals with a complete digestive tract?
    a. Arthropods and Nematoda
    b. Arthropods and Annelida
    c. Arthropods, Annelida, and Rotifera
    d. **Arthropods, Annelida, Rotifera, and Nematoda**
    e. Arthropods, Annelida, Rotifera, Nematoda, and Platyhelminthes

17. The main difference between a coelom and a pseudocoelom is that
    a. **a coelom is lined with tissue of mesodermal origin, whereas a pseudocoelom is not.**
    b. a coelom is a body cavity that is outside of the gastrovascular cavity, whereas a pseudocoelom is actually an outgrowth of the gastrovascular cavity.
    c. a coelom is the developmental end point of the blastocoel, whereas in the pseudocoelomate animals the blastocoel is obliterated.
    d. a coelom is lined with tissue of endodermal origin, whereas a pseudocoelom is lined with tissue of ectodermal origin.

18. In nematodes, or roundworms, the main form of locomotion is
    a. cilia that beat rhythmically to move the animal forward.
    b. antagonistic muscles that work against each other and the pseudocoelom to change the shape of the animal, causing it to move forward.
    c. **longitudinal muscles that pull against the pseudocoelom, causing whiplike movements.**
    d. a series of hairs that project backward and engage the substrate; back-and-forth movements cause the animal to move forward.
    e. expelling water through special ducts (nephridiopores), which causes the animal to move forward.

19. A major difference between the musculature of nematodes and the musculature of annelids is that
    a. nematodes have antagonistic muscles, whereas annelids do not.
    b. nematodes have only longitudinal muscles, whereas annelids have both longitudinal muscles and muscles that run around the body.
    c. nematode muscles contract for the entire length of the body, whereas annelid muscle contraction is subdivided into segments.
    d. nematodes have both longitudinal muscles and muscles that run around the body, whereas annelids have only longitudinal muscles.
    e. *b* **and** *c*

20. Which of the following statements is *not* true concerning nematodes?
    a. Nematode parasites infect many members of the animal kingdom, including many domestic animals.
    b. Nematode parasites infect many members of the plant kingdom, including many crop plants.
    c. Relatively few people in developed countries such as the United States have ever been infected by nematodes.
    d. Free-living nematodes are often extremely abundant.
    e. Much of what is known about roundworms has been stimulated by the desire to control parasitic species.

21. Which of the following classes of annelids has parapodia?
    a. Oligochaeta
    b. Polychaeta
    c. Hirudinea
    d. *a* and *b*
    e. All of the above

22. Which of the following groups has members with trochophore larvae?
    a. Annelida
    b. Arthropods
    c. Mollusca
    d. *a* and *b*
    e. *a* and *c*

23. Mollusks and annelids differ in that
    a. annelids are segmented, whereas mollusks are not segmented.
    b. annelids have a coelom, whereas mollusks do not have a coelom.
    c. annelids are protostomes, whereas mollusks are deuterostomes.
    d. annelids have an open circulatory system, whereas mollusks have a closed circulatory system.
    e. annelids have separate sexes, whereas mollusks are hermaphrodites.

24. Which of the following is *not* an advantage of an exoskeleton?
    a. It is a highly efficient means of anchoring muscles, providing more efficient movement.
    b. It provides protection from predators.
    c. It dominates the development of the animal after it emerges from the egg, since it must be shed for the animal to grow.
    d. It provides protection from water loss.
    e. It provides support for walking on dry land.

25. The class Arachnida includes scorpions, mites, spiders, and a host of other less well-known orders. Their bodies are divided into two major regions, the anterior of which bears two pairs of appendages modified to form mouth parts, and four pairs of walking legs. What subphylum do they belong to?
    a. Chelicerata
    b. Mandibulata
    c. Trilobita
    d. Araneida
    e. Xyphosura

26. Which of the following statements is *not* true concerning the tracheal system in insects?
    a. Tracheae are air sacs and tubular channels.
    b. Tracheae penetrate virtually every part of an insect's body.
    c. Tracheae work by providing oxygen to the blood, which carries it to the other tissues of the insect.
    d. Tracheae provide oxygen to the tissues of the insects.
    e. Tracheae extend from external openings inward to tissues throughout the body.

27. The main body parts common to all mollusks are
    a. the foot, the radula, and the mantle.
    b. the foot, the mantle, and the shell.
    c. the visceral mass, the radula, and the mantle.
    d. the visceral mass, the mantle, and the shell.
    e. the foot, the visceral mass, and the mantle.

28. Which of the following statements is *not* true concerning the mantle cavity of mollusks?
    a. It is lost altogether in octopuses.
    b. It has been modified into internal support in slugs and squids.
    c. It is used as a filtering device by bivalves.
    d. It is used as the basis of jet propulsion by cephalopods.
    e. It secretes the shell that provides external protection in most molluscan groups.

29. Cephalization is most commonly associated with
    a. spherical symmetry.
    b. radial symmetry.
    c. biradial symmetry.
    d. bilateral symmetry.
    e. sessile animals.

30. Deuterostomes and protostomes differ in a number of categories. From the following list, select one characteristic in which they do *not* differ.
    a. Cleavage type
    b. Ability to form a blastopore
    c. Origin of the mouth
    d. Derivation of mesoderm
    e. Manner in which the coelom is formed

31. Which of the following features is *not* characteristic of a true coelom?
    a. Located between the endoderm and ectoderm
    b. Derived from the mesoderm
    c. Must arise as an outpocketing of the gut tube
    d. A body cavity lined with peritoneum
    e. Found in both protostomes and deuterostomes

32. Which of the following events was probably *not* important in the evolution of animals?
    a. Increase in atmospheric oxygen concentration
    b. Evolution of division of labor among the cells of colonial protists
    c. Evolution of noncyclic photophosphorylation
    d. Success of cyanobacteria
    e. **Abundant zooplankton**

33. Which of the following is *not* characteristic of the body plan of members of the phylum Porifera?
    a. **Gastrovascular cavity**
    b. No distinct tissue layers or organs
    c. No separation between the different cell layers
    d. Organization around water tubes
    e. Asymmetry

34. Which of the following structures is *not* associated with a typical sponge?
    a. Choanocytes
    b. Porocytes
    c. Spicules
    d. **Mesoglea**
    e. Amoebocytes

35. Which of the following characteristics is *not* associated with members of the phylum Cnidaria?
    a. Alternation between polyp and medusa
    b. **Three distinct body layers**
    c. Nematocysts
    d. Gastrovascular cavity
    e. Planula larva

36. Which of the following terms is associated with both the medusa and polyp stage in a typical hydrozoan cnidarian?
    a. Asexual reproduction
    b. Sessile
    c. **Gastrovascular cavity**
    d. Produces planula larva
    e. Thick mesoglea

37. Select the following characteristic that is a feature unique to the cnidarians in the class Anthozoa.
    a. Most species are colonial
    b. Commonly known as jellyfish
    c. **Only the polyp form exists**
    d. Gametes produced directly by the medusa
    e. Possess nematocysts

38. Which of the following features is *not* considered to be a major factor in the overall success of the cnidarians?
    a. **Unidirectional movement of food through the gut**
    b. Symbiosis with dinoflagellates
    c. Low metabolic rate
    d. Nematocysts
    e. Ability to subdue large prey

39. In free-living flatworms, which of the following functions occurs by simple diffusion?
    a. Respiration
    b. Absorption of nutrients
    c. Excretion
    d. Distribution of nutrients within the body
    e. **All of the above**

40. Which of the following is *not* associated with a parasitic lifestyle as seen in tapeworms and flukes?
    a. **Complex digestive systems**
    b. Flattened body
    c. Extensive reproductive organs
    d. Life cycles with multiple hosts
    e. Reduced gastrovascular cavity

41. *Trichinella spiralis*, the causative agent of the disease trichinosis, is a member of the phylum
    a. Nemertea.
    b. Platyhelminthes.
    c. **Nematoda.**
    d. Rotifera.
    e. Annelida.

42. Which of the following pseudocoel-containing phyla has members that use cilia for both locomotion and feeding?
    a. Nemertea
    b. Platyhelminthes
    c. Nematoda
    d. **Rotifera**
    e. Annelida

43. You discover an animal with bilateral symmetry, a pseudocoelom, a tubular digestive system, and a thick multilayer cuticle. Select the phylum to which this animal most likely belongs.
    a. Nemertea
    b. Platyhelminthes
    c. **Nematoda**
    d. Rotifera
    e. Annelida

44. Which one of the following groups of arthropods includes the millipedes?
    a. Onychophora
    b. Trilobita
    c. Chelicerata
    d. Crustacea
    e. **Uniramia**

45. Which one of the following characteristics of insects is also found in other arthropods?
    a. Malpighian tubules
    b. Tracheae
    c. Wings
    d. **Molting**
    e. Incomplete metamorphosis

46. The evolution of an exoskeleton affected many aspects of arthropod evolution. From the following list, select the one aspect that was least affected.
    a. Division of labor among the body parts
    b. Mode of locomotion
    c. Pattern of growth
    d. Type of gas exchange system
    e. **Subdivision of the coelom**

47. In the molluscan body plan, select the body part that secretes the shell.
    a. **Mantle**
    b. Foot
    c. Visceral mass
    d. Radula
    e. Spicules

48. Which of the following features is characteristic of the molluscan class Gastropoda?
    a. A radula
    b. Torsion
    c. Complex eyes
    d. A single calcareous shell
    e. Gills

49. Which of the following is *not* a consequence of a sessile lifestyle?
    a. Decreased predation
    b. Increased competition for space
    c. Production of toxins
    d. Increased motility of the larval stages
    e. Coloniality

50. Select from the following list a characteristic that has *not* been a major theme in protostome evolution.
    a. Predation as a selective pressure
    b. Subdivision of the body cavity
    c. Switch from cilia to muscles for movement
    d. Mechanisms for ingesting large prey
    e. Development of hard, external body parts

51. Select from the following phyla the one that does *not* contain wormlike organisms.
    a. Platyhelminthes
    b. Ctenophora
    c. Annelida
    d. Nematoda
    e. Nemertea

52. Which one of the following phyla has a circulatory system, a complete digestive tract, and a reduction in the coelom?
    a. Rotifera
    b. Annelida
    c. Mollusca
    d. Nematoda
    e. Platyhelminthes

53. You discover an animal that has a true coelom and an exoskeleton that includes the modified polysaccharide chitin. Select the phylum to which this animal most likely belongs.
    a. Nematoda
    b. Annelida
    c. Arthropods
    d. Mollusca
    e. Platyhelminthes

54. You discover a dioecious marine animal that has parapodia. Select the phylum to which this animal most likely belongs.
    a. Nematoda
    b. Annelida
    c. Arthropods
    d. Mollusca
    e. Platyhelminthes

55. Clues to the evolutionary relationships of animals can be found in
    a. the fossil record.
    b. developmental patterns.
    c. anatomy and physiology.
    d. nucleotide sequence patterns.
    e. All of the above

56. The terms "acoelomate," "pseudocoelomate," and "coelomate" are used to describe
    a. cephalization.
    b. origin of the blastopore.
    c. ectoderm, mesoderm and endoderm.
    d. the origin of the body cavity of animals.
    e. the vertebrate body plan.

57. Corals, members of the class Anthozoa, flourish in nutrient-poor, clear, tropical waters because
    a. the symbiotic dinoflagellates that reside in that environment provide corals with carbohydrates.
    b. coral cnidoblasts and nematocysts are highly efficient.
    c. their digestive cells contain mammalian-like enzymes.
    d. *a* and *c*
    e. None of the above

58. Which of the following is *not* true of members of the phylum Platyhelminthes?
    a. Some are parasitic, like members of the classes Cestoda and Trematoda.
    b. They possess a mouth but no anus.
    c. They are diploblastic.
    d. They possess some cephalization.
    e. Some possess complex life cycles.

59. The body cavity of coelomate animals is called peritoneum and arises from
    a. endoderm tissue.
    b. ectoderm tissue.
    c. mesoderm tissue.
    d. the pseudocoel.
    e. mesenchyme.

60. Porifera and Placozoa differ from most other animals in that they
    a. have no distinct tissues and no cavity between layers.
    b. have no distinct tissues and are sessile.
    c. are sessile and are pseudocoelomate.
    d. are acoelomates and are triploblastic.
    e. are triploblastic and have no distinct tissues.

61. The Portuguese man-of-war is a deadly example of a
    a. poriferan.
    b. cnidarian.
    c. ctenophore.
    d. nematode.
    e. mollusk.

30 Deuterostomate Animals: Evolution of Larger Brains and Complex Behaviors

Fill in the Blank

1. An early vertebrate, the jawless fishes are called **Agnatha**.

2. The bony fishes are capable of regulating their buoyancy because of an organ called the **swim bladder**.

3. One group of Osteichthyes, the Crossopterygii, gave rise to the class **Amphibia**, which are capable of exchanging gases through their skin.

4. The first major animal group that can live completely out of water is the class **Reptilia**.

5. The reptiles were extraordinarily successful in part due to the evolution of the **amniotic egg**, which allows their offspring to develop in a protected watery environment.

6. The birds (subclass **Aves**) are in the class Dinosauria.

7. The class **Mammalia** includes monotremes, marsupials, and eutherians.

8. **Arrow worms** are members of the phylum Chaetognatha that swim in the open sea, have lateral and caudal fins, possess grasping spines, prey on protists and young fish, and have a tripartite body plan.

9. The evolutionary lineage leading to the **chordates** lost the lophophore and proboscis, replacing them with large pharyngeal slits as a feeding device.

10. Members of the phylum **Chordata** are bilaterally symmetrical and possess pharyngeal slits at some stage of their life cycle. They also possess an internal skeleton, a dorsal nervous system, a ventral heart, and a tail that extends beyond the anus.

11. The whale shark, with a length of over 15 meters and a weight of 9,000 kilograms, is considered the world's largest **fish**.

12. The evolution of primitive **lungs or lunglike sacs** in the early bony fishes (Osteichthyes) set the stage for the invasion of land, because they provided an alternative method of gaseous exchange.

13. In the evolutionary lineage leading to dinosaurs and birds, and in the lineage leading to mammals, two factors were important in enabling those organisms to run for long periods of time at fast rates. These factors were a more vertical position assumed by the legs and **special, ventilatory muscles** that could operate independently of locomotory muscles.

14. The large surface area of the lophophore makes it useful for both feeding and **gas exchange**.

Multiple Choice

1. The lophophorate animals clearly belong to the deuterostome lineage, yet have many resemblances to some protostomate animals. Which of the following is true concerning lophophorates?
   a. **They have a U-shaped ridge around the mouth with one or two rows of ciliated, hollow tentacles that are used for food collecting and gas exchange.**
   b. Adult animals of all lophophorate species are mobile.
   c. Most lophophorate species are terrestrial.
   d. They have a bipartite body plan, with the body divided into two parts, an anterior prosome and a posterior metasome.
   e. Only the prosome has a separate coelomic compartment, the protocoel.

2. Which of the following statements is *not* true concerning echinoderms?
   a. Echinoderms are radially symmetrical as adults.
   b. **Echinoderms have a tripartite body plan.**
   c. Echinoderms lack a brain.
   d. Echinoderms are radially symmetrical as larvae.
   e. Echinoderms have an extensive fossil record.

3. Which of the following phyla has a water vascular system?
   a. Mollusca
   b. Chordata
   c. Annelida
   d. **Echinodermata**
   e. *a* and *d*

4. The major change in the evolutionary lineage leading to the chordates was
   a. evolution of the water vascular system.
   b. **loss of the lophophore and enlargement of the pharyngeal gill slits for feeding.**
   c. calcification of an internal skeleton.
   d. development of the lophophore as an adaptation to predatory life.
   e. the ability to "suck mud" (extract food from mud or sand).

5. Which of the following is *not* found in the phylum Chordata?
   a. Bilateral symmetry
   b. A dorsal hollow nerve chord
   c. **An external skeleton**
   d. Gill slits at some stage during development
   e. A notochord at some stage during development

6. Which of the following is *not* a characteristic of the vertebrate body plan?
   a. A vertebral column
   b. Two pairs of appendages
   c. A large coelom in which body organs are found
   d. A well-developed circulatory system
   e. A large coelom that serves as a hydrostatic skeleton

7. The first vertebrates were in the class Agnatha, the jawless fishes. Which of the following was *not* found in the early Agnatha?
   a. Gill arches
   b. The ability to attach to carrion and suck fluids and partly decomposed tissues into the mouth
   c. Flexible, leathery skin that is not armored
   d. A notochord
   e. A dorsal nerve chord

8. The important evolutionary novelty that evolved from the Agnatha and is present in Placodermi, Chondrichthyes, and Osteichthyes is
   a. heavily armored skin.
   b. the jaw.
   c. fins.
   d. the ability to swim.
   e. All of the above

9. The jaw found in sharks is most likely evolved from
   a. a gill arch.
   b. dermal bone.
   c. the skull.
   d. the backbone.
   e. the mouth plates of echinoderms.

10. The cartilaginous fishes, including sharks, skates and rays, and chimaeras
    a. belong to the class Agnatha.
    b. have heavy external armor.
    c. have an open circulatory system.
    d. have a few bones, but mostly cartilage, in their skeletons.
    e. have less external armor and are faster swimmers than their ancestors.

11. A major ecological difference between bony fishes (class Osteichthyes) and Chondrichthyes is that bony fishes
    a. do not have swim bladders.
    b. have no cartilage in their skeletons.
    c. need more dissolved oxygen in the water than sharks do.
    d. evolved in fresh water.
    e. All of the above

12. Major differences between the classes Chondrichthyes and Osteichthyes include which of the following?
    a. Chondrichthyes have a lung or swim bladder, whereas Osteichthyes do not.
    b. Chondrichthyes have a cartilaginous skeleton, whereas Osteichthyes have a bony skeleton (with true bone).
    c. Chondrichthyes do not have paired fins, whereas Osteichthyes have paired fins.
    d. Chondrichthyes evolved in fresh water, whereas Osteichthyes evolved in salt water.
    e. *b* and *d*

13. The swim bladders of bony fishes
    a. evolved from lunglike sacs that supplemented the gills in respiration.
    b. are used for respiration in most contemporary species.
    c. are organs of buoyancy that help the fish control its depth in the water column.
    d. prevented bony fishes from existing in a marine environment.
    e. *a* and *c*

14. Lobe-finned fishes (Crossopterygians) had several adaptations that were instrumental in the transition to a life on land. Adaptations found in lobe-finned fishes that were important in the evolution of the amphibians include
    a. primitive lungs.
    b. jointed fins with strong muscular support.
    c. a watertight skin.
    d. *a* and *b*
    e. All of the above

15. Amphibians breathe air by which two means?
    a. Gills and swim bladders
    b. Gills and thin skin
    c. Lungs and gills
    d. Lungs and thin skin
    e. Swim bladders and lungs

16. Reptiles are the first group of animals to be completely liberated from a need to return to water for some portion of their life cycle. Adaptations first found in reptiles that are important in this freedom include which of the features below?
    a. A hard-shelled (amniotic) egg
    b. Parental care
    c. Watertight skin
    d. *a* and *c*
    e. All of the above

17. Which of the following statements comparing birds and mammals is *not* true?
    a. No mammals lay eggs, whereas all birds lay eggs.
    b. All mammals nurse their young (provide milk), whereas no birds provide milk.
    c. Both mammals and birds are homeothermic (warm-blooded), but they apparently evolved it independently.
    d. Mammals (generally) have hair for insulation, whereas birds have feathers for insulation (and flight).
    e. All of the above

18. The single most characteristic feature of birds is
    a. the ability to fly.
    b. the ability to lay eggs that will not dry out.
    c. feathers.
    d. the enormous amount of parental care they provide.
    e. All of the above

19. The birds, or Aves
    a. descended most recently from an amphibian ancestor.
    b. should be included in the class Mammalia.
    c. should include some modern reptiles.
    d. should include all modern reptiles.
    e. are in the class that includes dinosaurs and crocodiles.

20–22. Match the groups of organisms from the list below with their most important evolutionary novelty.
- a. Reptiles
- b. Birds
- c. Osteichthyes
- d. Chondrichthyes
- e. Echinoderms

20. Swim bladders *(c)*

21. Powered flight *(b)*

22. Jaws *(d)*

23. In several important ways, deuterostome and protostome evolution are similar. Which of the following is *not* true of both protostome and deuterostome evolution?
- a. Both lineages exploited the abundant food supplies buried in soft marine substrates.
- b. Many groups in both lineages developed elaborate structures for extracting prey from water.
- c. In both lineages, a coelomic cavity evolved and became divided into compartments that allowed better control of body shape and movement.
- d. Both lineages evolved locomotor abilities.
- e. **Both groups invaded land and evolved into very large terrestrial animals.**

24. The oldest fossil remains of members of our genus, *Homo*, suggest that early relatives of humans lived
- a. near the oceans, where fish are plentiful throughout the year.
- b. near rivers, where fish and fresh water are plentiful.
- c. in the Midwestern United States, where there is some of the most fertile soil in the world.
- d. in the American tropics, where there are long growing seasons and several species of fruits and berries.
- e. **in dry African savannas, eating roots, bulbs, tubers, and animals.**

25. Pterobranchs (class Pterobranchia) and acorn worms (class Enteropneusta) are two groups in the phylum Hemichordata. From the following, select a characteristic that these two classes do *not* share.
- a. A tripartite body plan
- b. A proboscis, collar, and trunk
- c. **A lophophore**
- d. Bottom-dwelling marine habitats
- e. Use of cilia and mucus to trap food

26. Which of the following features do arrow worms and the hemichordates share in common?
- a. A lophophore
- b. A similar diet
- c. Pharyngeal gill slits
- d. **Division of the body into three compartments**
- e. Motile larval stages

27. To which class of echinoderms do the brittle stars belong?
- a. Crinoidea
- b. **Ophiuroidea**
- c. Asteroidea
- d. Echinoidea
- e. Holothuroidea

28. From the following list of echinoderm classes, select the class that was much more abundant in the past than it is today.
- a. **Crinoidea**
- b. Ophiuroidea
- c. Asteroidea
- d. Echinoidea
- e. Holothuroidea

29. Select the following feature that does *not* characterize animals in the phylum Echinodermata.
- a. Bilaterally symmetric larvae
- b. Radially symmetric adults
- c. Presence of a water vascular system
- d. **An external skeleton of calcified plates**
- e. A tubular digestive system

30. Which class of echinoderms has tube feet at their anterior end modified into tentacles that are used for feeding?
- a. Crinoidea
- b. Ophiuroidea
- c. Asteroidea
- d. Echinoidea
- e. **Holothuroidea**

31. Which class of echinoderms are completely herbivorous and have no arms?
- a. Crinoidea
- b. Ophiuroidea
- c. Asteroidea
- d. **Echinoidea**
- e. Holothuroidea

32. The chordates
- a. all have a backbone.
- b. include some animals without a nervous system.
- c. **all pass through a developmental stage with gill slits.**
- d. are poorly represented in the fossil record.
- e. None of the above

33. Which of the following is *not* a characteristic unique to all members of the phylum Chordata?
- a. A notochord
- b. A ventral heart
- c. **Vertebrae**
- d. An endoskeleton
- e. A dorsal hollow nerve cord

34. Which of the following structures is seen in both the adult and tadpolelike larvae of the tunicates?
- a. **Pharyngeal gill slits**
- b. Notochord
- c. Dorsal hollow nerve cord
- d. A tunic
- e. An atrial siphon

35. The first true vertebrates were the
- a. **ostracoderms.**
- b. lampreys and hagfish.
- c. placoderms.
- d. tunicate larvae.
- e. cephalochordates.

36. Evolution of jaws first occurred in the group of fish known as the
- a. ostracoderms.
- b. **placoderms.**
- c. cartilaginous fishes.

d. Chondrichthyes.
e. Osteichthyes.

37. The sharks, skates, and rays are members of the vertebrate class known as
a. Osteichthyes.
b. Agnatha.
c. Placodermi.
d. Chondrichthyes.
e. Ascidiacea.

38. Select the following adaptation that was *not* first evolved in the group of fish known as the Osteichthyes.
a. An operculum
b. A swim bladder
c. Paired lateral fins
d. Bony skeletons
e. Lungs

39. Which of the following statements about the crossopterygian fishes is *not* true?
a. They are commonly known as the lobe-fin fishes.
b. *Latimeria chalumnae* is a well-known fossil crossopterygian fish.
c. Crossopterygian fishes have lungs, but no gills.
d. The bones in their pectoral and pelvic limbs have well-formed joints.
e. The crossopterygian fishes gave rise to the amphibians during the Devonian period.

40. Which of the following statements about the amphibians is *not* true?
a. The class Amphibia was more abundant in the past.
b. Waterproof coverings on the skin of amphibians permit them to be truly terrestrial.
c. Most amphibians must reproduce in or near water.
d. Amphibian eggs are very sensitive to drying.
e. Living amphibians belong to three major classes.

41. Select the following feature that was *not* a significant evolutionary advancement of the class Reptilia.
a. Water-impermeable skin
b. Maternal care of the young
c. Keratinized scales
d. More effective ventilation of the lungs
e. Better separation of oxygenated and deoxygenated blood in the heart

42. Which of the following subclasses of the class Reptilia has the fewest living members?
a. Squamata
b. Sphenodontida
c. Chelonia
d. Crocodilia
e. Aves

43. From the following list, select the vertebrate group that is *not* placed within the class Archosauria.
a. Turtles
b. Crocodiles and alligators
c. Dinosaurs
d. Birds
e. Plesiosaurs and ichthyosaurs

44. Select the following group of modern reptiles that includes entirely carnivorous species.
a. Lizards
b. Turtles and tortoises
c. Snakes
d. Caecilians
e. None of the above

45. Which of the following statements comparing Archaeopteryx with modern birds is *not* true?
a. The breastbone of Archaeopteryx lacked a keel.
b. Unlike modern birds, Archaeopteryx had a long tail.
c. Unlike modern birds, Archaeopteryx had long fingers on its forearms.
d. Although Archaeopteryx could fly, it lacked true feathers.
e. Unlike Archaeopteryx, no modern birds have teeth.

46. Which of the following characteristics is unique to the marsupial mammals?
a. They are the only egg-laying mammals.
b. Their mammary glands have no nipples.
c. They have no placentas.
d. Gestation is short, and the young complete development outside the uterus in a special pouch.
e. They are only found in Australia.

47. Which of the following hominids is oldest?
a. *Homo habilis*
b. *Homo erectus*
c. *Homo sapiens*
d. Neanderthals
e. Cro-Magnons

48. Which of the following was *not* a change that accompanied the transition from the australopithecines to *Homo habilis*?
a. Increase in brain size
b. A change in diet
c. An increase in body size
d. Use of tools
e. Dispersal of populations to Europe and Asia

49. Which of the following evolutionary themes did *not* occur in both the protostomes and deuterostomes?
a. Evolution of structures for filtering food from water
b. Evolution of wormlike burrowing forms
c. Evolution of adaptations needed for invasion of the land
d. Evolution of large terrestrial species
e. Evolution of jointed appendages for improved locomotion

50. Which of the following is *not* characteristic of deuterostomes?
a. Indeterminate or regulatory cleavage
b. Formation of mesoderm from an outpocketing of the embryonic gut
c. A blastopore that becomes the anus
d. Greater number of species than the protostomes
e. Triploblastic with well-developed coelomic body cavities

51. Members of the phylum _____ exhibit radial symmetry that evolved in association with a complex body plan.
    a. Porifera
    b. Chordata
    c. Annelida
    d. Cnidaria
    e. Echinodermata

52. The principal reason why we consider tunicates similar to ancestors of all chordates is that
    a. the body plan of adult tunicates parallels that of chordates.
    b. tunicate larvae reveal close evolutionary relationships with chordates.
    c. tunicates have a lophophore-style mouth.
    d. a and c
    e. None of the above

53. We sometimes refer to birds as "feathered dinosaurs" because
    a. birds are descendants of a lineage of dinosaurs.
    b. dinosaurs had feathers as well as scales.
    c. birds are warm-blooded, and preserve warmth via feathers, while dinosaurs were also warm-blooded.
    d. a and c
    e. None of the above

54. The calcification of an internal skeleton is characteristic of the phylum
    a. Chaetognatha.
    b. Echinodermata.
    c. Hemichordata.
    d. Brachiopoda.
    e. Phoronida.

55. The now-living species *Latimeria chalumae* belongs to a subclass that is an important link in evolution because
    a. it was the first subclass of fish with lungs.
    b. it was the first subclass of fish with paired fins.
    c. it was the first subclass of fish with lobed fins.
    d. it was the first subclass of bony fish.
    e. it was the first subclass of fish with cartilage.

56. A difference between amphibians and reptiles is
    a. amphibian eggs can survive out of water and reptile eggs cannot.
    b. amphibians have thin skins and reptiles have thick skins.
    c. amphibians have gills and lungs and reptiles have only lungs.
    d. a and b
    e. a, b, and c

57. Cartilaginous fishes
    a. rely on swimming to stay afloat.
    b. have swim bladders.
    c. belong to the class Osteichthyes.
    d. evolved in fresh water.
    e. are relatively poor swimmers.

58. Which of the following is *not* true concerning general characteristics of deuterostomate animal phyla?
    a. All have three body layers.
    b. All have closed circulatory systems.
    c. All have coelomic body cavities.
    d. All have complete digestive tracts.
    e. All of the above

59. Which of the following statements comparing members of the phyla Bryozoa and Brachiopoda is *not* true?
    a. Only bryozoans can extend and retract their lophophores.
    b. Only brachiopods are enclosed in shells.
    c. Bryozoans are mostly colonial; brachiopods are mostly solitary.
    d. Members of both phyla are sessile.
    e. Members of both phyla are mostly marine.

31 The Flowering Plant Body

Fill in the Blank

1. Unlike dicots, monocots have their vascular bundles in a **scattered** arrangement in their stems and possess a central region called the **pith** inside the xylem layer in their roots.

2. When a tree is cut down, the cut surface of the stump often shows **annual rings** due to variations in the size of the vessels.

3. The most common type of undifferentiated cells in plant bodies are **parenchyma** cells. **Sclerenchyma** cells have thickened secondary cell walls and actually function when dead, while in more flexible support tissue called **collenchyma** the elongate, living cells have a thickened primary cell wall.

4. Just inside the epidermis of the root is a region of unspecialized cells called the **cortex**.

5. In some plants a **mycorrhizal** association forms between root epidermal cells and a fungus; the plant benefits from this association because the fungi help in **water and mineral uptake**.

6. The **endodermis** is a layer of cells waterproofed by a layer of suberin in their cell walls.

7. Some vascular bundles contain a single layer of actively dividing cells known as the vascular **cambium**, which produces **xylem** cells toward the inside of the root or stem and **phloem** cells toward the outside of the plant.

8. In the epidermis of a leaf are pores called **stomata** whose opening is controlled by the action of a pair of **guard cells** surrounding the pore.

9. Division of cells that will form all of the plant's organs occurs from regions called **meristems**.

10. **Suberin** is the waxy substance that waterproofs the inner core of vascular tissues in the root and stem.

11. Potatoes are a portion of the **stem** of the plant and their eyes contain lateral **buds**.

12. Cork cambium, the bark on trees, arises from **phloem**.

Multiple Choice

1. Unlike primary cell walls, secondary cell walls have
   a. plasmodesmata.
   b. deposits of lignin or suberin.
   c. deposits of cellulose.
   d. pit pairs.
   e. permeability to small molecules.

2. In vessel elements and sieve tube elements, only sieve tube elements
   a. are dead at maturity.
   b. are stacked end-to-end.
   c. transport substances through the plant.
   d. often have companion cells.
   e. occur in all plant organs.

3. Sieve tube elements are unusual cells because they lack
   a. cell walls.
   b. cytoplasm.
   c. water.
   d. nuclei.
   e. plasma membranes.

4. Biennials are plants that set seed
   a. in their second year and then die.
   b. and then die in less than one year.
   c. every two years for many years.
   d. twice each year for many years.
   e. once a year for two years and then die.

5. A mycorrhiza is an association between a plant root and
   a. soil bacteria.
   b. earthworms.
   c. another plant root.
   d. a fungus.
   e. soil insects.

6. A mycorrhizal association benefits a plant by
   a. protecting its root hairs from damage.
   b. allowing it to use deeper soil layers.
   c. increasing its uptake of water and minerals.
   d. anchoring it more firmly in the soil.
   e. enlarging its capacity to store sugars.

7. Which of the following is characteristic of heartwood but *not* of sapwood?
   a. Lighter color
   b. Stores resin
   c. Conducts water and minerals
   d. Younger wood
   e. Has knots

8. Cacti thorns result from a modification of the same plant organ that produces
   a. coconut trunks.
   b. pea tendrils.
   c. strawberry runners.
   d. potato tubers.
   e. corn adventitious roots.

9. Tracheids, vessel elements, and sclereids are similar in that they all
   a. lack secondary cell walls.
   b. conduct water and minerals.
   c. harden seed coats.
   d. have open ends.
   e. function when dead.

10. Which cell type occurs in angiosperm xylem but *not* in gymnosperm xylem?
    a. Tracheids
    b. Cork
    c. Sieve tube elements
    d. Vessel elements
    e. Parenchyma

11. A root is called adventitious if it
    a. forms a mycorrhizal association.
    b. belongs to a fibrous root system.
    c. originates from a stem or leaf.
    d. is modified for storage.
    e. is actively growing.

12. The widening of a tree trunk is mostly due to the activity of its
    a. apical meristem.
    b. secondary phloem.
    c. phelloderm.
    d. vascular cambium.
    e. primary xylem.

13. In each vascular bundle, the tissue nearest the center of the stem is
    a. collenchyma.
    b. fibers.
    c. phloem.
    d. vascular cambium.
    e. xylem.

14. Moving from the center of a tree trunk outward, the order of vascular tissues is
    a. primary xylem, secondary xylem, vascular cambium, secondary phloem, primary phloem.
    b. secondary xylem, primary xylem, vascular cambium, primary phloem, secondary phloem.
    c. primary xylem, primary phloem, secondary xylem, secondary phloem, vascular cambium.
    d. primary xylem, primary phloem, vascular cambium, secondary phloem, secondary xylem.
    e. secondary xylem, secondary phloem, vascular cambium, primary xylem, primary phloem.

15. The branches of a flowering dogwood are held straight by
    a. compression wood on the branches' lower sides.
    b. tension wood on the branches' upper sides.
    c. tension wood on the upper sides and compression wood on the lower sides.
    d. tension wood in rings around the branches.
    e. compression wood in rings around the branches.

16. The pull of gravity is detected by a root's
    a. apical meristem.
    b. cap.
    c. endodermis.
    d. pericycle.
    e. region of elongation.

17. The collective term for phelloderm, cork cambium, and cork is
    a. pericycle.
    b. periderm.
    c. phloem.
    d. procambium.
    e. protoderm.

18. What are the major functions of plant roots?
    a. Absorption and support
    b. Absorption and reproduction
    c. Anchoring and support
    d. Anchoring and absorption
    e. Absorption and transport

19. Which of the following describes a fibrous root system?
    a. Deep-growing
    b. Thick roots
    c. Holds soil well
    d. Typical of many dicots
    e. Food storage organ

20. The meristem is
    a. the tip of the stem.
    b. the site on the stem where a bud forms.
    c. supporting tissue.
    d. growing tissue.
    e. the base of the leaves.

21–25. Match the following descriptions with one of the cell types from the list below.
    a. Parenchyma cells
    b. Sclerenchyma cells
    c. Collenchyma cells

21. Photosynthesis occurs in these cells *(a)*

22. These cells function when dead *(b)*

23. Stone cells of pears *(b)*

24. Cells that support growing organs *(c)*

25. Bulk of root cells *(a)*

26. The xylem tissue of advanced angiosperms typically is distinguished by its
    a. elongate tracheids.
    b. short stacked vessel elements.
    c. sieve tube elements.
    d. companion cells.
    e. thick-walled fiber cells.

27. Stems function in support and transport. Which type of cell accomplishes most of this function?
    a. Collenchyma
    b. Companion cells
    c. Parenchyma
    d. Sclerenchyma
    e. Vessel elements

28. The vascular tissue system of plants has the same function as what animal system?
    a. Circulatory
    b. Digestive
    c. Excretory
    d. Reproductive
    e. Respiratory

29. A layer of cells that protects the plant is the
    a. cuticle.
    b. endoderm.
    c. epidermis.
    d. ground tissue.
    e. pericycle.

30. The vascular cambium is located between which two tissues?
    a. Phloem and cork cambium
    b. Xylem and cork cambium
    c. Phloem and bark
    d. Xylem and phloem
    e. Phloem and ground tissue

31. Which is characteristic of secondary growth but *not* of primary growth?
    a. Growth in plant diameter
    b. Growth in plant height
    c. Typical of meristems
    d. Growth by cell elongation
    e. Occurs in all dicots

32. Which of the following is typical of cork cells?
    a. Interior to cork cambium
    b. Waxy suberin
    c. Water storage
    d. Active cell division
    e. Abundant in monocots

33. Root hairs are adaptations that
    a. increase surface area.
    b. defend the plant.
    c. reduce water loss.
    d. provide active growth.
    e. support the plant.

34. In the course of the development of an individual root cell, the typical sequence is
    a. elongation, division, differentiation.
    b. differentiation, elongation, division.
    c. division, differentiation, elongation.
    d. elongation, differentiation, division.
    e. division, elongation, differentiation.

35. Grasses can grow back after their tips are removed by clipping or grazing because
    a. their stems elongate at the tips.
    b. their stems elongate at the bases of internodes.
    c. their leaf primordia serve as a protective tip.
    d. their stems thicken with growth.
    e. their vascular bundles are scattered through the stem.

36. A root or stem increases in diameter when
    a. primary xylem cells divide.
    b. phloem cells divide.
    c. secondary xylem is deposited external to the vascular cambium.
    d. vascular cambium cells divide.
    e. phloem cells elongate.

37. Annual rings are seen in temperate-zone trees because
    a. heartwood cells alternate with sapwood cells.
    b. resin is deposited in rings in the stem.
    c. cork is deposited in rings in the stem.
    d. xylem activity varies with season.
    e. xylem cell size varies with season.

38. Lenticels are spongy regions on the surface of some woody stems that function in
    a. water uptake.
    b. water conservation.
    c. gas exchange.
    d. protection of growing layers.
    e. support of the plant.

39. What is the advantage of the spongy arrangement of mesophyll cells in the lower leaf layer?
    a. Maximum absorption of sunlight for photosynthesis
    b. Maximum diffusion of carbon dioxide in the leaf
    c. Maximum movement of water to leaf cells
    d. Minimum water loss from the leaf
    e. Minimum exchange of oxygen within the leaf

40. The veins of a leaf consist of
    a. mesophyll cells.
    b. guard cells.
    c. stomata.
    d. vascular cells.
    e. epidermal cells.

41. Cacti are plants with stems modified for water storage. Which type of tissue is well developed in cacti for this function?
    a. Cork
    b. Xylem
    c. Phloem
    d. Parenchyma
    e. Epidermis

42. The next time you sprinkle coconut on your ice cream, you can say, "What delicious _____."
    a. endosperm
    b. vascular bundles
    c. apical buds
    d. fermented sap
    e. fruit walls

43. Unlike fibrous root systems, taproot systems
    a. maximize surface area.
    b. are used as food-storage organs.
    c. anchor the plant.
    d. are better at holding soil.
    e. transport water and minerals to the stem.

44. The region of cell division in a primary root is located
    a. in the root cap.
    b. in the apical meristem.
    c. in the region of elongation.
    d. in the area containing differentiated tissues.
    e. throughout the root.

45. A common function of stems but *not* roots would be
    a. anchorage.
    b. transport.
    c. storage.
    d. support.
    e. absorption.

46. Which of the following is least likely to be formed from a lateral bud?
    a. A flower
    b. A branch
    c. A runner
    d. A leaf
    e. A root

47. Technically, simple tissues in plants
    a. are composed of only one type of cell.
    b. have only one function.
    c. only produce primary cell walls.
    d. can only be found in apical meristems.
    e. are only found in primitive plants.

48. "An organized group of plant cells, working together as a functional unit" best defines
    a. an organism.
    b. an organ.
    c. an organ system.
    d. a tissue.
    e. a tissue system.

49. Pit pairs allow plasmodesmata to travel through
    a. the primary cell wall.
    b. the secondary cell wall.
    c. both the primary and secondary cell walls.
    d. neither the primary nor the secondary cell walls.
    e. the secondary and sometimes the primary cell walls.

50. Which of the following is *not* true for parenchyma cells?
    a. They are the most common cell in the plant.
    b. They may contain chloroplasts.
    c. They are commonly used for food storage.
    d. They help support leaves in nonwoody plants.
    e. They usually have thick cell walls.

51. Compared to sclerenchyma, collenchyma cells
    a. have more secondary cell wall materials.
    b. are variously shaped.
    c. can be found in bundles.
    d. are used to support the plant.
    e. are more flexible.

52. Unlike tracheids, vessel elements
    a. function when dead.
    b. are spindle-shaped.
    c. are found primarily in gymnosperms.
    d. lose part or all of the end walls.
    e. evolved to be progressively longer.

53. In angiosperm phloem,
    a. both the sieve tube elements and the companion cells have nuclei.
    b. the sieve tube elements have nuclei but the companion cells do not.
    c. the companion cells have nuclei, but the sieve tube elements do not.
    d. neither the companion cells nor the sieve tube elements have nuclei.
    e. the sieve tube elements have nuclei but the companion cells may or may not have nuclei.

54. One primary function of the ground tissue in a plant is
    a. photosynthesis.
    b. to protect the plant.
    c. to anchor the plant.
    d. for water conduction.
    e. for conduction of sugars.

55. Oak trees commonly live for a few to many years. They are therefore called
    a. annuals.
    b. biennials.
    c. perennials.
    d. semiannuals.
    e. multiannuals.

56. In the development of a root, the protoderm gives rise to
    a. the cortex.
    b. the root hairs.

c. the endodermis.
d. the xylem and phloem.
e. the pith.

57. In a young root, you would observe xylem cells in the
    a. root cap.
    b. apical meristem.
    c. zone of cell division.
    d. zone of cell elongation.
    e. zone of cell differentiation.

58. Which of the following is *not* part of a monocot stele?
    a. Xylem
    b. Phloem
    c. Endodermis
    d. Pericycle
    e. Pith

59. The _____ is centermost tissue in a dicot stem.
    a. pith
    b. xylem
    c. phloem
    d. pericycle
    e. endodermis

60. Branch roots arise from the _____.
    a. epidermis
    b. pericycle
    c. endodermis
    d. cortex
    e. pith

61. In which of the following states would you be least likely to find annual rings in dicot tree trunks?
    a. Maine
    b. Washington
    c. Kansas
    d. Arizona
    e. Hawaii

62. The periderm of a tree functions primarily
    a. to transport sugars.
    b. to form branches.
    c. to absorb water.
    d. to protect the inner tissues.
    e. to support the leaves.

63. In a woody stem, gas exchange occurs through
    a. stomata.
    b. the waxy cuticle.
    c. lenticels.
    d. the cork cambium.
    e. bundle sheath cells.

64. In a typical dicot leaf, most of the chloroplasts would be found in the
    a. upper epidermal cells.
    b. palisade mesophyll cells.
    c. bundle sheath cells.
    d. phloem cells.
    e. guard cells.

65. The primary function of a typical leaf is
    a. photosynthesis.
    b. food storage.
    c. support.
    d. anchorage.
    e. absorption.

66. Guard cells
    a. protect the plant from herbivores.
    b. secrete a waxy cuticle to prevent evaporation.
    c. contain chemicals that poison insects.
    d. **control gas exchange.**
    e. inhibit germination of fungal spores.

67. The purpose of vascular rays is
    a. **to transport nutrients through the phloem to storage cells.**
    b. to transport water from the roots to the xylem.
    c. to transport water from the leaves to the phloem.
    d. to transport nutrients from the sclerenchyma to the phloem.
    e. to transport carbon dioxide from the leaves to the phloem.

68. Brown, or unmilled, rice contains _____, whereas milled, white rice does not.
    a. protein
    b. carbohydrates
    c. **B vitamins**
    d. secondary xylem
    e. secondary phloem

69. Which of the following is *not* a function of cork cambium?
    a. Protection from microorganisms
    b. Minimize water loss
    c. Secondary growth of stems and roots
    d. **Mineral uptake**
    e. To break off and allow expansion of tree trunks

32 Transport in Plants

Fill in the Blank

1. Minerals, taken up in their **ionic** form, are moved into root cells against their concentration gradient by the process of **active transport**.

2. When osmotic **potential** increases inside a cell, water moves **into** the cell passively by osmosis.

3. More negative water potential in the xylem of the root draws water into the xylem generating a force called **root pressure**.

4. Cell walls and spaces between cells make up a compartment called the **apoplast**, and the continuous meshwork of living cells connected by plasmodesmata is called the **symplast**.

5. Evaporative water is lost through pores in the leaf, which are called **stomata**.

6. Water in the xylem is pulled up to replace water lost by evaporation because of the **cohesion** of water molecules, a physical property of water due to the **hydrogen** bonding between water molecules.

7. In CAM plants, carbon dioxide is taken in through the **stomata** during the **night** but is *not* immediately used in photosynthesis. Instead, it is made into **organic acids** until photosynthesis can resume in the daytime. The advantage of CAM is that these plants can **conserve water**.

8. Water moves toward the region of more **negative** water potential.

9. Parenchymal cells known as **transfer** cells help transport mineral ions from the symplast into the apoplast.

10. Water transport through the xylem results from **evaporation** in the leaves and subsequent **tension** in the xylem, which pulls water up.

11. On a hot day, water is lost via evaporation through the shoot. This is called **transpiration**.

12. Endodermal cells are lined with waxy structures, called **Casparian strips,** that prevent water and ions from moving between cells.

Multiple Choice

1. Pumping protons (H^+) out of a cell can trigger these movements of potassium ion (K^+) and chloride ion (Cl^-):
   a. K^+ out of the cell and Cl^- into the cell.
   b. **K^+ into the cell and Cl^- out of the cell.**
   c. both K^+ and Cl^- out of the cell.
   d. both K^+ and Cl^- into the cell.
   e. K^+ into the cell and no movement of Cl^-.

2. Patch clamping is used to
   a. **monitor ion movement through a small number of carrier proteins.**
   b. measure the water potential in an isolated cell.
   c. determine the tension pressure of xylem sap.
   d. measure translocation rates in phloem tissue.
   e. monitor rates of loading sucrose into phloem.

3. A cell is prepared for patch clamping by
   a. killing it.
   b. reducing its water potential.
   c. giving it a supply of sucrose.
   d. **removing its wall.**
   e. exposing it to red light.

4. Water soluble substances that cannot cross plasma membranes would be stopped at what part of the root?
   a. Epidermis
   b. Cortex
   c. **Casparian strips**
   d. Pericycle
   e. Xylem

5. The fact that transfer cells have many mitochondria supports the statement that transfer cells move mineral ions by means of
   a. **active transport.**
   b. facilitated diffusion.
   c. simple diffusion.
   d. tension pressure.
   e. translocation.

6. Transfer cells are located in the root's
   a. epidermis and cortex.
   b. cortex and endodermis.
   c. endodermis and pericycle.
   d. **pericycle and xylem.**
   e. epidermis and xylem.

7. The theory that xylem sap moves by capillary action has been discounted because
   a. capillary action requires living cells, and tracheids were discovered to be dead.
   b. pumping mechanisms were discovered in xylem.
   c. **the diameters of tracheid cells would not have allowed water to reach the tops of tall trees.**
   d. root pressure was discovered to be the primary cause of sap movement.
   e. capillary action could not account for guttation.

8. The phenomenon of guttation is related to
   a. active transport.
   b. osmosis.
   c. **root pressure.**
   d. transpiration.
   e. translocation.

9. On moderately dry, hot days, recently watered plants have
   a. **high transpiration rates and low root pressures.**
   b. low transpiration rates and high root pressures.
   c. low transpiration rates and low root pressures.
   d. high transpiration rates and high root pressures.
   e. moderate transpiration rates and moderate root pressures.

10. The function of cellulose microfibrils in guard cell walls is to
    a. regulate the movement of K^+ in and out of the cell.
    b. **control the pattern of cell stretching.**
    c. sense blue light.
    d. act as carbon dioxide receptors.
    e. monitor the cells' water potentials.

11. To initiate stomatal opening, potassium ions
    a. passively diffuse into guard cells.
    b. passively diffuse out of guard cells.
    c. **are actively transported into guard cells.**
    d. are actively transported out of guard cells.
    e. bond to receptor sites on guard cell walls.

12. The advantage to CAM plants of forming organic acids from carbon dioxide rather than storing carbon dioxide directly is that
    a. concentrated carbon dioxide is toxic to cells.
    b. organic acids can be used directly in photosynthesis.
    c. more energy is required to store carbon dioxide than organic acids.
    d. the presence of organic acids keeps stomata closed.
    e. **storage as organic acids make it possible to take up higher concentrations of carbon dioxide into the cell.**

13. Which technique allowed researchers to conclude that the fibrous proteins in sieve tube elements are normally dispersed and only obstruct sieve plates in response to cell damage?
    a. Analyzing phloem sap extruded from aphid stylets
    b. **Freezing phloem tissue before cutting and examining it**
    c. Watering the plant before cutting and examining it
    d. Using patch clamping to examine sieve plates
    e. Exposing plants to blue light before examination

14. The active transport of sucrose molecules from a leaf's apoplast to its phloem requires the sucrose to be attached to
    a. **protons.**
    b. amino acids.
    c. potassium ions.
    d. chloride ions.
    e. water.

15. As a tree begins transpiring in the morning, tension pressure occurs first in
    a. **the leaves.**
    b. the branches.
    c. the trunk.
    d. the roots.
    e. all regions of the tree.

16. How do mineral nutrients enter the plant body directly from the environment?
    a. **Uptake through the roots**
    b. Uptake by the leaves
    c. Uptake from digested food molecules
    d. Uptake into vascular tissue
    e. Uptake through the stems

17. Water tends to move into a cell that has
    a. a high turgor pressure due to cell wall rigidity.
    b. a high, positive water potential.
    c. an interior solute concentration like that of distilled water.
    d. **a more negative water potential.**
    e. a high turgor pressure.

18. The wilting of plant tissue occurs when
    a. water potential is high.
    b. turgor pressure is high.
    c. interior solute concentration is high.
    d. osmotic potential is low.
    e. **turgor pressure is low.**

19. When the ion concentration inside root cells is higher than ions in the soil solution, such ions can enter root cells by what processes?
    a. **Active transport only**
    b. Simple diffusion only
    c. Simple diffusion and facilitated diffusion
    d. Facilitated diffusion and active transport
    e. Simple diffusion and active transport

20. The facilitated diffusion of ions from the soil solution into root cells requires
    a. that the concentration of ions at the root cells be the same outside and inside.
    b. that the concentration of ions outside the root cells be lower than that inside.
    c. the expenditure of ATP.
    d. **specific carrier proteins in the membranes.**
    e. cellular respiration.

21. When a large amount of water enters a plant cell, what happens?
    a. Entry of water increases as the water potential increases.
    b. **Entry of water is opposed by turgor pressure.**
    c. Water moves toward the region of more positive water potential.
    d. Entry of water reduces the turgor pressure.
    e. Entry of water causes increased active transport into the cell.

22. Which of the following is true about the apoplast (the transport route through intercellular spaces)?
    a. Osmosis of water is involved.
    b. Movement of materials is regulated by membranes.
    c. **Water and solutes can move by bulk flow.**
    d. Plasmodesmata are involved.
    e. Water and solutes enter the stele via this channel.

23. Which of the following is the Casparian strip?
    a. The layer of endodermal cells
    b. The layer of epidermal cells
    c. The apoplast
    d. The symplast
    e. **The waxy layer between endodermal cells**

24. Endodermal cells differ from other cells in the root in that they
    a. lack a symplast region.
    b. are nonselective with regard to solute uptake.
    c. have a high rate of water transport.
    d. are completely surrounded by a waxy layer.
    e. prevent water and ions from moving between them.

25. Water enters the xylem tissue from surrounding root cells via
    a. active transport.
    b. facilitated diffusion.
    c. osmosis.
    d. pressure pumping.
    e. guttation.

26. Which of the following causes the root pressure that moves water upward in plant xylem?
    a. Negative water potential in the xylem sap
    b. High pressure potential of water in the soil
    c. Movement of water from root cells into the soil
    d. Active transport of minerals from soil to root cells
    e. High atmospheric humidity

27. Which force accounts for the movement of water upward through a narrow tube?
    a. Cohesion of water molecules via hydrogen bonding
    b. Negative water potential in the xylem
    c. Active transport of water molecules
    d. Passive osmosis of water following ion movement
    e. Pumping of water into the phloem

28. The evaporation–tension–cohesion mechanism explains how
    a. water is lost from leaf openings.
    b. water is transported in the xylem.
    c. water and minerals enter the root.
    d. mineral ions move through the xylem.
    e. leaf epidermal cells minimize water loss.

29. Which of the following increases a plant's intake of carbon dioxide for photosynthesis?
    a. Thick waxy cuticle
    b. Loss of water vapor
    c. Darkness
    d. Electric imbalance
    e. Open stomata

30. Stomata begin to open when
    a. potassium enters guard cells and they become turgid.
    b. potassium leaves guard cells and they become less turgid.
    c. potassium enters guard cells and they become less turgid.
    d. potassium leaves guard cells and they become turgid.
    e. potassium reaches equilibrium between guard cells and their surroundings.

31. What structure of the leaf minimizes water loss?
    a. Stoma
    b. Epidermis
    c. Cuticle
    d. Phloem
    e. Xylem

32. Plants in habitats where excess water can be lost by transpiration might show which of the following adaptations to conserve water loss during gas exchange?
    a. Stomata that close at night
    b. Stomata concentrated on the upper sides of leaves
    c. A uniformly high density of stomata
    d. Stomata that close during the day
    e. Stomata that open only during photosynthesis

33. Succulent plants of the family Crassulaceae have unusual gas-exchange patterns in that they
    a. do not take in carbon dioxide through stomata.
    b. give off oxygen at night.
    c. accumulate carbon dioxide stored as organic acid.
    d. retain oxygen in the leaf.
    e. obtain gases mainly through their roots.

34. The value of CAM (crassulacean acid metabolism) plant reactions is that
    a. photosynthesis can proceed in darkness.
    b. carbon dioxide can be concentrated in the leaf for photosynthesis.
    c. stomata can remain open during daylight.
    d. sugar formation can occur during the night.
    e. excess water can be eliminated.

35. According to the pressure flow model for translocation,
    a. sugar concentration is highest near the sink area.
    b. water enters the sieve tube by osmosis.
    c. sugar is transported out of the sieve tubes near the source area.
    d. osmosis accomplishes the bulk flow of water and nutrients.
    e. little ATP expenditure is required.

36. When sugars are actively transported into a cell, what happens to the turgor pressure inside that cell as a result?
    a. No change; sugar concentration has no effect on turgor pressure
    b. Increases, because sugar concentration directly affects turgor pressure
    c. Increases, because water enters and affects turgor pressure
    d. Decreases, because water exits and affects turgor pressure
    e. Decreases, because sugar concentration directly affects turgor pressure

37. At the site where sugars are to be used, how do sugars move from the sieve tubes into the tissue?
    a. By diffusion
    b. Via the apoplast
    c. By active transport
    d. By osmosis
    e. Via translocation

38. Most water moving through the apoplast from the soil into the stele cells first crosses a plasma membrane in the cells of the _____.
    a. root hairs
    b. cortex
    c. endodermis
    d. pericycle
    e. tracheids

39. A transfer cell has knobby growths extending into the cell that facilitate movement of minerals
    a. from its cell wall into its cytoplasm.
    b. from its cytoplasm into its cell wall.
    c. from its cell wall into the next cell's cell wall.
    d. from its cytoplasm into the next cell's cytoplasm.
    e. from its cell wall into the next cell's cytoplasm.

40. Strasburger's experiments with the movement of poisons through trees demonstrated all of the following except
    a. the absence of "pumping" cells in xylem.
    b. the crucial role of leaves in water movement.
    c. the fact that water movement is *not* dependent upon root pressure.
    d. the requirement that water movement must be through narrow cells.
    e. the ability of xylem to function without living cells.

41. Guttation is most commonly observed under conditions of
    a. high atmospheric humidity and plentiful soil water.
    b. low atmospheric humidity but plentiful soil water.
    c. high atmospheric humidity but little soil water.
    d. low atmospheric humidity and little soil water.
    e. varying atmospheric humidity and plentiful soil water.

42. What process makes the water potential in a leaf more negative?
    a. The pressure placed on the leaf by the cuticle
    b. The evaporation of water from mesophyll cells
    c. The movement of water into the leaf by root pressure
    d. The increased potassium pumped out of guard cells
    e. The movement of water from the veins into the leaf

43. The evaporative loss of water from the shoot is called
    a. translocation.
    b. transformation.
    c. transportation.
    d. transpiration.
    e. transcention.

44. Cohesion is the tendency of water molecules to attract
    a. other water molecules by covalent bonds.
    b. other water molecules by hydrogen bonds.
    c. cellulose molecules by covalent bonds.
    d. cellulose molecules by hydrogen bonds.
    e. lignin molecules by covalent bonds.

45. If xylem sap in a stem is under tension, what will occur if you cut the stem?
    a. Xylem sap will spurt out.
    b. Xylem sap will stay at the cut surface.
    c. Air will be pulled into the xylem.
    d. The cut surface will form bubbles if placed under water.
    e. Xylem sap will run out if placed under water.

46. What happens to potassium ions to initiate stomatal closing?
    a. Ions are actively transported into the guard cells.
    b. Ions are actively transported out of the guard cells.
    c. Ions are actively transported from one guard cell to another.
    d. Ions diffuse into the guard cells.
    e. Ions diffuse out of the guard cells.

47. CAM plants have adapted to dry areas by
    a. only opening their stomata on cool days.
    b. having a more efficient Calvin–Benson cycle.
    c. producing organic acids at night.
    d. evolving an active pump for carbon dioxide.
    e. fixing carbon dioxide in one type of cell, then transferring the product to another cell for sugar formation.

48. Transport through both the xylem and the phloem
    a. stops if the tissue is killed.
    b. requires ATP.
    c. can occur simultaneously in both directions.
    d. requires negative pressure (tension).
    e. involves long, thin cells.

49. Plant physiologists can obtain pure phloem sap from individual phloem cells by
    a. using very tiny drills and capillary pipettes.
    b. collecting materials from aphid stylets.
    c. obtaining liquids oozing from cut stem surfaces.
    d. analyzing the contents of droplets formed by guttation.
    e. gathering the mycorrhizal fungi for subsequent analysis.

50. The pressure flow model of translocation depends entirely on the existence of mechanisms for loading sugars into phloem at the _____ regions and for unloading them at the _____ regions.
    a. sink; sink
    b. sink; source
    c. source; source
    d. source; sink
    e. source; source or sink

51. According to the pressure flow model, during fruit development, photosynthesizing leaves would be the _____ and the fruit would be the _____.
    a. sink; sink
    b. sink; source
    c. source; source
    d. source; sink
    e. source; source or sink

52. Sugars pass from cell to cell in the leaf, starting in the _____ of the mesophyll, through the _____ of other cells, and finally into the _____ of the sieve tube element.
    a. symplast; symplast; symplast
    b. apoplast; apoplast; apoplast
    c. symplast; apoplast; symplast
    d. apoplast; symplast; symplast
    e. apoplast; symplast; apoplast

53. Maple syrup is obtained from
    a. phloem sap.
    b. xylem sap.
    c. both xylem and phloem sap.
    d. cortex sap.
    e. both cortex and phloem sap.

54. Technically, water moves from the soil into the root by
    a. active transport.
    b. passive transport.
    c. facilitated transport.
    d. **simple diffusion.**
    e. facilitated diffusion.

55. Pure water under no applied pressure is defined as having a water potential of
    a. +10.
    b. +1.
    c. **0.**
    d. −1.
    e. −10.

56. A plant cell placed in distilled water will
    a. expand until the osmotic potential reaches that of distilled water.
    b. become more turgid until the osmotic potential reaches that of distilled water.
    c. become less turgid until the osmotic potential reaches that of distilled water.
    d. **become more turgid until the pressure potential of the cell reaches its osmotic potential.**
    e. become less turgid until the pressure potential of the cell reaches the outside water potential.

57. Water will move from the root hairs through the cortex to the xylem if the water potentials are
    a. root hairs = 0; cortex = 0; xylem = 0.
    b. root hairs = −1; cortex = −1; xylem = −1.
    c. root hairs = −2; cortex = −1; xylem = 0.
    d. root hairs = 0; cortex = +1; xylem = +2.
    e. **root hairs = 0; cortex = −1; xylem = −2.**

58. Which of the following parts of a plant cell are part of the apoplast?
    a. **Cell wall**
    b. Plasma membranes
    c. Plasmodesmata
    d. Cytoplasm
    e. Vacuole

59. Cell walls impregnated with water-repellent suberin are found in the cells of the _____.
    a. root hairs
    b. cortex
    c. **endodermis**
    d. pericycle
    e. tracheids

60. When cells in the leaf release abscisic acid, it is in response to
    a. **negative water potential.**
    b. not enough carbon dioxide.
    c. reduction of light (nighttime).
    d. increase of chloride ions in the stomata.
    e. increase in potassium ions in the stomata.

33 Plant Responses to Environmental Challenges

Fill in the Blank

1. Plants adapted to dry environments are called **xerophytes**.

2. Plants with long dormant periods interrupted by short periods of rapid growth and reproduction typically live in **dry** environments.

3. When cells accumulate the amino acid **proline**, their water potential becomes **more** negative.

4. Some swamp plants have root extensions called **pneumatophores** that grow into the air and deliver **oxygen** to the rest of the root system.

5. The major stress encountered by plants living in waterlogged soil is lack of soil **oxygen**.

6. In cacti, leaves are modified to form **spines**, and photosynthesis is carried out by the **stem** region.

7. Globally, the toxic substance that most restricts plant growth is **salt (sodium chloride)**.

8. Halophytes accumulate chloride and **sodium** ions and transport these substances to their **leaves**.

9. Plants with fleshy, water-storing leaves are called **succulents**, and can be either xerophytes or **halophytes**.

10. Plants produce small molecules called **phytoalexins** within hours of infection by fungi or **bacteria.**

11. **Secondary products** produced by plants are useful to humans as fungicides and pharmaceuticals.

12. Large molecules, called **pathogenesis-related** proteins, are produced as a defense against infection and in cleanup operations.

Multiple Choice

1. Which of the following is *not* an adaptation to dry environments?
   a. Thick cuticle
   b. Vertically hanging leaves
   c. Stomata in sunken cavities
   d. Epidermal hairs
   e. **Salt glands in leaves**

2. The typical environment for annual plants with a brief growing period and seeds capable of long dormant periods is a
   a. **desert.**
   b. salt marsh.
   c. freshwater marsh.
   d. heavy metals environment.
   e. grazed field.

3. Corn and related grasses roll up their leaves in response to
   a. excess water.
   b. **lack of water.**
   c. excess salt.
   d. heavy metals.
   e. herbivores.

4. Compared to plants in moderate environments, xerophytes carry out photosynthesis
   a. more slowly due to interference from the amino acid proline.
   b. more slowly because transpiration occurs more quickly.
   c. **more slowly because their adaptations minimize carbon dioxide uptake.**
   d. more quickly because their stems are also photosynthetic.
   e. more quickly because they have short periods of intense growth.

5. Swamp plants typically have root systems that
   a. grow quickly.
   b. penetrate deeply into the soil.
   c. **can carry out alcoholic fermentation.**
   d. alternate periods of growth and dying back.
   e. accumulate the amino acid proline.

6. The pneumatophores of swamp plants are modified
   a. flowers.
   b. leaves.
   c. **roots.**
   d. spines.
   e. stems.

7. Leaf parenchyma tissue with large spaces between cells is used by
   a. **aquatic plants to provide buoyancy.**
   b. aquatic plants to decrease transpiration rates.
   c. desert plants to store water.
   d. desert plants to form succulent leaves.
   e. halophytes to excrete salt.

8. By accumulating the amino acid proline, plants
   a. become toxic to most herbivores.
   b. can carry out alcoholic fermentation.
   c. **can extract more water from the soil.**
   d. can prevent toxic effects from sodium.
   e. attract animals that disperse seeds.

9. If a nonhalophyte and a halophyte are both placed in a salty environment, which will accumulate more sodium internally, and why?
   a. **The halophyte, because the nonhalophyte will not absorb much sodium.**
   b. The halophyte, because it requires sodium as a nutrient.

c. The halophyte, because its succulence gives it more volume for internal storage.

d. The nonhalophyte, because it cannot excrete sodium after absorption.

e. The nonhalophyte, because it needs sodium to create a negative water potential.

10. Which of the following adaptations is *not* found in both xerophytes and halophytes?
a. High root-to-shoot ratios
b. Sunken stomata
c. Large air spaces in the leaf parenchyma
d. Reduced leaf area
e. Thick cuticles

11. The reason that certain plants can grow in soils contaminated with high amounts of heavy metals is because the plants
a. do not take up the metals.
b. excrete the metals.
c. have a genetic tolerance to the metals.
d. use the metals for normal biochemical functions.
e. are toxic to herbivores due to the metals.

12. Which of the following is *not* true of plants that tolerate heavy metals?
a. The plants take up the heavy metals.
b. A population will tolerate all heavy metals equally.
c. Tolerant populations can evolve rapidly.
d. The plants' tolerance is due to their genotype.
e. The plants usually experience little competition.

13. Serpentine soils have a shortage of the nutrient
a. calcium.
b. carbon.
c. chromium.
d. magnesium.
e. nickel.

14. Compared to nonserpentine regions, the vegetation on serpentine soils is
a. smaller in size but otherwise the same.
b. more sparse but otherwise the same.
c. more sparse and less diverse.
d. more abundant and more diverse.
e. larger in size but less diverse.

15. Serpentine plants such as jewel flower
a. absorb calcium efficiently even in low concentrations.
b. grow without calcium.
c. use magnesium in place of calcium.
d. maximize their use of calcium by a high root-to-shoot ratio.
e. store calcium and excrete excess magnesium.

16. If tomatoes, a nonserpentine species, and jewel flowers, a serpentine species, both are grown in serpentine soil, what are the effects of supplying calcium to some of the plants?
a. The tomatoes grow better, and the jewel flowers are unaffected.
b. Both tomatoes and jewel flowers are unaffected.
c. Both tomatoes and jewel flowers grow better.
d. The tomatoes grow better, and the jewel flowers grow less well.
e. Both tomatoes and jewel flowers grow less well.

17. Grazing increases photosynthetic rates in certain plant species because
a. more light reaches younger, more active leaves.
b. the remaining leaves can transport sugars more slowly to the roots.
c. the roots will be able to take up more nitrogen.
d. older, dying leaves had acted as sugar sinks.
e. competition for atmospheric carbon dioxide is reduced.

18. The ability of grasses to grow from the base of the shoot and leaf is an adaptation to
a. dry environments.
b. soil fungi.
c. heavy metals.
d. grazing.
e. saline environments.

19. Grazed plants may exhibit increased productivity in all of the following ways, except
a. faster photosynthetic rates.
b. growing more stems.
c. greater seed distribution.
d. continuing growth later in the season.
e. growing more roots.

20. Secondary products
a. are essential for basic biological reactions.
b. are similar in all plants.
c. occur more often in animals than in plants.
d. attract or inhibit other organisms.
e. are usually of high molecular weight.

21. Alkaloids protect plants by
a. affecting herbivore nervous systems.
b. causing cancer in animals.
c. acting as fungicides.
d. mimicking animal hormones.
e. mimicking essential amino acids.

22. Terpenes protect plants by
a. affecting herbivore nervous systems.
b. causing cancer in animals.
c. acting as fungicides.
d. mimicking animal hormones.
e. mimicking essential amino acids.

23. Sulfur compounds protect plants by
a. affecting herbivore nervous systems.
b. causing cancer in animals.
c. acting as fungicides.
d. mimicking animal hormones.
e. mimicking essential amino acids.

24. One class of secondary plant products that attract pollinating and seed-dispersing animals is
a. alkaloids.
b. flavonoids.
c. phenolics.
d. quinones.
e. steroids.

25. The secondary plant products that prevent the normal development of insect herbivores are
a. alkaloids.
b. flavonoids.
c. phenolics.
d. quinones.
e. steroids.

26. Some plants produce the amino acid canavanine, which is toxic to many insects because
    a. insects lack a tRNA specific for canavanine.
    b. canavanine is a component of alkaloids.
    c. canavanine prevents cells from synthesizing the amino acid arginine.
    d. insects lack large vacuoles for the storage of secondary products.
    e. insect proteins that incorporate canavanine function poorly.

27. In defending their bodies against tissue damage caused by pathogens,
    a. animals repair tissues and plants seal off tissues.
    b. animals seal off tissues and plants repair tissues.
    c. both plants and animals sometimes seal off and sometimes repair tissues.
    d. both plants and animals only repair tissues.
    e. both plants and animals only seal off tissues.

28. Phytoalexins
    a. are always present in plants.
    b. occur in equal concentrations throughout plants.
    c. are toxic to many fungi and bacteria.
    d. have no effect on viral infections.
    e. cause the plants to seal off areas of damaged tissue.

29. A plant's hypersensitive reaction to infection may include all of the following, except
    a. cells around the infection site produce phytoalexins.
    b. the plant acquires long-term resistance to the infective agent.
    c. the plant synthesizes pathogenesis-related proteins.
    d. infected and nearby cells die.
    e. phytoalexins are transported to all parts of the plant.

30. Salicylic acid in plants does *not*
    a. increase resistance to pathogens.
    b. trigger the production of pathogenesis-related proteins.
    c. poison fungi and bacteria.
    d. protect against tobacco mosaic virus.
    e. have a role in the hypersensitivity response.

31. Non-water-soluble or hydrophobic poisons are stored in a plant's
    a. chloroplasts.
    b. epidermal waxes.
    c. Golgi bodies.
    d. mitochondria.
    e. vacuoles.

32. Plants can produce the respiratory poison cyanide without poisoning themselves because plants
    a. do not respire.
    b. store a cyanide precursor in one compartment and activating enzymes in a different compartment.
    c. store water-soluble cyanide in laticifers.
    d. have enzymes that are unaffected by cyanide.
    e. also produce proteins that bind and inhibit cyanide.

33. Which evidence best supports the hypothesis that the presence of toxic latex in leaves deters insects from feeding on the plant?
    a. Many insects do not feed on latex-producing plants.
    b. Latex plants release milky latex when the leaf is damaged.
    c. Beetles that drain latex out of part of the leaf can then feed on that part.
    d. Beetles that cut veins in the leaves can then feed on the latex released.
    e. Latex-producing plants have high survival rates.

34. Which of the following is *not* a special adaptation of leaves to dry environments?
    a. Leaves modified as spines
    b. Stomata in sunken cavities
    c. Dense epidermal hairs
    d. Fleshy leaves
    e. Horizontal leaves

35. The presence of pneumatophores in plants is an adaptation for success in which type of habitat?
    a. Desert
    b. Mountain
    c. Grassland
    d. Seashore
    e. Swamp

36. Some halophytic plants have salt glands that
    a. accumulate salt in the roots.
    b. serve as a barrier to salt intake.
    c. maintain high salt concentration in the plant.
    d. secrete salt onto the leaf surface.
    e. increase water loss from the plant.

37. What combination of adaptations is often seen in plants that live in saline environments?
    a. Salt glands and succulence
    b. Salt glands and broad leaves
    c. CAM metabolism and succulence
    d. Spines and thin cuticle
    e. Dense stomata and thick cuticle

38. Certain plants concentrate the harmless amino acid proline in their cells. What effect does this have on the plant?
    a. More negative water potential
    b. Increased rate of transpiration
    c. Increased positive water potential in the leaf
    d. Decreased salt loss
    e. Decreased water uptake

39. When some leaves are removed from a plant, what typically happens?
    a. Less light is available to the leaves.
    b. Less nitrogen is obtained from the soil.
    c. The plant dies.
    d. The remaining leaves photosynthesize faster.
    e. The transport of sugar from the remaining leaves decreases.

40. Once secondary compounds are formed by plants, where are those defensive compounds usually stored?
    a. Vacuoles
    b. Nucleus
    c. Cell walls
    d. Cytoplasm
    e. Bound to membrane proteins

41. Plants respond to conditions where water is plentiful but oxygen scarce by
    a. rapidly growing roots that penetrate deeply into the soil.
    b. fermenting sugars to lactic acid.
    c. inhibiting the production of ATP.
    d. forming aerenchyma tissue.
    e. producing oxygen from water.

42. All of the following are adaptations to saline environments, except
    a. accumulation and transport of sodium ions.
    b. sequestering sodium ions in the roots.
    c. salt glands in the leaves.
    d. fleshy, gummy leaves.
    e. smaller leaves with smaller cells.

43. One way that grazing increases the productivity of a plant is by
    a. supporting the food chains in nature.
    b. reducing the rate of photosynthesis in the remaining leaves.
    c. increasing the number of sinks for absorbed nitrogen.
    d. shading the younger leaves.
    e. causing the production of more replacement stems.

44. Which of the following is a "secondary product"?
    a. Proteins
    b. Lipids
    c. Alkaloids
    d. Carbohydrates
    e. Nucleic acids

45. Steroids produced by plants may function to
    a. attract pollinators and animals that disperse seeds.
    b. affect nervous systems of animals.
    c. inhibit fungal action.
    d. prevent normal development of insects.
    e. impair growth of competing plants.

46. Polysaccharides serve to
    a. store water in plants.
    b. defend against pathogens by antimicrobial activity.
    c. repel predators because they are toxic to animals.
    d. act as salt glands.
    e. strengthen cell walls to form a barrier against invasion.

47. Laticifers are
    a. specialized cells for containing sodium ions.
    b. latex-containing tubes for storing hydrophobic products.
    c. the waxy cells in the epidermis.
    d. cells that produce poisons such as alkaloids.
    e. cells in the roots that help to take up more water in dry environments.

48. In the gene-for-gene resistance mechanism, if a plant has a dominant resistance gene and a pathogen has a dominant avirulence gene
    a. the pathogen's gene will override the plant's gene and infect the plant.
    b. epistasis occurs and the plant is infected.
    c. epistasis occurs and the plant is resistant to the pathogen.
    d. the plant will be resistant to all pathogens whether they have the dominant avirulence gene or not.
    e. None of the above

49. Why aren't plants affected by their production of canavanine?
    a. Their tRNA molecules don't bind canavanine.
    b. They keep it sequestered in vacuoles.
    c. It isn't activated until it leaves the plant.
    d. Plants don't use amino acids like animals do.
    e. Plants are affected, but they produce it in such small amounts that is isn't detrimental to the plant.

50. Scientists discovered a protein called arcelin in wild bean plants. This protein is useful to the plant because
    a. it is toxic to herbivore predators.
    b. it helps the plants to grow in dry environments.
    c. it is a secondary product that is beneficial to humans.
    d. it is responsible for making the plant seeds resistant to predation by weevils.
    e. it makes the plant taste bad so weevils and other herbivores leave it alone.

34 *Plant Nutrition*

Fill in the Blank

1. Decomposed plant litter produces a dark-colored organic material called **humus**.

2. The source from which plants derive their carbon is **the atmosphere**.

3. Nitrogen fixation is catalyzed by the enzyme called **nitrogenase**, which cannot function in the presence of the element **oxygen**.

4. Plants acquire their essential mineral nutrients from **the soil**. The four elements used in greatest quantity are **carbon, hydrogen, oxygen,** and **nitrogen**.

5. Some essential elements that occur as positive ions in soils may be traded with H^+ ions in soil solutions by the process of **ion exchange**.

6. Because of leaching and crop production, soils may become depleted of nutrients and require addition of **fertilizer**.

7. The three elements most commonly added to agricultural soils are **nitrogen, phosphorus,** and **potassium**.

8. Nitrogen fixation is the conversion of atmospheric nitrogen to **ammonia** and is catalyzed by a single enzyme called **nitrogenase**.

9. All nitrogen fixation is carried out by **prokaryotes or bacteria**, and some of them live in symbiosis with other organisms.

10. The process that is the opposite of nitrogen fixation is called **denitrification**.

11. Some plants living in boggy areas of low pH have adaptations for carnivory in order to increase their intake of **nitrogen**. These carnivorous plants are considered to be **autotrophs** because they acquire energy from photosynthesis.

12. The large *Rhizobium* cells that live within plant root nodules are called **bacteroids**.

13. The mineral group required by plants in concentrations of less than 100 µg per gram of dry matter is called **micronutrients**.

14. Nutrients necessary for plant growth and reproduction, not replaceable by another nutrient, and directly required by the plant are called **essential elements**.

Multiple Choice

1. Which of the following describes how animals and plants acquire nutrients?
   a. Both animals and plants carry out bulk ingestion.
   b. Both animals and plants control their uptakes to match their biochemical needs.
   c. **Animals carry out bulk ingestion, and plants control their nutrient uptakes.**
   d. Animals control their nutrient uptakes, and plants carry out bulk ingestion.
   e. Both animals and plants carry out bulk ingestion and also control their nutrient uptakes to match their biochemical needs.

2. Which of the following elements does *not* reversibly attach to the surface of clay particles?
   a. Calcium
   b. **Chloride**
   c. Hydrogen
   d. Magnesium
   e. Potassium

3. Young soils exhibit a rapid increase in the amount of biomass, which is accompanied by a rapid
   a. **decline in available nutrients.**
   b. decline in the amount of humus.
   c. decline in the amount of clay.
   d. increase in available water.
   e. increase in soil fertility.

4. Which of the following does *not* have nitrogen-fixing root nodules containing *Rhizobium?*
   a. Alfalfa
   b. Beans
   c. Clover
   d. Peas
   e. **Rice**

5. Which mineral is deficient in a plant whose growth is stunted and whose oldest leaves turn yellow and die prematurely?
   a. Calcium
   b. Iron
   c. Magnesium
   d. **Nitrogen**
   e. Phosphorus

6. Which of the following is *not* a consequence of adding lime to soil?
   a. The soil's pH is raised.
   b. Nutrient availability increases.
   c. Calcium (Ca^{2+}) is added to the soil.
   d. Clay particles release hydrogen ions (H^+).
   e. **The soil's ability to retain water increases.**

7. One advantage of organic fertilizers over inorganic fertilizers is that organic fertilizers
   a. **improve the physical properties of the soil.**
   b. provide an almost instantaneous supply of soil nutrients.
   c. increase the soil's pH by liming.
   d. contain higher concentrations of essential nutrients.
   e. contain chemically active clay particles.

8. In general, soils of the northern temperate zone have higher nutrient concentrations than tropical soils because the northern soils are
   a. drier.
   b. subject to hot and cold seasons.
   c. less acidic.
   d. formed from more nutrient-rich bedrock.
   e. **younger.**

9. Which substance is actively transported from one cell to another in order to trigger the intercellular movement of water?
   a. Carbon
   b. Chloride
   c. Oxygen
   d. **Potassium**
   e. Sodium

10–14. Match the mineral nutrient in the list below with the following description. Each item may be used once, more than once, or not at all.
   a. Magnesium
   b. Nitrogen
   c. Phosphorus
   d. Potassium

10. When bonded to oxygen atoms, important in many energy-storing and energy-releasing pathways *(c)*

11. Pumped into cells to create osmotic potential *(d)*

12. Constituent of both proteins and amino acids *(b)*

13. Used as a cofactor by many enzymes *(a)*

14. Aids in maintaining the electric balance of cells *(d)*

15. What is the effect of the ionization of carbonic acid on soil?
   a. **It triggers the release of mineral ions from clay.**
   b. It lowers the pH of the soil.
   c. It reduces the leaching of phosphates and nitrates.
   d. It causes laterization of the A horizon.
   e. It increases the amount of soil carbon available to plants.

16. The capture and digestion of insects allows carnivorous plants to
   a. pollinate their flowers.
   b. **absorb nitrogen compounds.**
   c. disperse their fruits.
   d. overcome insect parasitism.
   e. neutralize acidic soils.

17. Although mistletoes are green, they are considered to be parasites because
   a. **they depend on other plants for water and minerals.**
   b. they cling to woody plants for physical support.

   c. they capture and digest insects.
   d. they have root nodules containing nitrogen-fixing *Rhizobium.*
   e. their chlorophyll is not functional.

18. The form of nitrogen that most plants prefer from the soil is
   a. found in amino acids.
   b. ammonia.
   c. dinitrogen.
   d. **nitrate.**
   e. nitrite.

19. Which of the following nitrogen compounds is used directly by plants to build proteins?
   a. **Ammonia**
   b. Dinitrogen
   c. Nitrate
   d. Nitrite
   e. Nitrous oxide

20. One example of a nutrient in reduced form is the
   a. carbon in carbon dioxide.
   b. hydrogen in water.
   c. **nitrogen in ammonia.**
   d. phosphorus in phosphate.
   e. sulfur in sulfate.

21. The most common gas in the atmosphere is
   a. carbon dioxide.
   b. **dinitrogen.**
   c. oxygen.
   d. ozone.
   e. water vapor.

22. Some bacteria function as denitrifiers; they
   a. oxidize ammoniun ions to nitrate.
   b. oxidize nitrate to nitrite.
   c. reduce dinitrogen to ammonia.
   d. reduce nitrates to ammonia.
   e. **oxidize ammonia to dinitrogen.**

23. Which of the following are autotrophs?
   a. Carnivorous organisms
   b. Herbivorous organisms
   c. **Chemosynthetic organisms**
   d. Organisms that require organic nutrients
   e. All of the above

24. Which of the following could serve as an energy source for chemosynthetic bacteria?
   a. Sugars
   b. Carbon dioxide
   c. **Hydrogen sulfide**
   d. Oxygen
   e. Protein

25. For a plant, an essential element is something that
   a. is used for reproduction.
   b. is synthesized in the plant.
   c. is a secondary plant toxin.
   d. is replaceable by another element.
   e. **must be taken in from the environment.**

26. Which of the following is more true of plant nutrient systems than of animal systems?
   a. They require an energy source.
   b. They release wastes.

c. **They control intake of nutrients.**
d. They do not use nutrients.
e. They do not need essential elements.

27. Which of these factors contributes least to the differences between soil types?
    a. Temperature
    b. Type of original parent rock
    c. Farming activities
    d. Rainfall
    e. **Depth**

28. In a soil profile, which is likely to be true of the horizon with the least mineral nutrients?
    a. **It contains the most organic material.**
    b. It contains the most original parent rock material.
    c. It is the thickest of the horizons.
    d. It generally does not support agriculture.
    e. It accumulates material from other horizons.

29. The process of ion exchange is the means by which
    a. carbonic acid is added to soils.
    b. **positive ion nutrients are replaced by H$^+$.**
    c. negative ion nutrients are incorporated into soils.
    d. positive ions are replaced by negative ions.
    e. neutral atoms become ions in soils.

30. The three elements most commonly added to agricultural soils in fertilizers are
    a. nitrogen, phosphorus, and iron.
    b. nitrogen, potassium, and iron.
    c. potassium, sulfur, and iron.
    d. **nitrogen, potassium, and phosphorus.**
    e. nitrogen, sulfur, and iron.

31. What is the advantage of adding organic fertilizers to soils instead of inorganic fertilizers?
    a. **Improved physical properties of soil**
    b. Use of specific nutrient formulas for specific problems
    c. More rapid increase in nutrients
    d. Leaching of the soils
    e. Increase of clay particles

32. Chemical weathering, an important part of soil formation, includes
    a. the splitting of clays.
    b. the effects of freezing and thawing.
    c. **the hydrolysis of rock.**
    d. the crushing of rock.
    e. the drying of soils.

33. Laterization and the formation of reddish soils involves all of the following except
    a. leaching of water-soluble nutrients.
    b. oxidation of iron and aluminum compounds.
    c. rapid movement of water through the soil.
    d. depletion of the A and B layers.
    e. **accumulation of nutrients in the soil.**

34. Which of the following is *not* true of semiarid regions?
    a. The B horizon is usually very hard.
    b. **Water moves very quickly downward through the soil.**
    c. Rainfall evaporates quickly.
    d. The soil is very fertile.
    e. Successful irrigated agriculture depends on these regions.

35. The process of nitrogen fixation is
    a. the uptake of atmospheric nitrogen by plants.
    b. **the conversion of atmospheric nitrogen into ammonia.**
    c. the production of nitrogen-bearing compounds in plants.
    d. the release of nitrogen into the atmosphere.
    e. the release of ammonia into the atmosphere.

36. Root nodules on plants of the legume family contain
    a. cyanobacteria.
    b. *Nitrosococcus* bacteria.
    c. ***Rhizobium* bacteria.**
    d. *Pseudomonas* bacteria.
    e. *Nitrobacter* bacteria.

37. Why do farmers plant clover or alfalfa alternately with their crops?
    a. These plants grow well because they have symbiotic nitrogen-fixing microbes.
    b. When these plants die and are plowed under they contribute fixed nitrogen to the soil.
    c. They release nitrogen-containing compounds as waste products.
    d. **Their root nodules release amino acids into the soil.**
    e. These plants have high rates of photosynthesis.

38. Which of the following statements about the chemical process of nitrogen fixation in cells is true?
    a. All three bonds between nitrogen atoms are broken simultaneously.
    b. **Hydrogen atoms are added in one reaction at a time.**
    c. Very little energy in the form of ATP is required.
    d. A different enzyme catalyzes each of the many reactions.
    e. It is enhanced by high oxygen concentrations.

39. How do nitrogen-fixing microbes first become symbiotic with their plants?
    a. They are carried in the seed.
    b. **They are attracted by chemicals on root hairs.**
    c. They move in via openings in the plant cell walls.
    d. They move into the vascular system and multiply.
    e. They enter root nodules previously produced by the plant.

40. Legume root nodules represent a symbiosis between the legume plant, which receives fixed nitrogen, and the bacteria, which receive
    a. sugars.
    b. oxygen.
    c. carbon dioxide.
    d. nitrogenase.
    e. **leghemoglobin**

41. The industrial production of nitrogen-containing fertilizer is currently limited by
    a. **its high energy expense.**
    b. the inability to insert nitrogenase genes into plants.
    c. the lack of nitrogenase for the industrial process.
    d. the limited supply of dinitrogen gas.
    e. the need to exclude free oxygen in the process.

42. The process that is the opposite of nitrogen fixation is
    a. nitrification.
    b. denitrification.
    c. aerobic breakdown of amino acids.
    d. release of ammonia.
    e. nitrate reduction.

43. The products of nitrogen-fixing organisms can be oxidized to form nitrites and nitrates by
    a. all living organisms.
    b. most living organisms that utilize oxygen.
    c. many types of microorganisms.
    d. a few specific genera of soil bacteria.
    e. only the nitrogen-fixing bacteria.

44. Bacteria that obtain their carbon from organic compounds and their energy from inorganic substances are called
    a. photosynthetic autotrophs.
    b. photosynthetic heterotrophs.
    c. chemosynthetic autotrophs.
    d. chemosynthetic heterotrophs.
    e. chemosynthetic organotrophs.

45. Most plants continue to obtain new sources of mineral nutrients by
    a. breaking down organic matter.
    b. growing longer roots.
    c. shading the plants below them.
    d. evolving more elaborate photosystems.
    e. absorbing minerals through the leaves.

46. One of the defining characteristics of an essential element is it
    a. is only necessary for early growth of the seedling.
    b. can be replaced by another element.
    c. has a direct function in the plant.
    d. may function by relieving the toxicity of another element.
    e. is found in relatively high concentrations in the environment.

47. Plants do *not* obtain which of the following elements from the soil?
    a. Carbon
    b. Nitrogen
    c. Potassium
    d. Sulfur
    e. Zinc

48. The five elements that comprise most proteins are
    a. carbon, oxygen, sulfur, phosphorus, and potassium.
    b. carbon, hydrogen, oxygen, nitrogen, and phosphorus.
    c. carbon, hydrogen, oxygen, nitrogen, and sulfur.
    d. carbon, hydrogen, nitrogen, sulfur, and potassium.
    e. carbon, hydrogen, oxygen, phosphorus, and potassium.

49. When potassium ions are actively transported from one cell to another, which ions follow passively, maintaining electric balance?
    a. Sodium
    b. Chloride
    c. Hydroxide
    d. Carbonate
    e. Magnesium

50. Which of the following is a micronutrient in plants?
    a. Potassium
    b. Sulfur
    c. Calcium
    d. Iron
    e. Magnesium

51. Special air filters were required to first demonstrate the essentiality of
    a. chlorine.
    b. nickel.
    c. oxygen.
    d. potassium.
    e. nitrogen.

52. The maximum diameter of a clay particle is
    a. 2,000,000 micrometers.
    b. 2,000 micrometers.
    c. 2 micrometers.
    d. 0.002 micrometers.
    e. 0.000002 micrometers.

53. Three major zones are recognized in the profile of a typical soil. Most of the earthworms can be found in
    a. the topsoil.
    b. the subsoil.
    c. the topsoil and subsoil.
    d. bedrock.
    e. subsoil and bedrock.

54. A 5-10-5 fertilizer has 5% _____, 10% _____, and 5% _____.
    a. nitrogen, calcium, phosphate
    b. nitrogen, phosphate, potassium
    c. nitrogen, potassium, calcium
    d. potassium, phosphate, calcium
    e. phosphate, potassium, nitrogen

55. The soils formed by laterization are very poor in nutrients because the clay particles only contain
    a. iron and aluminum.
    b. iron and silicon.
    c. aluminum and silicon.
    d. aluminum and zinc.
    e. iron and zinc.

56. Mor humus, which accumulates under conifer trees, is characterized as
    a. nutrient rich with an acidic pH.
    b. nutrient rich with a neutral pH.
    c. nutrient rich with a basic pH.
    d. nutrient poor with an acidic pH.
    e. nutrient poor with a neutral pH.

57. Most of the nitrogen on Earth is in the form of
    a. ammonia.
    b. nitrate ions.
    c. dinitrogen gas.
    d. amino acids.
    e. proteins.

58. Nitrogen fixers convert
    a. ammonia to dinitrogen.
    b. dinitrogen to ammonia.
    c. ammonia to nitrate.
    d. nitrate to ammonia.
    e. dinitrogen to nitrate.

59. All nitrogen fixers belong to the
    a. **kingdom Eubacteria.**
    b. kingdom Plantae.
    c. kingdoms Eubacteria and Plantae.
    d. kingdom Protista.
    e. kingdoms Protista and Plantae.

60. Cyanobacteria are known to fix nitrogen in association with all of the following except
    a. fungi in lichens.
    b. ferns.
    c. bryophytes.
    d. cycads.
    e. **wheat.**

61. Nitrogenase enzymes react with the substrate dinitrogen with _____ hydrogen atoms before releasing the product.
    a. two
    b. three
    c. four
    d. five
    e. **six**

62. Nitrogenase enzymes are extremely sensitive to _____ molecules.
    a. hydrogen
    b. **oxygen**
    c. water
    d. carbon dioxide
    e. calcium carbonate

63. During the formation of a nodule, at the stage when the infecting bacteria are called bacteroids, the bacteria can be found in which portion of the plant cell?
    a. Cell wall
    b. Nucleus
    c. Mitochondria
    d. **Vacuoles**
    e. Chloroplasts

64. Nodules that are actively fixing nitrogen are pink, demonstrating the presence of
    a. iron.
    b. chlorophyll.
    c. **leghemoglobin**
    d. anthocyanin.
    e. alkaloids.

65. Recombinant DNA technology is now being used in an attempt to "teach" new plants how to
    a. **produce their own nitrogenase enzymes.**
    b. fix nitrogen without using ATP.
    c. produce ammonia under anaerobic conditions.
    d. convert nitrates into ammonium ions.
    e. recycle their nitrogen instead of excreting it.

66. Denitrifying bacteria are part of nature's nitrogen cycle, converting
    a. ammonia to dinitrogen.
    b. dinitrogen to nitrate.
    c. ammonia to nitrate.
    d. nitrate to ammonia.
    e. **nitrate to dinitrogen.**

67. Plants take up sulfur in the _____ form and phosphorus in the _____ form.
    a. reduced; oxidized
    b. **oxidized; oxidized**
    c. oxidized; reduced
    d. oxidized; reduced or oxidized
    e. reduced; reduced or oxidized

68. The big drawback in recombinant DNA for making plants that produce their own nitrogenase is
    a. bacterial genes cannot be inserted into plants.
    b. the DNA degrades when inserted into a plant.
    c. plants can't exclude oxygen, so the enzyme is ineffective.
    d. plants can't take up dinitrogen, so the enzyme is ineffective.
    e. **the amount of energy needed for the reaction is very great.**

69. An advantage to using inorganic fertilizers is
    a. they also improve physical properties of the soil.
    b. **they can be taken up almost instantaneously.**
    c. their quality is better than organic fertilizers.
    d. the minerals are in their proper ionic form.
    e. None of the above

70. Which of the following is not a component of soil?
    a. Clay
    b. Air
    c. Water
    d. **Arthropods**
    e. Fungi

35 Regulation of Plant Development

Fill in the Blank

1. A germinating grass seedling embryo produces **gibberellins** that mobilize stored nutrients.

2. The hormone **auxin** is responsible for the phenomenon of apical dominance in plants.

3. It is thought that auxins control growth by **cell elongation**, while cytokinins cause growth by **cell division**.

4. The hormone that generally shuts down or inhibits plant activity is **abscisic acid**.

5. Application of **cytokinins** to leaves can keep them green and delay senescence.

6. Fruit shippers apply **ethylene** to speed up ripening.

7. Growth of a shoot tip toward the light is called **phototropism** and is thought to be mediated by the actions of the hormone **auxin**.

8. Bending toward the light occurs when the auxin moves to **the side away from the light** and has the effect of **loosening** the cell walls in that region, causing asymmetric growth and thus bending.

9. The process of seed coat modification to increase ease of germination is called **scarification**.

10. Molecules active at small concentrations that regulate plant development are called **hormones**.

11. Dicot seedlings form an **apical hook** to protect the stem while it grows through the soil.

Multiple Choice

1. Plant growth substances generally
   a. have a single specific role.
   b. affect mainly the cells that produce them.
   c. are species-specific.
   d. are produced in many parts of the plant.
   e. elicit rapid responses.

2. As a grass seed germinates, the embryonic plant secretes gibberellins that
   a. absorb light.
   b. cause elongation.
   c. mobilize stored foods.
   d. take up water.
   e. direct the shoot upwards.

3. The hormone responsible for phototropism is
   a. abscisic acid.
   b. auxin.
   c. ethylene.
   d. gibberellin.
   e. phytochrome.

4. Gibberellins were discovered by studying the "foolish seedling" disease of rice, in which seedlings
   a. grew unusually slowly.
   b. grew into tall, spindly plants.
   c. died after germination.
   d. produced seeds unusually early.
   e. had a harmful mutation.

5. Which of the following suggests that plants produce gibberellin growth hormones?
   a. Genetically dwarf corn seedlings grow tall with gibberellin treatment.
   b. Genetically tall corn plants grow even taller with gibberellin treatment.
   c. Gibberellins are produced by tall and dwarf varieties of corn.
   d. Gibberellins are produced by dwarf varieties of corn.
   e. Different varieties of corn produce different gibberellins.

6. In the Darwins' experiment, which part of the seedling was sensitive to light?
   a. The entire seedling
   b. The entire shoot above the roots
   c. The entire leaf sheath
   d. The sheath just below the tip
   e. The extreme tip of the leaf sheath

7. In plant tropisms
   a. an imbalance in ethylene concentration causes curvature.
   b. roots grow toward the light.
   c. one side of the root or shoot grows more rapidly than the other.
   d. DNA is the light receptor or gravity receptor.
   e. auxin is the light receptor or gravity receptor.

8. Leaf abscission is
   a. the separation of leaves from stems.
   b. the orientation of a leaf toward the light.
   c. the regeneration of a leaf after a wound.
   d. the maturation of leaf tissue.
   e. the initiation of growth of new tissue.

9. The phenomenon of apical dominance is strengthened most by
   a. removal of the tip.
   b. auxin production.
   c. removal of leaves.
   d. production of fruits.
   e. removal of fruits.

10. In order for plant cells to grow larger
   a. new cellulose is deposited on the outer surface of the cell wall.

b. the vacuole expands, thus rupturing the cell wall.

c. the cell wall is strengthened by deposits on its inside surface.

***d.* the cell wall is loosened by the breaking of bonds in the cell wall.**

e. cell division must occur just beforehand.

11. Which of the following observations provided evidence of the existence of a specific receptor protein for auxin in plant cells?
    a. The addition of H^+ ions causes rapid stem growth.
    b. Auxin binds with cellulose in cell walls.
    c. Cellulose linkages can be loosened by an enzyme.
    d. Synthetic auxins in herbicides harm dicot plants, but not monocots.
    ***e.* Certain mutants that cannot grow upright also cannot respond to auxins.**

12. Cytokinins are formed primarily in which area of the plant?
    a. Tips of the shoot
    b. Leaves
    c. Stems
    ***d.* Roots**
    e. Lateral buds

13. In the phenomenon of gravitropism the growth of a plant part is
    a. toward the center of Earth.
    ***b.* in a direction determined by gravity.**
    c. in a direction opposite that of the main light source.
    d. in a direction opposite that of the growth of the shoot.
    e. toward the darkest area.

14. Why are cell differentiation experiments often done with pith tissue cultures?
    ***a.* Pith cells are all unspecialized.**
    b. Only pith tissue grows rapidly.
    c. Pith tissue of stem does not differentiate to root cells.
    d. Pith tissue responds primarily to auxin.
    e. Pith tissue is found in all plants.

15. Which hormone stimulates lateral buds to grow into branches?
    a. Abscisic acid
    b. Auxin
    ***c.* Cytokinin**
    d. Ethylene
    e. Gibberellin

16. Which of the following hormones is a gas?
    a. Abscisic acid
    b. Auxin
    c. Cytokinin
    ***d.* Ethylene**
    e. Gibberellin

17. Which of the following is considered to be the plant's "stress hormone"?
    ***a.* Abscisic acid**
    b. Auxin
    c. Cytokinin
    d. Ethylene
    e. Gibberellin

18. If a shoot cutting is treated with auxin, which of the following is likely to result?
    ***a.* Extensive root production**
    b. Suppression of apical dominance
    c. Growth of lateral buds
    d. Bolting of the shoot
    e. Nothing will happen to the cutting

19. Which pair of hormones has opposing effects on senescence, in that the first promotes it and the second inhibits it?
    a. auxin; cytokinin
    ***b.* ethylene; cytokinin**
    c. ethylene; auxin
    d. cytokinin; auxin
    e. cytokinin; ethylene

20. Bolting, or rapid stem elongation, is induced by _____ and can be inhibited by _____.
    a. auxin; cytokinin
    b. abscisic acid; ethylene
    ***c.* gibberellin; abscisic acid**
    d. ethylene; auxin
    e. cytokinin; abscisic acid

21. When it is said that the movement of auxin in plants is "polar," this means that
    a. auxin is a chemically polar molecule.
    b. auxin is only produced at one part of the plant.
    ***c.* auxin is transported from tip to base of the plant.**
    d. auxin moves away from the light.
    e. auxin cannot move through gelatin.

22. Which of the following processes is *not* increased by ethylene?
    a. Breakdown of fruit cell walls
    b. Stimulation of leaf abscission
    c. Ripening of fruit
    d. Inhibition of stem elongation
    ***e.* Change in leaf color from green to red or yellow**

23. Which hormone has its highest concentrations in dormant (inactive) buds and seeds?
    ***a.* Abscisic acid**
    b. Auxin
    c. Cytokinin
    d. Ethylene
    e. Gibberellin

24. Plants utilize which cue to detect the onset of winter?
    a. Decreasing temperature
    b. Increasing precipitation
    c. Decreasing length of daylight
    ***d.* Increasing length of darkness**
    e. Height of the midday sun

25. In plants that germinate in response to a brief pulse of light,
    a. green light is most effective in triggering germination.
    b. far-red light is most effective in triggering germination.
    ***c.* far-red light reverses the effect of prior exposure to red light.**
    d. initiation of photosynthesis is the mechanism for germination.
    e. a rise in temperature also triggers germination.

26. A protein pigment called phytochrome is thought to monitor photoperiod because
    a. **it is converted between two forms by specific wavelengths of light.**
    b. a phytochrome that absorbs red light breaks down in darkness.
    c. the photoperiod response depends on how much light it absorbs.
    d. in darkness the pigment becomes inactive.
    e. the pigment response can only be observed in intact living plants.

27. Which of the following is most advantageous for a young plant seedling that has *not* yet been exposed to light?
    a. Increased production of chlorophyll
    b. **Rapid elongation of the shoot**
    c. Increased production of phytochrome
    d. Rapid photosynthesis
    e. Increased uptake of water

28. During germination of barley seeds, the _____ produces the signal that causes the _____ to synthesize digestive enzymes.
    a. **embryo; endosperm**
    b. endosperm; embryo
    c. seed coat; embryo
    d. seed coat; endosperm
    e. seed coat; embryo and endosperm

29. Gibberellins were first discovered by a biologist studying the "foolish seedling" disease of
    a. corn.
    b. wheat.
    c. **rice.**
    d. barley.
    e. millet.

30. Phinney reported the first evidence that gibberellins were produced by plants. In his studies with dwarf mutant strains of corn, he demonstrated that treatment with gibberellins caused the dwarf plants _____ while the normal tall plants _____.
    a. to grow taller; also grew taller
    b. **to grow taller; were virtually unaffected**
    c. to stop growing; also stopped growing
    d. to stop growing; were virtually unaffected
    e. were virtually unaffected; stopped growing

31. Why are there so many gibberellins?
    a. Each is needed for a different process.
    b. Each is produced by a different part of the plant.
    c. Some are no longer important to the plant's development.
    d. **Most are simply intermediates.**
    e. Most are produced by fungi, not the plants in which they are found.

32. Spraying biennials like cabbage with gibberellin causes them to bolt, a process first observed as an increase in
    a. leaf senescence.
    b. **stem elongation.**
    c. the number of flowers.
    d. the size of fruit.
    e. the number of seeds per fruit.

33. The discovery of auxin is traced back to the work of Charles and Francis Darwin. In their experiments, they studied
    a. photosynthesis.
    b. photorespiration.
    c. photophosphorylation.
    d. **phototropism.**
    e. photoperiodism.

34. In the Darwins' experiment with grass coleoptiles, they observed that the photoreceptor was _____; the actual bending took place _____.
    a. in the tip; at the tip also
    b. below the tip; below the tip also
    c. **in the tip; below the tip**
    d. below the tip; throughout the coleoptile
    e. at the tip; throughout the coleoptile

35. In Fritz Went's classic experiments, a gelatin block containing auxin was placed on one edge of a decapitated coleoptile. The result was that the coleoptile grew
    a. straight up if in the dark.
    b. **more on the side with the block if in the dark.**
    c. more on the side away from the block if in the dark.
    d. more on the side with the block if light was shining on the side with the block.
    e. more on the side with the block if light was shining on the side opposite the block.

36. Auxin transport is
    a. **from apex to base.**
    b. from base to apex.
    c. in either direction.
    d. primarily from apex to base, but a little in reverse.
    e. primarily from base to apex, but a little in reverse.

37. Your cat knocks over the coleus plant you were keeping in a dark closet. After a few days on its side, you notice that the shoots are growing upright again. You have just observed
    a. positive gravitropism.
    b. **negative gravitropism.**
    c. positive phototropism.
    d. negative phototropism.
    e. positive thigmotropism.

38. Removal of the auxin source demonstrates that leaf abscission is _____ by auxin, and apical dominance is _____ by auxin.
    a. promoted; promoted
    b. inhibited; inhibited
    c. promoted; inhibited
    d. **inhibited; promoted**
    e. promoted; unaffected

39. An ideal herbicide to kill weeds in a wheat field would kill _____ and break down _____ in soil.
    a. monocots and dicots; slowly
    b. monocots but not dicots; slowly
    c. dicots but not monocots; slowly
    d. monocots but not dicots; rapidly
    e. **dicots but not monocots; rapidly**

40. A cell wall is a network of crystalline _____ molecules in a jelly-like matrix of _____ molecules.
    a. cellulose; starch
    b. starch; other polysaccharide
    c. cellulose; other polysaccharide
    d. starch; nonpolysaccharide
    e. cellulose; nonpolysaccharide

41. The growth of a plant cell is driven primarily by
    a. the breakdown of ATP to ADP.
    b. the uptake of water into the vacuole.
    c. the strengthening of cell wall components.
    d. the deposition of new cell wall materials on the outside of the cell wall.
    e. the forces in transpirational pull.

42. The "wall-loosening factor" from the cytoplasm is now believed to be
    a. auxins.
    b. cellulose-digesting enzymes.
    c. starch-digesting enzymes.
    d. hydrogen ions.
    e. calcium ions.

43. Studies on diageotropica (*dgt*) mutants of tomato were the first to demonstrate that
    a. gravitropism is caused by auxins.
    b. positive and negative gravitropism involve different growth substances.
    c. hormone receptors occur in plants.
    d. auxins act upon the cell walls during gravitropism.
    e. plants use some of the same hormones as animals.

44. Undifferentiated cultures of tobacco pith tissue form roots when treated with _____ auxin and _____ cytokinin.
    a. low concentrations of; low concentrations of
    b. low concentrations of; high concentrations of
    c. high concentrations of; high concentrations of
    d. high concentrations of; low concentrations of
    e. zero; high concentrations of

45. Folke Skoog and Carlos Miller discovered cytokinins while looking for a growth substance that would
    a. retard cell elongation.
    b. regulate cell division.
    c. act as a selective herbicide.
    d. cause phototropism.
    e. cure a plant of "foolish seedling" disease.

46. Cytokinins are believed to form primarily in the plant's
    a. roots.
    b. stems.
    c. leaves.
    d. vegetative apical meristem.
    e. floral apical meristem.

47. Unlike other plant hormones, ethylene
    a. is not produced by plants.
    b. exerts a number of effects.
    c. inhibits plant development.
    d. acts either as an inhibitor or a promoter.
    e. is a gas.

48. Leaf senescence is important for the survival of a plant. Which of the following statements about senescence is *not* true?
    a. The delicate leaves could be a liability during winter.
    b. Amino acids in the leaves are exported to the stems.
    c. Leaf senescence is a reversible process.
    d. The hormone ethylene promotes leaf senescence.
    e. Controlled leaf abscission costs the plant little and benefits it greatly.

49. Abscisic acid concentrations are _____ in some dormant seeds and _____ in buds during winter dormancy.
    a. high; high
    b. high; low
    c. low; high
    d. low; nonexistent
    e. low; low

50. Which of the following is true for phytochrome P_r?
    a. It absorbs green light.
    b. It looks red in a test tube.
    c. It is the active form of phytochrome.
    d. It spontaneously converts to the other form in the dark.
    e. It controls a variety of plant responses.

51. When seeds germinate below the soil surface, the young etiolated seedlings
    a. grow very slowly.
    b. produce chlorophyll.
    c. elongate so that the apical meristem is the first part of the shoot to break through the soil surface.
    d. have small, unexpanded leaves.
    e. keep the cotyledons enclosed in a coleoptile.

52. Which of the following does *not* serve to break dormancy and initiate seed germination?
    a. Scarification by abrasion
    b. Scarification by fire
    c. Exposure to water
    d. Abscisic acid
    e. Growth promoters

53. Which of the following characterizes most seeds that germinate only when well-buried?
    a. Germination is triggered by light.
    b. They are large and nutrient-rich.
    c. Germination is triggered by fire.
    d. Freedom from competition with other seedlings
    e. Germination occurs only after scarification.

54. A shift away from rapid vegetative growth often accompanies which of the following phenomena?
    a. Repeated mitotic division
    b. Production of new leaves
    c. Production of flowers
    d. Increased root development
    e. Increased rate of photosynthesis

55. Which of the following is *not* a way to break seed dormancy?
    a. Soaking with water
    b. Singeing with flame
    c. Passing through an animal's gut
    d. Tumbling with stones
    e. Absorption of carbon dioxide

56. Seed dormancy is usually an adaptation to ensure that
    a. the embryo is mature.
    b. germination occurs at a favorable time.
    c. seeds germinate near the parent plant.
    d. levels of abscisic acid are high enough.
    e. plenty of other seeds are ready to germinate.

57. For plants whose seedlings require bright sunlight, what condition for breaking dormancy would you expect?
    a. Adequate rainfall
    b. Deep burial in soil
    c. Warm temperatures
    d. Longer day lengths
    e. High light intensity

58. All of the following occur during the earliest stages of germination, except
    a. intake of water.
    b. activation of enzymes.
    c. cell division.
    d. lengthening of the root.
    e. mobilization of food reserves.

59. What changes do gibberellins trigger in the aleurone layers of the seed?
    a. Manufacture of enzymes
    b. Cell division
    c. Intake of water
    d. Loss of water
    e. Release of abscisic acid

60. Stratification is used in agriculture to overcome what mechanism of dormancy?
    a. An impermeable seed coat that prevents the entry of water
    b. A seed coat that mechanically restrains the embryo
    c. The need of the seed to be underground at a certain level
    d. The requirement of natural promoters for seed germination
    e. The need for a cold period

61. The adaptive advantage of seed dormancy includes all of the following, except
    a. it may result in germination at a favorable time.
    b. it may increase the probability of a seed germinating in the right place.
    c. it may be a way to avoid competition.
    d. it may result in an increase in the likelihood of dispersal.
    e. it may ensure the success of a population by germinating while still attached to the parent plant.

62. During the initial stages of seed germination, all of the following increase, except
    a. cell size.
    b. respiration.
    c. RNA synthesis.
    d. DNA synthesis.
    e. protein synthesis.

63. Which of the following allows more energy to be stored in a smaller space?
    a. Protein
    b. Lipid
    c. Starch
    d. Sugar
    e. Amino acid

64. During germination of barley seeds, the aleurone layer
    a. imbibes water.
    b. produces gibberellins.
    c. synthesizes digestive enzymes.
    d. digests starch reserves.
    e. transports sugars to the embryo.

65. As a seed germinates, DNA synthesis begins
    a. when gibberellins are secreted by the embryo.
    b. when the embryonic root, or radicle, begins to grow.
    c. as imbibition takes place.
    d. when the endosperm starts metabolizing starches, proteins, and lipids.
    e. when the aleurone layer assembles enzymes, proteases, and ribonucleases.

66. Treatment of some plants with gibberellin or auxin causes parthenocarpy; this is
    a. flowers with petals in multiples of four.
    b. the shoot dividing into two separate shoots in one plant.
    c. fruit with only one seed instead of many.
    d. fruit formation without fertilization.
    e. formation of an embryo without fertilization.

67. Skoog and Miller, in their experiments, discovered two cytokinins, kinetin and zeatin. What is the difference between these two?
    a. Kinetin is found only in aged DNA, and zeatin is found only in fresh DNA.
    b. Kinetin is the active form of zeatin.
    c. Zeatin is the active form of kinetin.
    d. Zeatin is a naturally occurring plant cytokinin; kinetin is not.
    e. Zeatin is a synthetic cytokinin, and kinetin is naturally occurring.

68. Why do florists use silver thiosulfate?
    a. To delay abscission of petals caused by ethylene action
    b. To delay abscission of leaves caused by abscisic acid action
    c. To keep the petals from turning brown because of ethylene action
    d. To keep the leaves green longer
    e. To promote flower fertilization

69. Abscisic acid promotes the formation of bud scales; these are
    a. for retaining more water in dry areas.
    b. for waterproofing the leaf primordia and stem during winter.
    c. for cell elongation at the apical meristem.
    d. for helping leaf drop during autumn.
    e. for trapping more carbon dioxide in the leaves.

36 Reproduction in Flowering Plants

Fill in the Blank

1. Angiosperm plants are characterized by double fertilization, in which one sperm nucleus fertilizes the **ovum** to begin the embryo and the other fertilization results in production of **endosperm** tissue.

2. The male gametophyte in seed plants is the **pollen grain**, and the mature female gametophyte is an embryo sac with **eight** haploid nuclei.

3. The process of transfer of pollen grains to the stigma is called **pollination**.

4. The major role of the fruit of a flowering plant is to facilitate **seed dispersal**.

5. The production of progeny all having identical genotypes to the parent is called, in general, **asexual** reproduction, while **vegetative** reproduction is the modification of a vegetative part of the plant to produce new individuals.

6. The asexual production of seeds is called **apomixis**.

7. In order for grafting to be successful, the **cambium** tissue of the two parts must meet and fuse.

8. One agricultural industry in which grafting is an important technique is **fruit** production.

9. Instead of germinating immediately after release, many seeds undergo an inactive period of **dormancy**.

10. Experiments on the effect of light cues on flowering have shown that the significant cue that is sensed or measured by plants is **night length**.

11. The physiological mechanism by which plants measure photoperiod involves a pigment called **phytochrome**. This pigment alternates between two forms, which absorb **red and far-red** light.

12. Most plants do not rely on light cues for inducing flowering and are called **day-neutral** plants.

13. An opening in the ovule called the **micropyle** is where the pollen tube grows toward the egg.

14. Pollination before the flower bud opens is an example of **self-fertilization**.

15. Angiosperms are unique in that the endosperm is **triploid**.

16. Floral meristems produce flowers; this is **determinate** growth.

17. Organ identity genes that produce flower parts are analogous to **homeotic** genes of developing animals.

18. A northern tree that is placed too far south will not flower well; this is an example of **vernalization**.

Multiple Choice

1. An apomictic seed contains an embryo that is
   a. produced when two sperm fertilize one egg.
   b. developed from one egg alone.
   c. the result of parental self-fertilization.
   d. genetically identical to its parent.
   e. homozygous for most genetic traits.

2. Within a flower's style, a chemical gradient of calcium ions or other substances is necessary to
   a. guide the pollen tubes' growth.
   b. direct the swimming sperm.
   c. orient the egg near the micropyle.
   d. align the pollen grain on the stigma.
   e. trigger meiosis within the ovule.

3. A plant's transition to a flowering state is often marked by
   a. an increased rate of photosynthesis.
   b. a decrease in vegetative growth.
   c. an increase in root development.
   d. a decreased rate of respiration.
   e. an increase in lateral bud growth.

4. After pollination, which of the following events is crucial for fertilization?
   a. Sperm swim to the egg and the polar nuclei.
   b. Petals close around the reproductive parts.
   c. Meiosis occurs within the pollen grain.
   d. A pollen tube grows from the stigma to the ovule.
   e. An insect delivers pollen to the stigma.

5. Two modifications of vegetative parts used for asexual reproduction are short, vertical stems called
   a. rhizomes and tubers.
   b. tubers and corms.
   c. bulbs and corms.
   d. bulbs and rhizomes.
   e. corms and rhizomes.

6. After fertilization of the egg, the flower's integument develops into the
   a. cotyledons.
   b. embryo.
   c. endosperm.
   d. fruit.
   e. seed coat.

7. In some dicots, no distinct endosperm can be seen. Why?
   a. The embryo has digested the endosperm.
   b. The cotyledons have absorbed the endosperm.
   c. The seeds never produced endosperm.
   d. The endosperm has become the seed coat.
   e. The fruit has incorporated the endosperm.

8. The loss of water from a developing seed causes it to
   a. produce a root.
   b. die.
   c. be released from the fruit.
   d. become dormant.
   e. be protected from animal predators.

9. The following experiment supports the theory that a specific flower-initiating hormone is produced in plants: When a short-day plant (SDP) and a long-day plant (LDP) are grafted together and the SDP is exposed to a photoperiod that causes it to flower, the LDP flowers as well. The proper control for this experiment would be to repeat the same design and
   a. omit the grafting.
   b. omit the LDP.
   c. omit the SDP.
   d. omit the SDP photoperiod.
   e. include an LDP photoperiod.

10. Which of the following is a gametophyte of a flowering plant?
    a. Flower
    b. Egg
    c. Pollen grain
    d. Anther
    e. Entire plant

11. The megagametophyte of flowering plants consists of
    a. the pollen grain.
    b. the pollen tube.
    c. the eight-nucleate embryo sac.
    d. the ovule.
    e. the megasporangium and cells within it.

12. From megasporocyte to egg cell, what processes are required?
    a. Meiosis followed by mitosis
    b. Mitosis followed by meiosis
    c. Several meiotic divisions only
    d. Several mitotic divisions only
    e. Several nuclear fusion events

13. Within which of the following structures does meiosis occur?
    a. Petal
    b. Ovule
    c. Stigma
    d. Sepal
    e. Pollen grain

14. The advantage of self-fertilization in plants is
    a. increased genetic recombination.
    b. that meiosis can occur.
    c. greater efficiency of pollination.
    d. that no flowering is needed.
    e. that only asexual reproduction is necessary.

15. The advantage of cross-fertilization in plants is
    a. increased genetic recombination.
    b. that meiosis can occur.
    c. greater efficiency of pollination.
    d. that no flowering is needed.
    e. that only asexual reproduction is necessary.

16. Plants with wind-pollinated flowers tend to have
    a. colorful petals.
    b. smooth stigmas.

   c. **large quantities of pollen.**
   d. large quantities of nectar.
   e. pollination before the bud opens.

17. Where does fertilization occur in flowering plants?
    a. Where pollen lands on the stigma
    b. Where the pollen tube germinates
    c. Inside the pollen tube
    d. At the base of the embryo sac
    e. Inside the seed

18. The egg can be fertilized by
    a. one tube nucleus.
    b. one sperm nucleus.
    c. two sperm nuclei.
    d. one generative nucleus.
    e. two synergid nuclei.

19. What is the fate of the seven cells of the embryo sac?
    a. All but one disintegrate upon fertilization.
    b. Two become fertilized; the others disintegrate.
    c. Two become fertilized; the others fuse to form endosperm.
    d. All are involved in nuclear fusion events.
    e. They all become part of the seed tissue.

20. What nuclei fuse to form the endosperm?
    a. Egg and sperm nucleus
    b. Egg and two sperm nuclei
    c. Synergid nuclei and sperm nucleus
    d. Polar nuclei and generative nucleus
    e. Polar nuclei and sperm nucleus

21. Which of the following describes the ploidy of the components of a seed?
    a. Diploid embryo, triploid endosperm, diploid seed coats
    b. Haploid embryo, triploid endosperm, diploid seed coats
    c. Diploid embryo, triploid endosperm, haploid seed coats
    d. Diploid embryo, diploid endosperm, diploid seed coats
    e. Haploid embryo, diploid endosperm, haploid seed coats

22. The term "suspensor" applies to which of the following?
    a. The pollen tube
    b. The organ supporting the ovule
    c. The base of the flower
    d. The narrow part of the embryo
    e. The stalk of the stamen

23. The fruit generally develops from which part of the flower?
    a. Petals
    b. Sepals
    c. Ovary
    d. Stamens
    e. Pedicel

24. What process early in development initiates cell specialization in the embryo?
    a. Mitotic division of the zygote
    b. Uneven distribution of cell contents
    c. Absorption of food from endosperm
    d. Seed germination
    e. Maturation of the fruit

25. Which of the following is a distinguishing characteristic of all angiosperms?
    a. Double cotyledons
    b. Fleshy cotyledons
    c. Seed with nutrients
    d. Double fertilization
    e. Pollen production

26. What is the function of the nutritious flesh of many fruits?
    a. To nourish the embryo
    b. To attract seed eaters
    c. To attract pollinators
    d. To attract seed dispersers
    e. To ensure that the seeds fall close to the parent plant

27. Asexual reproduction is the best strategy for plants
    a. that are well adapted to their stable environment.
    b. as winter approaches.
    c. when new genes must be introduced.
    d. that have underground stems.
    e. that have low seed production that season.

28. What is necessary for successful grafting to occur?
    a. Each section must be able to form roots.
    b. The grafted section must be able to form seeds.
    c. Fusion of the two vascular tissues must occur.
    d. Fusion of the two cambial tissues must occur.
    e. Each section must be from the same species.

29. Fruit-eating bats tend to feed extremely rapidly on fruits and have relatively inefficient digestion (sometimes defecating seeds as early as an hour after feeding on them). Why are they good seed dispersal agents?
    a. Seed survival in bat guts is low.
    b. Undigested seeds are deposited in a heap at the bats' roost site.
    c. Undigested seeds are deposited at various bat feeding sites.
    d. Undigested seeds are deposited near the same plant that produced them.
    e. Digested seeds are dispersed in bat waste products.

30. Which of the following is a characteristic of wind-dispersed seeds?
    a. Abundant pollen
    b. Hooked extensions
    c. Fleshy fruit
    d. Air chambers
    e. Flat, winged extensions

31. Which of the following probably involves sexual reproduction?
    a. A water hyacinth population fills a pond in midsummer.
    b. A potato plant grows from the "eye" of a potato.
    c. Genetically identical, interconnected aspen trees produce flowers.
    d. Apomictic seed production in dandelions
    e. A bamboo plant gives rise to a forest of plants.

32. The process of grafting involves which of the following?
    a. Allowing a piece of one plant to grow onto the root of another
    b. Allowing cross fertilization between two plants
    c. Preparing several cuttings from a plant to grow into individual plants
    d. The production of xylem and phloem from the same cambium layer
    e. Interbreeding of two species of plants

33. Artificial seeds have been developed that contain
    a. an embryo surgically removed from a seed and packaged in a gel.
    b. a tissue culture product packaged in a gel.
    c. a tissue culture product implanted into a seed coat.
    d. extra amounts of nutrients added into the seed.
    e. an embryo removed from its seed coats and nutrients.

34. Flowering plants produce _____ by meiosis and _____ by mitosis.
    a. sperm nuclei; spores
    b. sperm nuclei; sperm nuclei
    c. spores; sperm nuclei
    d. integuments; sperm nuclei
    e. sperm nuclei; integuments

35. In a flower, the microsporangia are found in the
    a. anther.
    b. filament.
    c. stigma.
    d. ovule.
    e. ovary.

36. Which is the correct order of events?
    a. Megagametophyte–megasporocyte–megaspore
    b. Megagametophyte–megaspore–megasporocyte
    c. Megasporocyte–megaspore–megagametophyte
    d. Megaspore–megasporocyte–megagametophyte
    e. Megaspore–megagametophyte–megasporocyte

37. In the mature embryo sac, the cells closest to the micropyle are the
    a. polar nuclei.
    b. synergids.
    c. eggs.
    d. egg and polar nuclei.
    e. egg and synergids.

38. In flowering plants, the pollen is transferred to the
    a. stigma.
    b. style.
    c. ovary.
    d. ovule.
    e. micropyle.

39. A flower that is wind-pollinated would be least likely to
    a. have numerous anthers.
    b. have sticky or feather-like stigmas.
    c. produce large numbers of pollen grains.
    d. have a colorful corolla.
    e. have smooth wall sculpturing on its pollen.

40. Pollination is
    a. the fusion of the egg and sperm nuclei.
    b. the transfer of pollen from the anther to the stigma.
    c. the development of the two-celled pollen grain.
    d. the growth of the pollen tube after pollen germination.
    e. the division of the generative nucleus to produce two sperm nuclei.

41. The three nuclei in a mature pollen grain are formed by
    a. one meiotic division and one mitotic division.
    b. two meiotic divisions and one mitotic division.
    c. one meiotic division and two mitotic divisions.
    d. two meiotic divisions and two mitotic divisions.
    e. one meiotic division in which one of the four cells degenerates.

42. The "embryo sac" is also called the
    a. megaspore.
    b. megasporangium.
    c. megasporocyte.
    d. megasporophyll.
    e. megagametophyte.

43. Double fertilization results in the formation of
    a. two diploid embryos.
    b. one diploid embryo and a diploid endosperm.
    c. two diploid embryos and a haploid endosperm.
    d. one diploid embryo and a triploid endosperm.
    e. two diploid embryos and a diploid seed coat.

44. The integuments of the ovule develop into the _____ while the carpels ultimately become the wall of the _____.
    a. cotyledons; endosperm
    b. seed coats; fruit
    c. cotyledons; seed coats
    d. endosperm; seed coats
    e. cotyledons; fruit

45. The function of a fleshy fruit is to
    a. feed the new embryo.
    b. attract pollinators.
    c. disperse the seeds.
    d. protect the immature embryo.
    e. keep the rest of the plant from being eaten.

46. Coconut fruits are dispersed by
    a. monkeys.
    b. wind.
    c. fruit bats.
    d. water.
    e. birds.

47. In instances where dispersal is extensive,
    a. paternal genes travel farther than maternal genes.
    b. maternal genes travel farther than paternal genes.
    c. paternal and maternal genes travel the same distance.
    d. paternal and maternal genes travel the same distance, but only the paternal genes survive.
    e. paternal and maternal genes travel the same distance, but only the maternal genes survive.

48. Self-pollination results in progeny that
    a. are identical to the parent.
    b. are somewhat different because mutations are common.
    c. may express a recessive gene if the parent is heterozygous.
    d. may be heterozygous in a locus where the parent is homozygous.
    e. may be as varied as from cross-pollination.

49. A clone of white potatoes may be derived from underground _____.
    a. stolons
    b. tubers
    c. rhizomes
    d. bulbs
    e. root suckers

50. The major advantage of asexual reproduction is
    a. it results in no genetic variation.
    b. it results in increased dispersal.
    c. production of more progeny.
    d. the ability to invade new environments.
    e. only observed when the habitat is unstable.

51. The ability of plants to measure night length was determined in what sort of experiment?
    a. Growing plants in 12 hours of darkness alternating with 12 hours of light
    b. Growing plants in continuous light
    c. Interrupting the dark period with a brief pulse of light
    d. Measuring flowering in plants of different ages placed in the same light/dark schedule
    e. Keeping a 24-hour cycle but increasing the relative length of light

52. Which of the following photoperiods would induce flowering in a short-day plant with a critical day length of 15 hours?
    a. Twelve hours of light alternating with 12 hours of darkness
    b. Sixteen hours of light alternating with 8 hours of darkness
    c. Fourteen hours of light alternating with 8 hours of darkness
    d. Eight hours of light alternating with 8 hours of darkness
    e. Fifteen hours of light alternating with 9 hours of darkness, interrupted by one short burst of white light.

53. Vernalization refers to the requirement for which of the following before flowering can occur?
    a. Exposure to cold
    b. Availability of soil calcium
    c. Minimal day length
    d. Sufficient moisture for a minimum period
    e. One full year of growth

54. Short-day annuals usually flower
    a. in the spring.
    b. in midsummer.
    c. in late summer.
    d. in midsummer and late summer.
    e. throughout the summer.

55. Technically, short-day plants flower when
    a. the light period exceeds a critical period.
    b. the light period is less than a critical period.
    c. the light period equals the critical period.
    d. the dark period exceeds a critical period.
    e. the dark period is less than a critical period.

56. Long-day plants will not flower if exposed to
    a. a long day.
    b. a long day interrupted by a dark period.
    c. a long day interrupted by a period of red light.
    d. a long night interrupted by a light period of far-red light.
    e. A long day interrupted by a light period of red light.

57. Circadian rhythms
    a. are always exactly 24-hour cycles in nature.
    b. are only found in multicellular organisms.
    c. have a period that is remarkably sensitive to temperature.
    d. do not continue when placed in complete darkness.
    e. can be made to coincide to the light–dark regime.

58. Although it has never been discovered, there is evidence that the mysterious "flowering hormone" is synthesized in the
    a. floral meristem.
    b. vegetative apical meristem.
    c. stems.
    d. leaves.
    e. roots.

59. A benefit of sexual reproduction in plants is
    a. greater number of progeny.
    b. pollination is easier.
    c. better adaptation to new environments.
    d. the haploid plant becomes diploid.
    e. progeny are dispersed farther.

60. Plants, and some protists, are unique in that the gametophyte
    a. is haploid, while in all other organisms it is diploid.
    b. is an alternate generation from the vegetative plant.
    c. is only found in the male plants.
    d. develops from the seed.
    e. does not undergo mitosis.

61. *In vitro* fertilization of plant eggs has only recently been accomplished. Which is *not* a means of overcoming difficulties with this process?
    a. Treating gametes with electricity
    b. Using feeder cells
    c. Digesting embryo sac cell walls with enzymes
    d. Uniting the sperm and egg with osmotic treatments
    e. Microdissection to release the egg from the embryo sac

62. Long-short-day plants bloom
    a. in the spring.
    b. in the fall.
    c. in the summer.
    d. only after a cold winter.
    e. only after the second year of the plant's life.

63. Initial resistance of the French wine grape *Vitis vinifera* to plant lice was accomplished by
    a. self-fertilization of resistant plants.
    b. grafting the scion onto resistant plants' roots.
    c. recombinant DNA techniques using the resistance gene.
    d. planting seeds in California where there are no plant lice.
    e. taking cuttings, or slips, of the plants and inserting them into California soil, without lice, where they formed roots and grew.

64. Nutrients for the seedling are generally stored in what form in a seed?
    a. As monomers in solution
    b. As monomers in fat storage
    c. As macromolecules
    d. As cellular enzymes
    e. As cellular organelles

65. As a pollen tube grows into the female organ, the nucleus that enters the synergid first is called the
    a. sperm nucleus.
    b. generative nucleus.
    c. tube nucleus.
    d. pollen nucleus.
    e. microspore.

37 *Physiology, Homeostasis, and Temperature Regulation*

Fill in the Blank

1. The largest organ of the body is the **skin**.

2. In certain "hot" fish such as bluefin tuna, heat is exchanged between blood vessels carrying blood in opposite directions. This adaptation is called **countercurrent heat exchange**.

3. The metabolic rate of a resting animal at a temperature within the thermoneutral zone is called the **basal metabolic rate**.

4. Small endotherms can extend the period over which they can survive without food by dropping body temperature. This adaptive hypothermia is called **daily torpor**.

5. The maintenance of more or less constant physiological conditions within an organism is called **homeostasis**.

6. The process of physiological and behavioral regulation of body temperature is **thermoregulation**.

7. Animals whose temperature fluctuates to match that of the environment are termed **poikilotherms**.

8. Animals that can affect their body temperature by generating metabolic heat are called **endotherms**.

9. In the vertebrate animal, the thermostat is located in the **hypothalamus**.

10. During the condition called **hypothermia**, when an endotherm's body temperature is far below normal, metabolic rates become **lower**, and relatively **less** oxygen is used by the tissues. Although this condition can be harmful to endotherms, it becomes a useful strategy for some overwintering mammals called **hibernators**.

11. Hibernating mammals are likely to rely on the metabolism of **brown fat** tissue for release of heat to wake up.

12. The linings of most tissues consist of **epithelial** cells.

13. The most abundant protein in our bodies is **collagen**, found in connective tissue.

14. **Control** is the ability to change the rate of a reaction, whereas **regulation** is the ability to maintain the rate in a physiological system.

15. Bacteria release **pyrogens** when they infect, which causes you to have a fever.

Multiple Choice

1. Which of the following is a difference between the lymphatic and the circulatory systems?
   a. The lymphatic system does not form a complete circular system.
   b. The lymphatic system transports extracellular fluid only.
   c. The lymphatic system lacks a pumping organ.
   d. The lymphatic system transports mainly plasma.
   e. **All of the above**

2. Nitrogenous waste products principally result from the metabolism of
   a. carbohydrates.
   b. lipids.
   c. **proteins and nucleic acids.**
   d. nitric acid.
   e. respired nitrogen in the air.

3. Homeostasis
   a. **favors a constant internal physiological environment regardless of the changes in the external environment.**
   b. keeps vital organs working at their maximum potential.
   c. keeps all cells working at the same metabolic rate.
   d. keeps the body's metabolic rate constant in varying environmental temperatures.
   e. keeps the body's temperature constant in varying environmental temperatures.

4. Most organisms relying on behavioral rather than metabolic mechanisms to regulate body temperature are
   a. homeothermic.
   b. endothermic.
   c. heterothermic.
   d. **ectothermic.**
   e. None of the above

5. Compared to an ectothermic organism's body temperature response to a 10°C rise in environmental temperature, an endotherm's body temperature will
   a. rise at a constant rate.
   b. fall at a constant rate.
   c. fall to a point, then become stable.
   d. rise to a point, then become stable.
   e. **remain relatively constant.**

6. Active muscles of some ectotherms can produce sufficient energy to raise body temperatures above that of the ambient surrounding if the rate of heat loss to the environment is not greater than the rate of heat production. Which of the following adaptations would *not* favor an increase in an ectotherm's body temperature?
   a. Isometric muscle contractions
   b. Cluster or huddling behavior
   c. Decreased surface-to-surface contact with the cold environment
   d. Circulatory changes to maintain core or internal temperatures greater than the animal's peripheral temperatures
   e. Metabolic compensation

7. Readjustment of an organism's metabolic rate to compensate for seasonal thermal change is termed
   a. homeostasis.
   b. negative feedback.
   c. metabolic compensation.
   d. acclimatization.
   e. regulation.

8. Which term best describes an organism that depends upon the mobilization of metabolic mechanisms for heat production and a repertoire of active mechanisms for heat loss to thermoregulate?
   a. Ectotherm
   b. Homeotherm
   c. Heterotherm
   d. Poikilotherm
   e. All of the above

9. A desert lizard slowly crawls from its burrow after cool nighttime temperatures and postures on a rock warmed by the mid-morning sun. Which one of the following is most likely *not* part of the lizard's thermoregulatory responses?
   a. Orientation to the sun to maximize exposure to solar radiation
   b. Increased peripheral circulation via dilation of surface blood vessels
   c. Increased metabolic heat production
   d. Increased physical contact between the warm rock and the relatively cool lizard
   e. Orientation of the lizard on the rock into more suitable microenvironments to avoid cold air currents

10–13. The endocrine and nervous systems are both responsible for integration of information and control of organs and organ systems. This control differs with each system. Fill in the blanks using the following answer choices.
   a. Hormones
   b. Neurons
   c. Blood
   d. Ductless glands

10. The principal organ(s) of the endocrine system are _____. *(d)*

11. The nervous system routes control to specific targets by way of _____. *(b)*

12. The chemical messages of the endocrine system are called _____. *(a)*

13. The nervous system principally receives sensory input about the body from _____. *(b)*

14. What are the normal value(s) for most biological Q_{10}'s?
   a. 1
   b. 1–2
   c. 2–3
   d. 2–10
   e. 0–45

15–21. Use the following answer choices to correctly indicate the probable metabolic response of a mammal exposed to a 3–5°C environmental temperature change.
   a. Increased metabolic rate
   b. Decreased metabolic rate
   c. No change in the metabolic rate
   d. Death

15. The environmental temperature increases above the upper critical temperature. *(a)*

16. The environmental temperature increases above the lower critical temperature. *(c)*

17. The environmental temperature decreases far below zero. *(d)*

18. The environmental temperature decreases below the lower critical temperature. *(a)*

19. The environmental temperature decreases below the upper critical temperature. *(c)*

20. The environmental temperature fluctuates between the upper and the lower critical temperature. *(c)*

21. The environmental temperature increases above zero. *(b)*

22. An endotherm's thermoneutral zone can be variously described. Which of the following does *not* describe the limits of the thermoneutral zone?
   a. A range of environmental temperatures between the upper critical and lower critical temperature
   b. A range of environmental temperatures over which the organism exhibits a basal metabolic rate
   c. A range of body temperatures at which the metabolic rate is maximum
   d. A range of environmental temperature over which the organism's metabolic rate neither increases nor decreases for thermoregulation
   e. A range of environmental temperatures over which the organism produces minimal metabolic activity to support itself

23. Increased heat for thermoregulation or thermogenesis is produced either by shivering or by nonshivering mechanisms. Nonshivering thermogenesis is dependent upon all of the following, except
   a. brown fat.
   b. thermogenin.
   c. production of ATP.
   d. uncoupled oxidative phosphorlyation.
   e. increased mitochondria.

24. The hypothalamus in part serves as an integrated thermoregulatory center defining an organism's response to changes in its thermal environment. Since the hypothalamus normally serves to produce metabolic changes to reverse the direction of environmental temperature change, the control is termed
    a. positive feedback.
    b. metabolic compensation.
    c. **negative feedback.**
    d. feedforward.
    e. metabolic torpor.

25. Which of the following is *not* true of hibernation?
    a. Controlled by an endogenous biological clock with a circannual rhythm
    b. Is a form of regulated hypothermia
    c. Body temperature is turned down to a low level
    d. Metabolic rate is reduced to only a fraction of the basal metabolic rate
    e. **Is an adaptive mechanism widely found in many diverse species of birds and mammals**

26. Which of the following is an organ system that processes information?
    a. Urinary system
    b. Nervous system
    c. Endocrine system
    d. **b and c**
    e. a, b, and c

27. The temperature sensitivity of a reaction or process can be called its Q_{10}. Most biological Q_{10}'s are between
    a. 0.2–0.3.
    b. **2–3.**
    c. 20–30.
    d. 200–300.
    e. None of the above

28. _____ is the process of physiological and biochemical change that an animal undergoes in response to seasonal changes in climate.
    a. **Acclimatization**
    b. Homeostasis
    c. Sublimity
    d. Metabolic compensation
    e. Hybridization

29. Adaptations used by endotherms to reduce heat loss include which of the following?
    a. Rounder body shapes
    b. Shorter appendages
    c. Increased thermal insulation
    d. **a, b, and c**
    e. b and c

30. The vertebrate thermal regulatory center ("thermostat") is located within the central nervous system in the
    a. pons.
    b. cerebellum.
    c. hypophysis.
    d. medulla.
    e. **hypothalamus.**

31. Homeostasis refers to the tendency to keep the body systems
    a. matched to the external environment.
    b. the same relative to one another.
    c. **at a steady state over time.**
    d. under the control of the brain.
    e. at the same specific temperature.

32. Which type of animal tissue is responsible for secreting digestive enzymes?
    a. Connective
    b. **Epithelial**
    c. Matrix
    d. Muscle
    e. Nervous

33. To which type of animal tissue does bone belong?
    a. **Connective**
    b. Epithelial
    c. Matrix
    d. Muscle
    e. Nervous

34. The cellular components of blood belong to which type of animal tissue?
    a. **Connective**
    b. Epithelial
    c. Matrix
    d. Muscle
    e. Nervous

35. Connective tissues differ from each other mostly in their
    a. cellular structure.
    b. function in support.
    c. **matrix properties.**
    d. location in the body.
    e. cell packing.

36. In regulatory systems, the phenomenon of negative feedback
    a. **provides a means of control.**
    b. provides information about the system.
    c. increases the error signal in the system.
    d. increases the pace of the system.
    e. decreases the pace of the system.

37. Which of the following would serve to increase the set point of a regulatory system?
    a. Negative feedback
    b. Positive feedback
    c. **Feedforward**
    d. Insensitivity to information
    e. None of the above

38. Cellular functions are generally limited to what temperature range (in °C)?
    a. 20–100
    b. 20–45
    c. 0–20
    d. **0–45**
    e. 0–100

39. The upper temperature limit at which cells can function is determined by
    a. the boiling point of water.
    b. the melting point of water.
    c. the melting point of fats.
    d. the denaturation point of nucleic acids.
    e. **the denaturation point of proteins.**

40. The Q_{10}, which describes the sensitivity of a reaction to temperature, is calculated as
    a. the rate of a process at a certain temperature divided by its rate at 10°C.
    b. the rate of a process at a certain temperature divided by its rate at a temperature 10° lower.
    c. the rate of a process at a certain temperature divided by its rate at a temperature 10° higher.
    d. the temperature at which the rate of a certain process doubles.
    e. the temperature at which the rate of a certain process becomes insignificant.

41. The rate of a particular biological function is X at 15°C. Which of these statements is true about the rate of that function at 25°C?
    a. If the Q_{10} were 1, the rate would be X.
    b. If the Q_{10} were 1, the rate would be 25.
    c. If the Q_{10} were 2, the rate would be X.
    d. If the Q_{10} were 3, the rate would be 2X.
    e. If the Q_{10} were 3, the rate would be 30.

42. If an animal exhibits metabolic compensation, this means that
    a. it adapts its physiology to local environmental conditions.
    b. its metabolic rate is always slightly higher in summer.
    c. its Q_{10} stays constant despite seasonal temperature change.
    d. its body temperature and its metabolic rate are independent of each other.
    e. it uses the same reaction pathways at all temperatures.

43. Poikilothermic animals are those whose body temperature
    a. is maintained at a constant level some of the time.
    b. is maintained at a constant level all of the time.
    c. fluctuates with environmental temperature.
    d. is kept at a higher temperature than the environment.
    e. is kept at a lower temperature than the environment.

44. Which of the following is a thermoregulatory strategy that depends largely on external sources of heat?
    a. Ectothermy
    b. Endothermy
    c. Heterothermy
    d. Homeothermy
    e. Poikilothermy

45. Which of the following animals would be most successful with the part-time temperature regulation strategy of heterothermy?
    a. A desert lizard that basks in the sun
    b. A deep ocean fish in an environment with little temperature change
    c. An insect that is warm-blooded in flight and cold-blooded at rest
    d. An amphibian that is sometimes in water and sometimes on land
    e. A hummingbird that drops its body temperature at night

46. Which of the following animals is behaving as an endotherm in order to warm its body?
    a. A moth that shivers its wings before flight
    b. A black beetle that absorbs solar radiation
    c. A snake that lies on a warm blacktop road
    d. A fish that moves to a warm, shallow part of a pond
    e. An insect that positions its body for maximum exposure to sunlight

47. A certain desert lizard thermoregulates as an ectotherm. Which of the following phenomena should *not* be a part of its thermoregulatory behavior?
    a. Staying in a burrow when the surface temperature is below 10°C
    b. Basking in the sun during the early morning hours
    c. Moving into the shade during the midday hours
    d. Pressing its body to the earth during the midday hours
    e. Consuming prey that is ectothermic

48. In which case would a poikilothermic animal have a higher body temperature than a homeotherm?
    a. When resting in an underground burrow
    b. At midday on the desert floor
    c. When swimming in the cold ocean
    d. When basking in the early morning sun
    e. When the poikilotherm is much larger

49. Which of the following exemplifies adaptive thermoregulatory behavior in humans?
    a. A cowboy wearing a broad-brimmed hat
    b. An Eskimo wearing lightweight white clothing
    c. A desert home with large windows facing the sun
    d. Swimming in winter weather
    e. Humans are endothermic and don't exhibit thermoregulatory behavior.

50. Which physiological control mechanism is a response to a rise in body temperature?
    a. Slower heart rate
    b. Increased blood flow to the skin
    c. Constriction of blood vessels in the skin
    d. Contraction of muscles
    e. Retention of water

51. What is the adaptive advantage of honeybee workers clustering to maintain warm temperatures in the hive?
    a. Protection of the queen bee
    b. Protection of the comb structure
    c. Optimal pollen collection
    d. Optimal digestion of honey
    e. Optimal brood development

52. Within a range of environmental temperatures called the thermoneutral zone, the metabolic rate of an endotherm is
    a. constant.
    b. low and independent of temperature.
    c. high and independent of temperature.
    d. near the basal metabolic rate.
    e. dependent upon the temperature.

53. In a mammal, metabolic rate is highest in which of the following situations?
    a. The animal is at rest; the temperature is within the thermoneutral zone.

b. **The animal is active; the temperature is below the thermoneutral zone.**

c. The animal is active; the temperature is within the thermoneutral zone.

d. The animal is at rest; the temperature is above the thermoneutral zone.

e. The animal is at rest; the temperature is below the thermoneutral zone.

54. Which of the following is *not* true of brown fat?
    a. It contains abundant mitochondria.
    b. **It provides most energy for shivering.**
    c. It produces less ATP in metabolism.
    d. It converts more fuel to heat.
    e. It is more abundant in hibernating animals.

55. The mechanism of heat production in brown fat depends on
    a. inefficient use of ATP in metabolism.
    b. more rapid breakdown of protein.
    c. more rapid breakdown of fatty acids.
    d. additional shivering of skeletal muscles.
    e. **uncoupling oxidative phosphorylation.**

56. Which of the following is an important adaptation of animals to cold climates?
    a. Increased tendency to shiver
    b. Thinner layers of body fat
    c. Reduced density of fur or feathers
    d. **Reduced surface area-to-volume ratio**
    e. Increased flow of blood to surface

57. The elephant is better adapted to tropical habitats than to cold climates because of its
    a. **sparse hair.**
    b. large size.
    c. stocky appendages.
    d. vegetarian diet.
    e. thick skin.

58. Why is evaporative cooling used only as a last resort by animals in dry environments?
    a. It is ineffective at dissipating heat.
    b. **Water may be a limiting resource.**
    c. Sweating requires little energy expenditure.
    d. It requires an insulating layer in the skin.
    e. Resetting the thermostat is required.

59. When the temperature of the hypothalamus rises, which thermoregulatory responses result?
    a. Increased metabolic heat production
    b. Resetting of the thermostat higher
    c. **Dilation of blood vessels in the skin**
    d. Overall increase in body temperature
    e. Initiation of shivering movements

60. Which of the following accurately describes the thermoregulatory set point?
    a. The set point for shivering is the same as the set point for panting.
    b. In a given individual, the set points are relatively constant.
    c. The set points are the same for all members of the same species.
    d. **Information on skin temperature can change the metabolism set point.**
    e. Temperature information serves as a positive feedback signal.

61. What controls the hibernation season?
    a. The nighttime environmental temperature
    b. The daytime environmental temperature
    c. The environmental day length
    d. The completion of reproduction
    e. **The circannual rhythm**

62. Single-celled and simple multicellular animals meet their needs by having every cell exposed to the environment; more complex animals have an internal environment. Which of the following is *not* an advantage of an internal environment?
    a. **Each cell is capable of performing every function.**
    b. Cells can be specialized for one function.
    c. Cells can be more efficient.
    d. It can be maintained independently of the external environment.
    e. Complex animals can occupy otherwise inhospitable habitats.

63. Which is true of smooth muscle?
    a. It controls the pumping of blood through the heart.
    b. It is responsible for our behavior.
    c. **It is not under conscious control.**
    d. It is what connects bones to bones.
    e. It makes up the ciliated lining of the gut.

64. When scientists measured the Q_{10} of the fish in the lab, their prediction of the metabolic rate differed from the observed rate because
    a. **the fish acclimated to the temperature change in the lake.**
    b. metabolic rate varied in different parts of the fish.
    c. the fish do not produce metabolic compensation.
    d. the internal temperature of the fish is higher than the surrounding environment.
    e. it is impossible to figure the Q_{10} of a poikilotherm.

65. "Hot" fish, such as bluefin tuna, keep a higher temperature difference between their body and the surrounding water than cold fish because
    a. these fish are actually endotherms.
    b. **they use a countercurrent heat exchange system of veins and arteries.**
    c. they use shivering to create more heat.
    d. they have many brown fat tissues.
    e. they have a large dorsal aorta which keeps them warmer.

66. Your body responds to an infection by producing _____, which cause(s) many of the symptoms of sickness such as fever and feeling crummy.
    a. pyrogens
    b. interleukins
    c. prostaglandins
    d. *a* and *b*
    e. **b and c**

67. As the environmental temperature increases, up to 25°C, the metabolic rate of an ectotherm _____ and an endotherm _____.
    a. increases; increases
    b. **increases; decreases**
    c. decreases; increases
    d. decreases; decreases
    e. stays the same; decreases

38 *Animal Hormones*

Fill in the Blank

1. In an emergency, the adrenal glands produce **epinephrine** immediately and then **cortisol** for prolonged response to stress.

2. A prominent second messenger that responds to hormones is **cyclic AMP, or cAMP.**

3. Troponin and calmodulin are membrane proteins that trigger many responses in cells when they bind with **calcium** ions.

4. Hormonal and other systems in which production is shut off when the product becomes abundant are known as **negative** feedback systems.

5. **Prolactin** is a pituitary hormone that stimulates the secretion of milk.

6. A lobed gland located around the windpipe is called the **thyroid.**

7. Sweat glands and other glands with ducts are called **exocrine** glands.

8. Overproduction or underproduction of **growth hormone** in children can cause gigantism or dwarfism.

9. Pituitary peptide hormones that affect other target glands are called **tropic** hormones.

10. Local or circulating hormones are capable of acting on target cells only if target cells are equipped with response components termed **receptors.**

11. The hormone in *Rhodnius* produced by the corpora allata that prevents metamorphosis into an adult is called **juvenile hormone.**

12. The **pituitary gland** is an important endocrine gland derived embryonically from an outpocketing of the mouth region of the digestive tract and a downgrowth of the floor of the brain.

13. An experiment was run in which cortisol-induced cells of the hippocampi of aging rats appear less able to inhibit the secretion of adrenocorticotropin-releasing hormone by the hypothalamus than those of younger rats. Essentially, this is an example of a **negative** feedback loop.

14. Protein hormones do not cross the cell membrane, but **steroid** hormones do.

15. When a hormone's receptor is on the secreting cell, it is called a(n) **autocrine** message.

16. When people fail to get adequate amounts of iodine in their diet, they get goiter. The hormone that is nonfunctional is called **thyroxine.**

17. If the circulating levels of calcium are high, **calcitonin** stimulates the osteoblasts to resorb it.

18. The steroid hormones are synthesized from **cholesterol.**

Multiple Choice

1. Which of the following is a local hormone?
   a. Adrenaline
   b. Estrogen
   c. **Histamine**
   d. Insulin
   e. Thyroxine

2. The hormones of invertebrates
   a. differ in structure from vertebrate hormones.
   b. are mostly involved in controlling growth.
   c. are mostly produced in the animal's head.
   d. **have different functions from those in vertebrates.**
   e. require large quantities to have an effect.

3. Which of Wigglesworth's observations of *Rhodnius* bugs helped describe the role of insect hormones?
   a. A blood meal triggers molting in these bugs.
   b. If decapitated immediately following a blood meal, the bug will molt.
   c. When two bugs are connected, they molt simultaneously.
   d. **When two bugs are connected, the feeding status of one can trigger molting in the other.**
   e. When two bugs are decapitated and connected, they will never molt into adults.

4. Insect brain hormone serves to
   a. **stimulate the prothoracic gland to release molting hormone.**
   b. stimulate the corpora cardiaca to release molting hormone.
   c. directly stimulate molting if food reserves are adequate.
   d. inhibit molting until the insect is a certain size.
   e. have a general inhibitory effect on insect growth.

5. Which of the following hormones directly stimulates an insect larva to molt?
   a. Brain hormone
   b. **Ecdysone**
   c. Juvenile hormone
   d. Moltin
   e. Prolactin

6. Why can juvenile hormone be used by humans as an effective control of insect populations?
   a. It causes juvenile insects to die.
   b. It causes the insects to fail to molt.

c. **It causes the insects to fail to pupate.**
d. It causes even tiny insects to pupate.
e. It causes insects to molt too quickly.

7. The neurohormones vasopressin (antidiuretic hormone) and oxytocin are
   a. produced by the anterior pituitary and released by the posterior pituitary.
   b. **produced by the hypothalamus and released by the pituitary.**
   c. produced by the pituitary and signal to the hypothalamus.
   d. produced by the hypothalamus and signal to the brain.
   e. produced by the pituitary and signal to the reproductive organs.

8. Which of the following is an effect of oxytocin?
   a. **Stimulation of uterine contractions at birth**
   b. Increased reabsorption of water in the kidney
   c. Stimulation of tropic hormone release
   d. Increased productivity of the hypothalamus
   e. Increased rate of ovulation in the ovary

9. Which of these pairs is a correct match of a pituitary hormone and its target organ?
   a. Oxytocin; breast tissue
   b. Melanocyte-stimulating hormone; kidney
   c. **Endorphins; brain**
   d. Growth hormone; skin and hair
   e. Luteinizing hormone; thyroid

10. The best source of pituitary hormones for medical uses today is
    a. from the slaughter of sheep or cattle.
    b. by extraction from human cadavers.
    c. by synthesis from amino acids in the laboratory.
    d. **from genetically engineered bacteria.**
    e. by extraction from human blood samples.

11. Under which of these conditions would a mammal need to increase thyroxine levels?
    a. In a female following childbirth
    b. During illness and fever
    c. When blood glucose levels are high
    d. During sleep and rest
    e. **When exposed to cold**

12. Which of the following would signal a reduction in thyrotropin release?
    a. Increased levels of thyrotropin
    b. Decreased levels of thyrotropin
    c. **Increased levels of thyroxine**
    d. Decreased levels of thyroxine
    e. Decreased activity of the thyroid

13. Which type of goiter (enlarged thyroid) can be reduced by addition of iodine to the diet?
    a. Hyperthyroid goiter, in which the thyroxine does not turn off the pituitary
    b. Hyperthyroid goiter, in which the thyroid is activated
    c. **Hypothyroid goiter, involving low functional thyroxine and high thyrotropin**
    d. Hypothyroid goiter, involving high functional thyroxine and low thyrotropin
    e. All types of goiter

14. In addition to thyroxine, the mammalian thyroid gland also produces
    a. adrenaline.
    b. **calcitonin.**
    c. iodine.
    d. prolactin.
    e. thyrotropin.

15. The parathyroid glands are involved in regulation of blood levels of
    a. **calcium.**
    b. glucose.
    c. iodine.
    d. sodium.
    e. thyroxine.

16. Why does a lack of insulin cause diabetes mellitus?
    a. Insulin is required for excretion of glucose.
    b. Insulin is required for glucose breakdown.
    c. **Insulin is required for glucose uptake.**
    d. Insulin is required for converting glucose to glycogen.
    e. Insulin is required for synthesizing glucose.

17. Which of the following is *not* produced by the pancreas?
    a. **Cortisol**
    b. Glucagon
    c. Insulin
    d. Somatostatin
    e. Digestive enzymes

18. The adrenal medulla develops from nervous tissue and produces the hormone
    a. **epinephrine.**
    b. adrenocorticotropin.
    c. aldosterone.
    d. cholesterol.
    e. cortisol.

19. Cortisol, the stress hormone, has all of the following effects, except
    a. metabolizing fats for energy.
    b. increasing blood pressure.
    c. slowing down digestion.
    d. **stimulating the immune response.**
    e. slowing down protein synthesis.

20. Muscle-building anabolic steroids are known to cause all of the following effects, except
    a. irregular menstrual periods in women.
    b. increase in body and facial hair in women.
    c. **increase in body and facial hair in men.**
    d. breast enlargement in men.
    e. kidney disease.

21. The hormones secreted by the gonads are synthesized from
    a. complex carbohydrates.
    b. amino acids.
    c. **cholesterol.**
    d. hemoglobin.
    e. nucleic acids.

22. What determines whether a developing mammalian gonad will become an ovary or a testis?
    a. Any gonad with cells containing Y chromosomes will become a testis.
    b. A steady high level of estrogens cause a gonad to become an ovary.
    c. The absence of estrogens causes a gonad to become a testis.
    d. The absence of androgens causes a gonad to become an ovary.
    e. Androgen release at a critical fetal stage causes a testis to develop.

23. Which of the following is the earliest event in puberty?
    a. The pituitary produces more gonadotropins
    b. The level of circulating androgens rises in males
    c. Initiation of the menstrual cycle in females
    d. Differentiation of the gonads into testes or ovaries
    e. Increased subcutaneous fat in males

24. Why do some hormones (first messengers) need to trigger a "second messenger" to activate a target cell?
    a. The first messenger requires activation by ATP.
    b. The first messenger is not a water-soluble molecule.
    c. The first messenger binds to too many types of cells.
    d. The first messenger cannot cross a plasma membrane.
    e. There are no specific cell surface receptors for the first messenger.

25. Which event leads to cyclic AMP (cAMP) production and release in a cell?
    a. A hormone activates adenylate cyclase, which enables cAMP production.
    b. A hormone activates cAMP release from the plasma membrane.
    c. An activated membrane receptor triggers cAMP release.
    d. An activated GTP is converted to cAMP by adenylate cyclase.
    e. GTP activates adenylate cyclase to produce cAMP.

26. How does cyclic AMP (cAMP) behave as a second messenger?
    a. It activates one specific enzyme present in all target cells.
    b. It interacts with various protein kinases in the target cells.
    c. It triggers one reaction pathway specific to each target cell.
    d. It makes the cell membrane more receptive to the hormone.
    e. One cAMP leads to production of two product molecules.

27. After the hormone signal is removed from the cell's environment, what happens?
    a. The reactions initiated by the hormone continue until reactants are used up.
    b. cAMP begins to inactivate the enzymes it previously activated.
    c. cAMP is broken down and removed by enzymes.
    d. cAMP is taken back into the membrane until more hormone is present.

    e. More cAMP is produced in order to make up for lack of the hormone.

28. Steroid hormones initiate the production of target cell substances in which manner?
    a. They initiate second messenger activity.
    b. They bind with membrane proteins.
    c. They initiate DNA transcription.
    d. They activate enzyme pathways.
    e. They bind with membrane phospholipids.

29. A _____ stimulus releases hormones that produce and coordinate major developmental, physiological, and behavioral changes in the male cichlid fish of Lake Tanganyika. The hormone transforms "wimpy" males into big, brightly colored, aggressive, sexually attractive, "macho" males.
    a. chemical
    b. physical
    c. behavioral
    d. a, b, and c
    e. a and b

30. Hormones are secreted by
    a. endocrine glands like the thyroid gland.
    b. individual cells like those lining portions of the digestive tract.
    c. cells in the nervous system (neurohormones).
    d. a and c
    e. a, b, and c

31. Which of the following is the chronological order of events in the molting of *Rhodnius*, as determined by Sir Vincent Wigglesworth?
    a. Blood meal, brain hormone release, and ecdysone release
    b. Blood meal, ecdysone release, and brain hormone release
    c. Ecdysone release, blood meal, and brain hormone release
    d. Brain hormone release, blood meal, and ecdysone release
    e. None of the above

32. Hormones belong to a number of distinct chemical groups. Which of the following is *not* a chemical group to which hormones belong?
    a. Steroids
    b. Proteins
    c. Amino acids and peptides
    d. Carbohydrates
    e. Modified fatty acids

33. Which of the following hormones is *not* produced within the islets of Langerhans of the pancreas?
    a. Somatostatin
    b. Insulin
    c. Glucagon
    d. Calcitonin
    e. None of the above

34. Sweat and saliva are secretions from
    a. the prothoracic gland.
    b. target glands.
    c. endocrine glands.
    d. ductless glands.
    e. exocrine glands.

35. What was the target for the brain hormone demonstrated by Sir Wigglesworth's experiments with *Rhodnius*?
    a. **Prothoracic gland**
    b. Pupal stage
    c. Larval stage
    d. Corpora cardiaca
    e. Corpora allata

36. Hormones released by the posterior pituitary are produced in the
    a. anterior pituitary.
    b. **hypothalamus.**
    c. pineal.
    d. thymus.
    e. thyroid.

37. Which of the following hormones is produced by the posterior pituitary gland?
    a. Prolactin
    b. **Oxytocin**
    c. Endorphins
    d. Growth hormone
    e. Luteinizing hormone

38. Tropic hormones control other endocrine glands. Which of the following is *not* a tropic hormone?
    a. Thyrotropin
    b. Adrenocorticotropin
    c. Luteinizing hormone
    d. **Enkephalins**
    e. Follicle-stimulating hormone

39. Which of the following is an effect of increased levels of vasopressin in the blood?
    a. Stimulates the letdown of milk from mammary tissues
    b. Stimulates uterine contractions during birth
    c. **Stimulates water conservation by the kidney**
    d. Stimulates the liver to produce somatomedins
    e. Stimulates the hypothalamus to produce enkephalins

40. Which of the following hormones are *not* correctly paired with their target organ?
    a. **Oxytocin; endocrine adrenal cortex**
    b. Prolactin; endocrine mammary tissue
    c. Endorphin; endocrine spinal cord neurons
    d. Vasopressin; endocrine kidneys
    e. Luteinizing hormone; endocrine gonads

41. Which of the following maladies is a result of too much growth hormone in the human adult?
    a. Gigantism
    b. Cretinism
    c. Goiter
    d. Dwarfism
    e. **Acromegaly**

42. The 1972 Nobel prize in medicine was awarded to Roger Guillemin and Andrew Schally for discoveries relating to
    a. the cause of hypopituitary dwarfism.
    b. **hypothalamic releasing factors.**
    c. production of growth hormone using genetically engineered bacteria.
    d. the cause of acromegaly.
    e. the discovery of a cure for hypothyroidism.

43. Malfunctioning of the thyroid gland in the human adult can result in
    a. cretinism.
    b. diabetes.
    c. dwarfism.
    d. **goiter.**
    e. hermaphroditism.

44. Which of the following hormones is produced by the parathyroid glands?
    a. **Parathormone**
    b. Thyroxine
    c. Calcitonin
    d. Oxytocin
    e. Pitressin

45–55. Associate the hormones from the list of answer choices below with the correct function or action.
    a. Insulin
    b. Glucagon
    c. Epinephrine
    d. Somatostatin
    e. Cortisol

45. First extracted by Frederick Banting and Charles Best *(a)*

46. Produced by the adrenal medulla *(c)*

47. Released in response to rapid rises of glucose and amino acids in the blood *(d)*

48. Responsible for stimulation of cells to use glucose as a metabolic fuel and to convert excess glucose into fat and glycogen storage *(a)*

49. Responsible for the conversion of glycogen into glucose when serum glucose levels fall *(b)*

50. Inhibits the release of insulin and glucagon *(d)*

51. Slowly released in response to short-term stress *(e)*

52. Rapidly released in response to immediate stress *(c)*

53. Insufficient levels result in diabetes mellitus *(a)*

54. Inhibits the release of growth hormone *(d)*

55. Steroidal hormone synthesized from cholesterol *(e)*

56. Which of the following best describes the function of the corticosteroid hormones?
    a. Depresses the immune response
    b. Stimulates sexual and reproductive activity
    c. Influences blood glucose concentrations
    d. Influences ionic and osmotic concentration of the blood
    e. **All of the above**

57. Which of the following does *not* describe a major difference between water-soluble and lipid-soluble hormones?
    a. Lipid-soluble hormones pass readily through the plasma membrane.
    b. Lipid-soluble hormones act by stimulating the synthesis of new kinds of proteins through gene activation.
    c. Many water-soluble hormones normally require compounds such as cAMP and cGMP to function properly.

d. Water-soluble hormones exert their effect by altering the activity of enzymes normally present in the target cell.
e. **Lipid-soluble hormone action is characterized by a cascade of regulatory steps resulting in an amplification of the effect of a single hormone molecule.**

58. Hormones produced by the adrenal cortex are
    a. water-soluble.
    b. proteins.
    c. **steroids.**
    d. carbohydrates.
    e. tropic hormones.

59–62. Associate the appropriate hormone from the list of answer choices given below with the correct gland, target organ, function, or action.
    a. Cholecystokinin
    b. Melatonin
    c. Aldosterone
    d. Testosterone
    e. Atrial natriuretic hormone

59. Increases sodium ion excretion *(e)*

60. Increases reabsorption of sodium ions *(c)*

61. An androgen *(d)*

62. Secreted by the pineal *(b)*

63. Which of the following is *not* true of a hormone?
    a. One hormone can have different effects on different cells.
    b. **Its actions are as fast as a neural impulse.**
    c. It can act on the same cell that secretes it.
    d. It can enter a cell's nucleus.
    e. It is usually present in very small amounts.

64. Portal blood vessels connect the _____ to the _____.
    a. hypothalamus; brain
    b. hypothalamus; posterior pituitary
    c. **hypothalamus; anterior pituitary**
    d. anterior pituitary; posterior pituitary
    e. pancreas; liver

65. The hormone responsible for releasing calcium from bone into the bloodstream is
    a. insulin.
    b. calcitonin.
    c. **parathormone.**
    d. thyroxine.
    e. somatostatin.

66. In the signal transduction pathway, after a hormone binds a receptor on the cell surface,
    a. GTP converts to cAMP.
    b. protein kinase converts to cAMP.
    c. **GTP activates adenylate cyclase.**
    d. adenylate cyclase converts cAMP to ATP.
    e. adenylate cyclase activates the G protein.

67. The action of a protein kinase is to
    a. **phosphorylate other proteins.**
    b. dephosphorylate other proteins.
    c. convert ATP to cAMP.
    d. activate G protein.
    e. be the receptor for a hormone.

68. Inositol triphosphate and diacylglycerol are
    a. hormones.
    b. intracytoplasmic receptors for lipid-soluble hormones.
    c. cell surface receptors for water-soluble hormones.
    d. **second messengers in signal transduction.**
    e. molecules that bind DNA and stimulate transcription.

69. In order for a cell to be responsive to a lipid-soluble hormone, it must have
    a. a specific cell surface receptor.
    b. **a specific receptor in the cytoplasm or nucleus.**
    c. cAMP.
    d. G protein.
    e. a specific DNA sequence that the hormone binds to.

70. Lipid-soluble hormones' actions are _____ and _____ than water-soluble hormones.
    a. faster; longer lasting
    b. faster; shorter
    c. slower; shorter
    d. slower; last the same amount of time
    e. **slower; longer lasting**

39 *Animal Reproduction*

Fill in the Blank

1. A behavior that transfers sperm into the female's reproductive tract is **copulation**.

2. The uterus opens into the vagina at a muscular neck region called the **cervix**.

3. A **follicle** consists of one egg cell and the surrounding ovarian cells.

4. In humans, fertilization typically occurs when the egg is located in the **oviduct**.

5. Females of most mammals have a periodic sexual receptivity called the **estrous** cycle.

6. The corpus luteum secretes the hormones **progesterone** and **estrogen**.

7. Following fertilization, the human chorionic gonadotropin secreted by the **blastocyst or embryo** keeps the corpus luteum functional.

8. Birth control pills contain synthetic hormones that prevent the ovarian cycle through negative feedback to the **pituitary**.

9. A major evolutionary step allowing the first reptiles to succeed on land was the **shelled egg**.

10. In addition to establishing the diploid nucleus of the zygote and initiating the first stages of development, **fertilization** also activates the sperm and the egg, blocks the entry of additional sperm into the egg nucleus, and stimulates the final meiotic division of the egg nucleus.

11. As opposed to viviparous animals, **oviparous** animals lay eggs in the environment, and their offspring go through embryonic stages outside the body of the mother.

12. Very early in the life of a new mammalian embryo, a population of cells arises during the first few cell divisions. These special cells, called **germ cells**, ultimately end up in the region of the developing gonads where they populate and become gametes.

13. The **ejaculate or semen** of the human male contains sperm and secretions from accessory glands, seminal vesicles, and the prostate gland.

14. **Parthenogenesis** is a specialized type of asexual reproduction in which offspring develop from unfertilized eggs.

15. The time from conception to birth, or the period of pregnancy, is called **gestation**.

16. The male sexual response includes a **refractory** period following orgasm, in which he cannot achieve a full erection.

Multiple Choice

1. What is a disadvantage of asexual reproduction?
   a. Only mitosis is necessary for cell division.
   b. Populations can grow until limited by resources.
   c. Single individuals can produce offspring.
   d. All the offspring are identical in a changing environment.
   e. No energy expenditure is required for mating and fertilization.

2. In order for an organism to reproduce by regeneration, what must be true?
   a. Each cell must be able to give rise to a new organism.
   b. A fragment must be broken off at a particular place.
   c. A fragment must contain all essential tissues.
   d. Gamete formation must occur.
   e. From one individual, two are developed.

3. Parthenogenetic reproduction involves
   a. meiosis but not fertilization.
   b. neither meiosis nor fertilization.
   c. fertilization but not meiosis.
   d. both meiosis and fertilization.
   e. identical copies of the same fertilized egg.

4. In populations that alternate between periods of asexual and periods of sexual reproduction, when would asexual reproduction be most advantageous?
   a. When adults of both sexes are numerous
   b. During the most stressful season
   c. When there is a threat of hybridization
   d. When conditions are favorable and stable
   e. When the population is most dense

5. In male gametogenesis, the second meiotic division produces four haploid
   a. germ cells.
   b. primary spermatocytes.
   c. secondary spermatocytes.
   d. spermatids.
   e. spermatogonia.

6. Unlike spermatogenesis, oogenesis in humans
   a. is continuous over the life of the woman.
   b. begins when the woman is a fetus.
   c. produces four haploid gametes.
   d. occurs at a more rapid rate.
   e. results in swimming cells.

7. Hermaphrodites are organisms that
   a. **possess both male and female reproductive systems.**
   b. breed for a time as one sex, then change to the other.
   c. develop offspring from unfertilized eggs.
   d. usually fertilize themselves.
   e. have abnormal reproductive organs.

8. Reproductive systems with external fertilization are most common in
   a. terrestrial animals.
   b. populations with many more males.
   c. **animals that are sessile.**
   d. animals producing few gametes.
   e. animals that are widely dispersed.

9. For what organisms is a penis necessary?
   a. For all male animals
   b. For all but hermaphrodites
   c. For species with external fertilization
   d. **For species with internal fertilization**
   e. For species whose courtship precedes mating

10. What is the advantage of mammalian testes being located in a sac outside the body cavity?
    a. A shorter distance for semen to be ejaculated
    b. A shorter distance for sperm to swim
    c. The testes can be held at a constant temperature
    d. **The body temperature is too high for sperm production**
    e. It allows testis enlargement with sexual maturity

11. How does the penis become erect during arousal in human males?
    a. A bone is moved into place.
    b. **Spongy tissue is engorged.**
    c. Smooth muscles contract.
    d. By emission of glandular fluid.
    e. Intercellular fluid accumulates.

12. Which part of the testis produces the male sex hormones?
    a. Epididymis
    b. **Leydig cells**
    c. Scrotum
    d. Seminiferous tubules
    e. Vas deferens

13. About what proportion of the seminal fluid is made up of sperm?
    a. Over 90%
    b. 75–80%
    c. 50%
    d. 30–40%
    e. **Less than 10%**

14. A human female has the maximum number of primary oocytes in her ovaries
    a. **at birth.**
    b. just prior to puberty.
    c. early in her fertile years.
    d. midway through her fertile years.
    e. at menopause.

15. Following ovulation, if the egg is not fertilized, what becomes of the follicle cells?
    a. They move with the egg in the oviduct.
    b. **They degenerate.**
    c. They grow into a corpus luteum.
    d. They stop secreting hormones.
    e. They begin to develop a new egg.

16. By what means does the egg move through the oviduct?
    a. It is propelled by cilia on its surface.
    b. It is propelled by its flagellum.
    c. It is propelled by oviduct contractions.
    d. **It is propelled by cilia lining the oviduct.**
    e. It is propelled by its amoeboid motion.

17. In which way is the human female different from females of other mammalian species?
    a. She cycles at the slowest rate per year.
    b. She reabsorbs the uterine lining.
    c. She must copulate in order to ovulate.
    d. **She can be continuously sexually receptive.**
    e. Her cycles are under hormonal control.

18. Female reproductive events are coordinated by two pituitary gonadotropins,
    a. follicle-stimulating hormone and estrogen.
    b. estrogen and progesterone.
    c. follicle-stimulating hormone and progesterone.
    d. luteinizing hormone and progesterone.
    e. **luteinizing hormone and follicle-stimulating hormone.**

19. In the early half of the menstrual cycle, before the events of ovulation, which hormone is at its highest level in the blood?
    a. **Estrogen**
    b. Follicle-stimulating hormone
    c. Luteinizing hormone
    d. Progesterone
    e. Testosterone

20. What triggers ovulation and corpus luteum formation?
    a. A peak in estrogen
    b. **A peak in luteinizing hormone**
    c. A peak in progesterone
    d. Presence of sperm in the reproductive tract
    e. Readiness of the endometrium lining the uterus

21. Why do new follicles not mature as long as the corpus luteum is maintained?
    a. **Corpus luteum hormones inhibit gonadotropin release.**
    b. Corpus luteum hormones inhibit ovarian secretion.
    c. The ovary receives negative feedback from the uterine endometrium.
    d. The ovary receives negative feedback as long as the egg is viable.
    e. New follicle development depends upon corpus luteum hormones.

22. In the female sexual response, the structure that is most analogous to the male's penis is the
    a. breast.
    b. **clitoris.**
    c. labium.
    d. uterus.
    e. vagina.

23. With a few exceptions, mammals are
    a. oviparous.
    b. ovoviviparous.

c. **viviparous.**
d. multiparous.
e. marsupial.

24. Which of the following is *not* true of the umbilical cord?
    a. It contains blood vessels from the embryo.
    b. **It contains blood vessels from the mother.**
    c. It is derived from the allantois.
    d. It joins to the placenta.
    e. It carries nutrients and wastes.

25. What is the source of high levels of estrogen and progesterone during the latter half of pregnancy?
    a. Corpus luteum
    b. Chorion
    c. Follicle cells
    d. **Placenta**
    e. Uterus

26. The intrauterine device is a means of birth control that prevents
    a. ovulation.
    b. fertilization.
    c. **implantation.**
    d. ejaculation.
    e. sperm reaching the egg.

27. Which of the following is *not* a part of asexual reproduction?
    a. Identical Genetics
    b. **Fertilization**
    c. Parthenogenesis
    d. Regeneration
    e. Budding

28. Sexual reproduction has an evolutionary advantage since it
    a. promotes both males and females of a species.
    b. does not allow for rapid reproduction.
    c. **promotes genetic variability to cope with changes in the environment.**
    d. is controlled by many hormonal mechanisms.
    e. protects and nurtures the embryo.

29. Which of the following is the correct order of structures that sperm passes through during copulation?
    a. Vas deferens, seminiferous tubules, epididymis, urethra
    b. Epididymis, seminiferous tubules, vas deferens, urethra
    c. Seminiferous tubules, vas deferens, epididymis, urethra
    d. **Seminiferous tubules, epididymis, vas deferens, urethra**
    e. Vas deferens, urethra, epididymis, seminiferous tubules

30. Semen, which is the fluid matrix for sperm during emission and ejaculation, contains all of the following except
    a. seminal fluid from the seminal vesicles.
    b. **glucose to serve as an energy source for the sperm.**
    c. alkaline secretions from the prostate gland.
    d. hormone-like substances termed prostaglandins.
    e. clotting enzymes.

31. Which of the following is a probable advantage of hermaphroditism?
    a. **For a given species, every organism is a potential mate.**

b. The reduction of cultural strife resulting from societies that contain individuals of only one sex.
    c. Simpler behavior patterns are required for mate selection.
    d. More rapid reproduction of genetically successful individuals is possible through asexual reproduction.
    e. Complex hormonal control and feedback mechanisms are *not* required.

32–37. Choose the appropriate cell stage from the list below to respond to questions 32–37. There may or may not be more than one correct answer.
    a. Oogonia
    b. Spermatogonia
    c. Spermatid
    d. Primary spermatocyte
    e. Secondary spermatocyte

32. Germ cell from the female gonad *(a)*

33. Diploid germ cell of the male gonad *(b)*

34. Cell resulting from the initial mitotic division of a germ cell *(d)*

35. First haploid germ cell produced in spermatogenesis *(e)*

36. Cell resulting from the second meiotic division *(c)*

37. The terminal cell type that, at the end of spermatogenesis, differentiates into a sperm cell *(c)*

38–42. Choose the appropriate cell stage from the list below to respond to questions 38–42. There may or may not be more than one correct answer.
    a. Second polar body
    b. Primary oocyte
    c. Secondary oocyte
    d. First polar body
    e. Ootid

38. Stage during which the energy, raw materials, and RNA needed for the first cell divisions after fertilization is acquired *(b)*

39. First essential haploid germ cell produced during oogenesis *(c)*

40. Daughter cell to the secondary oocyte *(a and e)*

41. Largest haploid cell resulting from second meiotic division of oogenesis *(e)*

42. The terminal cell type that, at the end of oogenesis, differentiates into the mature ovum *(e)*

43. After spermatogenesis, sperm cells are generally stored in the
    a. spermatophore.
    b. prostate gland.
    c. vas deferens.
    d. seminiferous tubule.
    e. **epididymis.**

44. Progeny inherit all the characteristics of a single parent through
    a. copulation.
    b. **asexual reproduction.**
    c. gametogenesis.
    d. sexual reproduction.
    e. fertilization.

45. Generally, among vertebrates
    a. cells are haploid during embryological development, becoming diploid at birth.
    b. cells are diploid during embryological development, becoming haploid at birth.
    c. cells are diploid during adult life, haploid as a blastocyst.
    d. only gametic cells are haploid.
    e. only gametic cells are diploid.

46. Progesterone and estrogen are hormones produced by the
    a. anterior pituitary.
    b. posterior pituitary.
    c. Sertoli cells.
    d. hypothalamus.
    e. corpus luteum.

47. The blastocyst normally implants within the
    a. vas deferens.
    b. ovarian follicle tissue.
    c. endometrium of the uterus.
    d. myometrium of the vagina.
    e. chorion.

48. Which of the following describes the principal difference between menstrual and estrous cycles?
    a. The estrous cycle occurs if the female is fertilized, and the menstrual cycle occurs if the female isn't fertilized.
    b. The menstrual cycle occurs if the female is fertilized, and the estrous cycle occurs if the female isn't fertilized.
    c. The estrous cycle lacks menstruation.
    d. The menstrual cycle occurs in mammals; the estrous cycle occurs in birds.
    e. Only the menstrual cycle is controlled by estrogen.

49. Which of the following is the most effective form of birth control?
    a. Vasectomy
    b. Condom
    c. Intrauterine device
    d. Rhythm method
    e. "The pill"

50. A vasectomy is a minor operation which involves removing a small section of the _____ and tying off the loose ends.
    a. epididymis
    b. urethra
    c. oviduct
    d. vas deferens
    e. seminiferous tubules

51. In most male mammals, what anatomical feature is shared by both the reproductive and excretory systems?
    a. Prostate gland
    b. Seminiferous tubules
    c. Vas deferens
    d. Urethra
    e. Epididymis

52. In mammals, which is the correct sequence of structures that sperm passes through before fertilizing an egg?
    a. Uterus, vagina, cervix, oviduct, ovary
    b. Vagina, cervix, uterus, oviduct
    c. Oviduct, uterus, vagina, ovary, body cavity
    d. Vagina, cervix, oviduct, uterus
    e. Vagina, uterus, oviduct, body cavity

53. The hormone that normally triggers mammalian ovulation is
    a. progesterone.
    b. follicle stimulating hormone.
    c. luteinizing hormone.
    d. estrogen.
    e. prolactin.

54. The formation of the zygote is synonymous with
    a. copulation.
    b. fertilization.
    c. intercourse.
    d. recombination.
    e. capacitation.

55. The birth control method that prevents ovulation is
    a. a vasectomy.
    b. an intrauterine device.
    c. a condom.
    d. the rhythm method.
    e. "the pill."

56. Most methods of birth control focus on
    a. preventing oogenesis.
    b. preventing spermatogenesis.
    c. preventing fertilization.
    d. decreasing libido.
    e. decreasing testosterone levels.

57. The end result of spermatogenesis is the production of four haploid spermatids. The end result of oogenesis is the production of
    a. four haploid ootids.
    b. two haploid ootids.
    c. one diploid ootid.
    d. two diploid ootids.
    e. one haploid ootid with two polar bodies.

58. The terms "monoecious" and "dioecious" refer to
    a. patterns of inheritance.
    b. circulatory physiology.
    c. patterns of cleavage in developing embryos.
    d. the evolutionary origin of species.
    e. None of the above

59. At birth, a female has about one million primary oocytes in each ovary. During a woman's fertile years, about _____ of these oocytes will mature completely into eggs and be released at ovulation.
    a. 100
    b. 200
    c. 300
    d. 450
    e. None of the above

60. Which of the following is *not* true of asexual reproduction?
    a. It occurs mostly in invertebrates.
    b. Asexually reproducing species tend to live in variable environments.
    c. Asexually reproducing species tend to be sessile.
    d. Asexually reproducing species live in sparse populations.
    e. One method is parthenogenesis.

61. The contraceptive method that also helps prevent the spread of sexually transmitted disease is
    a. the diaphragm.
    b. the cervical cap.
    c. the condom.
    d. the IUD.
    e. "the pill."

62. The contragestational pill, RU-486, blocks _____ receptors, preventing implantation of the embryo in the uterine lining.
    a. luteinizing hormone
    b. progesterone
    c. estrogen
    d. follicle-stimulating hormone
    e. gonadotropin-releasing hormone

63. Which of the following is *not* true of the first trimester of pregnancy?
    a. There is rapid cell division.
    b. It is when the embryo is the most sensitive to radiation and drugs.
    c. The fetus grows rapidly in size.
    d. The mother has symptoms of nausea and mood swings.
    e. There are high levels of estrogen and progesterone circulating in the mother's blood.

64. During labor, uterine contractions are stimulated by
    a. estrogen.
    b. oxytocin.
    c. progesterone.
    d. *a* and *b*
    e. *a*, *b*, and *c*

65. During the second trimester of pregnancy,
    a. the blastocyst implants in the uterine lining.
    b. the mother goes through noticeable hormonal responses.
    c. the fetal digestive system starts to function.
    d. the fetal brain undergoes cycles of sleep and waking.
    e. the fetus grows fingers and toes.

66. Some species of mites and scorpions use indirect fertilization, excreting a gelatinous container of sperm called a(n)
    a. spermatophore.
    b. spermatogonium.
    c. spermatid.
    d. cloaca.
    e. amplexus.

67. The human sexual response consists of four phases in the order:
    a. excitement, refractory, plateau, orgasm
    b. plateau, excitement, orgasm, resolution
    c. plateau, excitement, refractory, orgasm
    d. excitement, plateau, orgasm, resolution
    e. excitement, plateau, refractory, orgasm

68. Gamete intrafallopian transfer is
    a. removing eggs from a woman's oviduct and fertilizing them in a culture medium.
    b. moving eggs from the ovaries to the oviduct when the ovaries are blocked.
    c. injecting eggs and sperm into the oviduct when the oviduct entrance is blocked.
    d. injecting an embryo into the oviduct.
    e. removing an embryo form the oviduct and injecting it into the uterus when the oviducts are blocked.

40 Animal Development

Fill in the Blank

1. Although we often stress the embryo in discussing animal development, it is a process that continues through all stages of life, ceasing only with **death**.

2. The primitive streak of bird and mammalian embryos is analogous to the **blastopore** of amphibians and sea urchins.

3. The cells that surround unfertilized mammalian eggs when they first arrive in the oviducts are **cumulus** cells.

4. The protein shell that surrounds the eggs of many mammalian species is called **zona pellucida**.

5. The membrane of an egg is sometimes called the **vitelline** envelope.

6. The **acrosome** contains enzymes that assist the sperm in penetrating the zona pellucida.

7. The region of the frog egg that is darkly pigmented is known as the **animal** pole.

8. **Spina bifida** is a birth defect caused by the failure of the closure of the posterior region of the neural tube.

9. The process in development by which germ layers form and take specific positions relative to each other is **gastrulation**.

10. The rod of cartilage that forms from the chordomesoderm and gives structural support to the developing embryo is called the **notochord**.

11. In chicks, the first extraembryonic membrane to form is the **yolk sac**.

12. In mammals, the first extraembryonic membrane to form is the **trophoblast**.

Multiple Choice

1. Development occurs
   a. only during growth of the organism.
   b. **throughout the life of the organism.**
   c. only in nondividing cells.
   d. in ectoderm and endoderm, but not in mesoderm.
   e. only in animals.

2. Bird eggs have large amounts of yolk. This results in
   a. gradual metamorphosis.
   b. complete metamorphosis.
   c. **incomplete cleavage.**
   d. incomplete mitosis during cleavage.
   e. bicoid larvae.

3. If one of the blastomeres is removed from a developing mouse embryo, the remaining cells will go on to develop into a normal mouse. This is an example of
   a. **regulated development.**
   b. the zone of polarizing activity.
   c. cleavage.
   d. gastrulation.
   e. ingression.

4. During cleavage, the number of cells in a developing frog embryo increases. The cytoplasm in these new cells
   a. **comes from the egg cytoplasm.**
   b. is synthesized by the blastomeres.
   c. does not contain any yolk.
   d. is the vegetal pole.
   e. undergoes mitosis.

5. The mesoderm
   a. is located on the outside of the embryo.
   b. **lies between the endoderm and the ectoderm.**
   c. is found in blastula-stage embryos.
   d. gives rise to the linings of the gut.
   e. is formed during cleavage.

6. The formation of the endoderm during gastrulation in frogs results from
   a. **movement of cells from the surface layer to the interior.**
   b. migration of cells within the blastocoel.
   c. formation of columnar cells at the vegetal pole.
   d. rapid cell division.
   e. migration of secondary mesenchyme cells.

7. You are studying gastrulation in frog embryos. You take embryos at the beginning of gastrulation and stain a patch of cells on the surface of the animal end, just above the dorsal lip of the blastopore. At the end of gastrulation, where will these cells be located?
   a. In the endoderm
   b. In the mesoderm
   c. In the ectoderm
   d. **a and b**
   e. a and c

8. In bird eggs, cells migrate into the interior of the embryo through the primitive streak. In nonmammalian embryos with smaller amounts of yolk, this involution occurs at the
   a. archenteron.
   b. **blastopore.**
   c. notochord.
   d. mesenchyme.
   e. endoderm.

9. Adult humans do not grow because
   a. human growth is discontinuous.
   b. human growth does not follow an S-shaped curve.
   c. cell division does not occur in adults.
   d. growth is most rapid during early embryonic development.
   e. **in adults, the rate of cell division equals the rate of cell loss.**

10. If cells from the neural tube of a frog embryo are transplanted onto the ventral surface of a second embryo, the transplanted tissue will still go on to develop into tissues of the nervous system. The transplanted cells are
    a. differentiated.
    b. totipotent.
    c. discontinuous.
    d. **determined.**
    e. endodermal.

11. When the blastopore dorsal lip is grafted from one frog embryo onto a second embryo, the second dorsal lip will
    a. change the polarity of the adjacent segments.
    b. block gastrulation.
    c. **change the developmental fate of the surrounding cells.**
    d. change the prospective potency of the surrounding cells.
    e. cause rapid cell division.

12. Because of the differences between regulative and mosaic development, which of the following animals would *not* develop as twins?
    a. Humans
    b. Frogs
    c. Sea urchins
    d. **Insects**
    e. Birds

13. Invagination
    a. requires unique cell movement.
    b. creates unique positions for new cell interactions.
    c. occurs in sea urchins.
    d. forms the archenteron in echinoderms.
    e. **All of the above**

14. Gastrulation is the stage of development
    a. when neural tube formation begins.
    b. when sea urchins begin to form primary mesenchyme.
    c. when new embryonic tissue begins to form in the frog embryo.
    d. that precedes cleavage.
    e. **b and c**

15. The gray crescent
    a. **is observable in the zygote and two-celled frog embryo.**
    b. is a homeobox gene.
    c. controls cellular affinities.
    d. induces the optic cup to form.
    e. can be mimicked by retinoic acid.

16. An organism (such as the chick) with extensive yolk distribution
    a. has complete cleavage.
    b. **has incomplete cleavage.**
    c. forms a blastoderm but no blastocoel.
    d. has yolk evenly distributed in the egg.
    e. fails to synthesize DNA during cleavage.

17. Because the human embryo is able to split at the 64-cell level of organization to produce two viable progeny, it is said to exhibit _____ development
    a. mosaic
    b. determinative
    c. definitive
    d. classical vertebrate
    e. **regulated**

18. The gray crescent experiments of Hans Spemann provided experimental evidence for
    a. the cell theory of life.
    b. the concept of determination.
    c. **the unequal distribution of cytoplasmic determinants.**
    d. the basis of genetic engineering.
    e. None of the above

19. Place the following developmental events in their proper chronological sequence: (1) formation of the neural tube, (2) movement of neural folds, and (3) thickening of neural ectoderm.
    a. 1, 2, and 3
    b. 2, 1, and 3
    c. 3, 1, and 2
    d. **3, 2, and 1**
    e. None of the above

20. The gray crescent is the region of the egg
    a. that is opposite the site of sperm penetration.
    b. that later is where the gastrulation begins.
    c. that was pigmented before the cytoplasm rearranged.
    d. where the dorsal lip of the blastopore will be.
    e. **All of the above**

21. The sea urchin differs from insects in early cleavage. The sea urchin undergoes _____, whereas the insects undergo _____
    a. **complete cleavage, superficial cleavage.**
    b. complete cleavage, incomplete cleavage.
    c. superficial cleavage, incomplete cleavage.
    d. incomplete cleavage, complete cleavage.
    e. superficial cleavage, complete cleavage.

22. When frog eggs are placed in water, which end is up?
    a. It would be random.
    b. **The animal pole would face up.**
    c. The vegetal pole would face up.
    d. The eggs would lay on their side.
    e. None of the above

23. The embryos of complex multicellular animals must establish spatial coordinates in order for development to progress. An important reference coordinate for developing frogs is _____
    a. the location of the vegetal pole.
    b. the location of the animal pole.
    c. **the point of sperm penetration.**
    d. the location where gastrulation begins.
    e. the direction of the sun.

24. Which of the cells below are the ones that actually develop into the fetus?
    a. Trophoblast
    b. Extraembryonic
    c. Inner cell mass
    d. Cumulus cells
    e. All of the above

25. The initial formation of the nervous system occurs during neurulation. The important inducing structure is (are)
    a. the somites.
    b. imaginal discs.
    c. Hensen's node.
    d. the notochord.
    e. spinal column.

26. The structure in birds and mammals that is most analogous to the dorsal lip of the blastopore found in frogs is the
    a. primitive streak.
    b. the archenteron.
    c. yolk plug.
    d. Hensen's node.
    e. notochord.

27. The trophoblast cells in frogs are important
    a. to form the placenta.
    b. to protect the egg from sunlight.
    c. for gas exchange.
    d. to help with implantation.
    e. None of the above

28. Gastrulation in birds and frogs differs quite a lot. The one main feature that both have in common is
    a. that three germ layers are established.
    b. the gut is formed from the migration of cells from the surface of the blastocyst.
    c. that the neural tube forms.
    d. that somites are created.
    e. All the above

29. The fate of the mesodermal cells after gastrulation is to contribute predominantly to the developing
    a. brain and nervous system.
    b. skeletal system and muscles.
    c. inner lining of the gut and respiratory tract.
    d. sweat glands.
    e. liver and pancreas.

30. The fate of the ectodermal cells after gastrulation is to contribute predominantly to the developing
    a. brain, nervous system, and sweat glands.
    b. skeletal system and muscles.
    c. inner lining of the gut and respiratory tract.
    d. liver and pancreas.
    e. None of the above

31. The fate of the endodermal cells after gastrulation is to contribute predominantly to the developing
    a. brain, nervous system, and nails.
    b. skeletal system and muscles.
    c. inner lining of the gut and respiratory tract.
    d. sweat glands and milk secretory glands.
    e. None of the above

32. An interesting difference between identical twins formed prior to the formation of the trophoblast versus after is
    a. if prior to trophoblast formation, each developing embryo has its own chorion.
    b. if after trophoblast formation, each developing embryo has its own chorion.
    c. if prior to trophoblast formation, each developing embryo has its own amnion.
    d. if after to trophoblast formation, each developing embryo has its own amnion.
    e. All the above

33. Identical twins can form either before or after trophoblast formation. Identical twins that form after the trophoblast has formed are thought to arise from
    a. the egg splitting in the womb.
    b. one embryo having two inner cell masses.
    c. the embryo splitting during hatching.
    d. each of the first two blastomeres forming separate blastocycts.
    e. fertilization with more than one sperm.

34. If the mouth forms from the region of the blastopore, the species would be members of the
    a. deuterostomes.
    b. yokalstomes.
    c. epistomes.
    d. echinoderms.
    e. protostomes.

35. If the anus forms from the region of the blastopore, the species would be members of the
    a. deuterostomes.
    b. yokalstomes.
    c. epistomes.
    d. echinoderms.
    e. protostomes.

36. Humans are
    a. deuterostomes.
    b. yokalstomes.
    c. epistomes.
    d. echinoderms.
    e. protostomes.

37. Mollusks are
    a. deuterostomes.
    b. yokalstomes.
    c. epistomes.
    d. echinoderms.
    e. protostomes.

38. Bottle cells are
    a. cells of the trophoblast.
    b. cells of the inner cell mass.
    c. fluid filled cells used for storage.
    d. cells with the shape of a bottle found around the blastopore.
    e. another name for granulosa cells.

39. The movement of cells toward the blastopore is called
    a. mass cellular migration.
    b. embryonic cellular integration migration.
    c. fulfilling cellular destiny.
    d. Spemania.
    e. epiboly.

40. One of Hans Spemann's important experiments involved
    a. dividing human embryos into equal halves.
    b. killing a single cell in a sea urchin embryo to study the effects.
    c. dividing a frog embryo in two using a human hair.
    d. removing cytoplasm from muscle cells and transplanting it into a fertilized egg.
    e. All the above

41. The epiblast and the hypoblast are structures found during
    a. human development.
    b. frog development.
    c. human and frog development.
    d. chicken development.
    e. human and chicken development.

42. In humans the amniotic cavity begins to form
    a. below the hypoblast.
    b. above the epiblast.
    c. below the epiblast.
    d. from the trophoblast.
    e. from the yolk sac.

43. The _____ is the structure from which the ectoderm, endoderm, and mesoderm cells are derived in mammals and birds.
    a. trophoblast
    b. cumulus cells
    c. blastocyst
    d. epiblast
    e. hypoblast

44. The cells that form the neural tube come from the
    a. notochord.
    b. mesoderm.
    c. endoderm.
    d. ectoderm.
    e. neuroderm.

45. The _____ is an important structure for waste storage in birds and some mammals including pigs, but not in humans.
    a. allantois
    b. yolk sac
    c. placenta
    d. umbilical cord
    e. amnion

46. Eggs are triggered into activity by
    a. sperm fusion.
    b. sodium influx.
    c. ovulation.
    d. ionic diequilibrium.
    e. actozymes.

47. The fast block to polyspermy includes the rise in
    a. intracellular potassium ions.
    b. extracellular potassium ions.
    c. intracellular sodium ions.
    d. release of cortical vesicles.
    e. All the above

48. In many mammalian species, before the sperm can penetrate the zona pellucida the sperm must
    a. capacitate.
    b. activate.
    c. decapitate.
    d. swim for hours.
    e. penetrate the oviducts.

49. In sea urchins, slow block takes
    a. a day.
    b. a few seconds.
    c. 1 hour or more.
    d. 20 seconds or more.
    e. None of the above

50. _____ contain enzymes that are released during the slow block to polyspermy.
    a. Cortical vesicles
    b. Exocytotic granules
    c. Acrosome
    d. Endoplasmic reticulum
    e. Mitochondria

51. The sperm contribute _____ to the embryo.
    a. a nucleus
    b. half of the mitochondria and a nucleus
    c. a centriole and nucleus
    d. cilium and a nucleus
    e. All of the above

52. In frogs, the sperm penetrates
    a. in the region of the vegetal pole.
    b. in the region of the animal pole.
    c. in the region of the gray crescent.
    d. anywhere.
    e. somewhere on the border of the animal and vegetal pole.

53. If a cell is removed from an 8-cell embryo and a part ends up missing in the organism, the development is termed
    a. regulative.
    b. controlled.
    c. banished.
    d. irreversible.
    e. mosaic.

54. _____ are used by mesenchyme cells to move along extracellular matrix molecules.
    a. Amoeboids
    b. "Walking feet"
    c. Filopodia
    d. Sliding cell receptors
    e. None of the above

55. The segmented characteristic of human embryonic development is evident from these bricklike structures that form along the notochord.
    a. Neural tubes
    b. Mesoderm
    c. Ectoderm
    d. Somites
    e. Somatomeres

41 Neurons and Nervous Systems

Fill in the Blank

1. The part of the neuron specialized for receiving impulses is the **dendrite**.

2. The initial membrane event of an action potential is the flow of **sodium** ions across the membrane.

3. Following an action potential, a neuron has a **refractory period** during which it cannot be stimulated.

4. In myelinated axons of vertebrate neurons, breaks in the insulation occur at points called the **nodes of Ranvier**.

5. Deadly nerve gases are examples of inhibitors of the enzyme **acetylcholinesterase**.

6. The central nervous system process of learning to ignore a repeated stimulus is called **habituation**.

7. The information that flows through the nervous system consists of chemical and **electrical** messages.

8. When a neuron contacts another neuron or a muscle or gland, special junctions called **synapses** transmit the message carried by the incoming neuron.

9. Special glial cells, called **astrocytes**, surround the smallest, most permeable blood vessels in the brain, thereby participating in the formation of the blood–brain barrier.

10. The nicotinic receptors of acetylcholine are found in **skeletal** muscles.

11. The depolarization of a neuron must rise above the **threshold** before an action potential is achieved.

12. A nervous system has three components: sensory input, **integration,** and output.

13. Neurons that make connections between afferent and efferent neurons are known as **interneurons**.

Multiple Choice

1. A human being has the maximum number of functioning neurons
   a. as a fetus.
   b. **just after birth.**
   c. at about college age.
   d. at middle age.
   e. just before death.

2. The functions of glial cells include all of the following except
   a. supporting developing neurons.
   b. supplying nutrients.
   c. **conducting nerve impulses.**
   d. consuming foreign particles.
   e. insulating nerve tissue.

3. About how many neurons are in the human brain?
   a. 100 thousand
   b. A million
   c. 100 million
   d. A billion
   e. **100 billion**

4. Each of the body's peripheral nerves
   a. carries information from the brain to the body.
   b. carries information from the brain to the spinal cord.
   c. carries a particular type of sensory information.
   d. **consists of a bundle of axons.**
   e. consists of glia and cell bodies.

5. Which of the following describes the resting potential of the neuronal cell membrane?
   a. The inside is 60 millivolts more positive than the outside.
   b. **The outside is 60 millivolts more positive than the inside.**
   c. The inside is 30 millivolts more positive than the outside.
   d. The outside is 30 millivolts more positive than the inside.
   e. The inside has about the same charge as the outside.

6. Which of the following can carry electric charges across the cell membrane?
   a. Electrons
   b. Protons
   c. Water
   d. **Ions**
   e. Proteins

7. What generally maintains the electric charge across the neuronal membrane?
   a. **Sodium–potassium pump**
   b. Action potential
   c. Resting potential
   d. Voltage-gated channels
   e. Negative ion pump

8. Which of the following describes the mechanism of voltage-gated channel proteins?
   a. Depending upon the membrane voltage, ions are pumped through.
   b. **Depending upon the membrane voltage, ions can diffuse through.**
   c. Ions can move through only if the overall membrane voltage stays the same.
   d. Ions are pumped through in order to maintain same membrane voltage.
   e. When the gates close, membrane voltage changes.

9. What happens if Na⁺ channels open and sodium ions diffuse into the cell?
   a. The cell becomes hyperpolarized.
   b. Other sodium ions will move out of the cell.
   c. Voltage-gated channels will remain closed.
   d. The charge across the nearby membrane will change.
   e. Action potentials will be triggered.

10. Following depolarization, the neural membrane potential is restored when
    a. Na⁺ ions rush outward through the membrane.
    b. K⁺ ions rush outward through the membrane.
    c. Cl⁻ ions rush inward through the membrane.
    d. a pump moves ions to their original concentrations.
    e. the membrane becomes freely permeable to many ions.

11. Which of the following would an electrode record as an action potential moves along an axon membrane?
    a. The inside membrane voltage becomes negative.
    b. The membrane voltage permanently changes.
    c. The action potential may reverse its direction.
    d. The action potential moves at a constant speed.
    e. The height of the action potential does not change.

12. Which of the following nerves would have the most rapid action potentials?
    a. Those with the thinnest axon diameters
    b. Those with sheaths of myelin
    c. Those in invertebrate animals
    d. Those with a greater membrane potential
    e. Those with the most ion channels

13. Saltatory conduction results when
    a. continuous propagation of the nerve impulse speeds up.
    b. a nerve impulse jumps from one neuron to another.
    c. the threshold for an action potential is suddenly increased.
    d. action potentials spread from node to node down the axon.
    e. the direction of an action potential suddenly changes.

14. When an action potential arrives at the axon terminal, the voltage-gated calcium channels there
    a. release calcium into the synaptic cleft.
    b. actively transport neurotransmitter into the synaptic cleft.
    c. cause vesicles to release neurotransmitter into the synaptic cleft.
    d. depolarize the membrane at the axon terminal.
    e. cause the membrane receptors to bind neurotransmitter.

15. When the neurotransmitter diffuses across the synaptic cleft,
    a. it automatically causes depolarization of the postsynaptic membrane.
    b. it can be excitatory or inhibitory depending upon the type of postsynaptic membrane.
    c. a single molecule is sufficient to trigger activation of the postsynaptic membrane.
    d. only a few molecules make it to the postsynaptic membrane.
    e. it must move through nodes in the myelin sheath.

16. Poisons like curare and atropine are nervous system poisons because
    a. they bind to neurotransmitter receptors and inactivate them.
    b. they bind to neurotransmitter receptors and activate them.
    c. they bind to neurotransmitters and inactivate them.
    d. they cause increased release of neurotransmitter.
    e. they inhibit the release of neurotransmitter.

17. What is the effect of inhibiting acetylcholinesterase?
    a. Neurotransmitter release from the presynaptic membrane is inhibited.
    b. Synthesis of neurotransmitter in cells is inhibited.
    c. Breakdown of neurotransmitter in the synapse is inhibited.
    d. Stimulation of the postsynaptic membrane is inhibited.
    e. Cholinergic receptors are inhibited.

18. Which of the following is true about monosynaptic reflex loops?
    a. They make up most of the body's neural circuits.
    b. They contain two or three synapses.
    c. The postsynaptic membrane is on a motor neuron.
    d. The sensory neuron dendrites are in the spinal cord.
    e. The main synapse is located within the brain.

19. Experimental monitoring of the siphon-withdrawal response of sea slugs indicates that
    a. sensitization is caused by increased sensitivity to neurotransmitter.
    b. habituation is caused by decreased neurotransmitter release.
    c. sensitization is caused when more circuits are activated.
    d. habituation results when polysynaptic circuits are used.
    e. sensitization is stimulated by the reticular activating system.

20. The two primary cell types of the nervous system are
    a. fibroblasts and chondrocytes.
    b. neurons and glial cells.
    c. epithelial cells and glandular cells.
    d. neurons and epithelial cells.
    e. neuromuscular cells and epithelial cells.

21. Synaptic clefts can be cleansed of neurotransmitters in which of the following ways?
    a. Enzymatic degradation
    b. Simple diffusion
    c. Active transport
    d. *a* and *c*
    e. *a, b,* and *c*

22. Which of the following ions is most responsible for generating an action potential?
    a. Na⁺
    b. K⁺
    c. Cl⁻
    d. H⁺
    e. OH⁻

23. Most nerve cells communicate with others through
    a. electric signals that pass across at synapses.
    b. chemical signals that pass across at synapses.
    c. bursts of pressure that "bump" the postsynaptic cell membrane.
    d. Na+ ions, released from one cell and entering the next.
    e. None of the above

24. The resting potential of a neuron is produced by
    a. voltage-gated channels in the membrane.
    b. chemically gated channels in the membrane.
    c. permanently open potassium channels in the membrane.
    d. the concentration difference in Na+ across the membrane.
    e. blocking the sodium–potassium pump.

25. The myelin sheath that surrounds some axons in the peripheral nervous system is formed by
    a. neurons.
    b. Schwann cells.
    c. bacteria that have invaded the nervous system.
    d. synapses.
    e. None of the above

26. Which of the following limits the frequency at which a single neuron can "fire" action potentials?
    a. The number of synapses that the neuron forms
    b. The number of other cells that the neuron contacts
    c. The refractory period for the neuron's Na+ channel
    d. The length of the axon of the neuron
    e. The number of dendrites on the neuron

27. Which of the following can be said to be involved in triggering synaptic transmission?
    a. The action potential
    b. The opening of Ca2+ channels at the synaptic terminal
    c. The entry of Ca2+ into the presynaptic terminal
    d. Fusion of synaptic vesicles with the presynaptic membrane
    e. All of the above

28. Neurons
    a. have a uniform shape throughout the nervous system.
    b. are more numerous than glial cells in the nervous system.
    c. are found only in mammals and birds.
    d. communicate with other cells at synapses.
    e. All of the above

29. Curare
    a. blocks acetylcholine release from synapses.
    b. binds to nicotinic acetylcholine receptors and blocks them.
    c. stimulates calcium release from neurons.
    d. blocks ATP production.
    e. stimulates acetylcholinesterase.

30. The spinal cord
    a. conducts information between the brain and the organs of the body.
    b. has a central area of gray matter surrounded by an area of white matter.
    c. has motor neurons whose axons exit through the ventral root.
    d. gains information through sensory fibers that enter the spinal cord through dorsal roots.
    e. All of the above

31. Which of the following statements about the spinal cord is correct?
    a. The spinal cord is a simple relay station to the brain and does little information processing or integration.
    b. The spinal cord does not contain cell bodies, only fiber tracts.
    c. Motor output leaves the spinal cord via the ventral root.
    d. The spinal cord does not contain interneurons.
    e. All of the above

32. The action potential
    a. begins with an increased permeability to potassium.
    b. returns to resting when the sodium channels open.
    c. can be triggered in very rapid succession, with no delay.
    d. involves voltage-gated channels in the membrane.
    e. propagates only because chloride ions move through the membrane.

33. The action potential
    a. travels along all axons at the same speed.
    b. is slowed down if a nerve cell has myelin around it.
    c. is blocked at the nodes of Ranvier.
    d. causes a brief depolarization of the membrane potential.
    e. triggers a change in potential simultaneously along the entire axon.

34. Monosynaptic reflexes
    a. only involve one synapse in the entire body.
    b. only involve one synapse on a single neuron.
    c. involve a number of synapses, but the signal travels directly from sensory neurons to motor neurons.
    d. involve a number of synapses, but only one in the brain itself.
    e. None of the above

35. Choose the correct statement about vertebrate nervous systems.
    a. The nervous system consists only of brain and spinal cord.
    b. They can carry out many tasks at the same time.
    c. All nervous system functions are voluntary.
    d. All neuronal circuits consist of monosynaptic reflexes.
    e. Ions are equally distributed across nerve cell membranes.

36. Choose the statement about acetylcholine that is *not* true.
    a. Acetylcholine is a neurotransmitter.
    b. Acetylcholine is found at mammalian neuromuscular junctions.
    c. Both smooth muscles and skeletal muscles respond to acetylcholine.

d. Acetylcholine is degraded by acetylcholinesterase.
e. Acetylcholine binds to curare and atropine.

37. Atropine
    a. binds selectively to nicotinic acetylcholine receptors.
    b. binds selectively to muscarinic acetylcholine receptors.
    c. binds well to both nicotinic and muscarinic receptors.
    d. binds to norepinephrine receptors in the brain.
    e. binds to norepinephrine receptors in the peripheral nervous system.

38. Neurons
    a. fire action potentials only on the basis of the number of excitatory inputs they receive.
    b. sum excitatory and inhibitory postsynaptic potentials.
    c. make the "decision" to fire in the dendrites of the neuron.
    d. make a spatial summation, but not a temporal one.
    e. never form synapses on synapses.

39. Patch clamping is used to
    a. fix a break in a cell membrane.
    b. record electrical activity inside a cell.
    c. record ion movements through a single channel.
    d. record ion movement through the entire neuron.
    e. record an action potential through a single channel.

40. When an action potential arrives at an axon terminal, it causes the opening of _____ channels, which triggers fusion of a neurotransmitter vesicle with the cell membrane.
    a. calcium
    b. sodium
    c. potassium
    d. chloride
    e. acetylcholine

41. Muscarinic receptors of acetylcholine are
    a. found in heart muscle.
    b. tend to be inhibitory.
    c. bound by curare.
    d. *a* and *b*
    e. *a*, *b*, and *c*

42. Which of the following is *not* true of atropine?
    a. It binds to muscarinic acetylcholine receptors.
    b. It constricts the pupils of the eyes.
    c. It increases heart rate.
    d. It decreases digestive juices.
    e. It comes from the plant *Atropa belladonna*.

43. Which of the following neurotransmitters has been shown to be involved with sleep/wake cycles?
    a. Acetylcholine
    b. Norepinephrine

c. GABA
d. Serotonin
e. Dopamine

44. Which of the following neurotransmitters is a peptide?
    a. Acetylcholine
    b. Norepinephrine
    c. Serotonin
    d. Glycine
    e. Endorphin

45. When a postsynaptic cell adds together information from synapses at different sites, this is called a(n)
    a. excitatory postsynaptic potential.
    b. inhibitory postsynaptic potential.
    c. spatial summation.
    d. temporal summation.
    e. action potential.

46. The most critical area in a neuron for "decision making" is the
    a. axon hillock.
    b. presynaptic terminal.
    c. postsynaptic terminal.
    d. cell body.
    e. synapse.

47. Some glial cells communicate with each other through
    a. a myelin sheath.
    b. axons.
    c. dendrites.
    d. gap junctions.
    e. tight junctions.

48. Anesthetics and alcohol can permeate the blood–brain barrier because
    a. they are small molecules.
    b. they are water-soluble.
    c. they are fat-soluble.
    d. they pass through gated channels.
    e. there are receptors for them on blood vessels.

49. The sodium–potassium pump
    a. needs energy to work.
    b. expels three sodium ions for every two potassium ions imported.
    c. works against a concentration gradient.
    d. *a* and *b*
    e. *a*, *b*, and *c*

50. A polysynaptic reflex
    a. is responsible for the knee jerk reaction.
    b. is responsible for the coordinated movement of opposing muscles.
    c. involves only motor neurons.
    d. sends messages to the brain about sensory input.
    e. does not involve interneurons.

42 Sensory Systems

Fill in the Blank

1. Equilibrium organs called **statocysts** use hair cells to respond to gravity.

2. Fish can detect pressure waves in water through their **lateral line** sensory system.

3. The three tiny bones of the middle ear transmit vibrations from the **tympanic membrane** to the oval window.

4. The inner ear is a long, coiled structure called the **cochlea.**

5. The molecule **rhodopsin** is the basis for photosensitivity.

6. A dense layer of photoreceptor cells at the back of the eye forms the **retina.**

7. Besides vertebrates, the other group of animals that has eyes that form images like cameras is the **cephalopod mollusk** group.

8. Mammals can alter the shape of the lens by contracting the **ciliary muscles.**

9. Just prior to transmission to the brain via the optic nerve, visual signals are processed by **ganglion cells.**

10. The phenomenon of emitting sounds and creating images from reflections of those sounds is **echolocation.**

11. **Action potentials** arriving in the visual cortex are interpreted as light; in the auditory cortex as sound; and in the olfactory cortex as smell.

12. An important characteristic of many sensors is that they can stop being excited by a stimulus that initially caused them to be active. This ability to ignore background or unchanging conditions while remaining sensitive to changes or new information is called **adaptation.**

13. The snake's forked tongue fits into cavities in the roof of its mouth that are richly endowed with olfactory sensors. Thus the snake is really using his tongue to **smell** his environment.

14. A common cause of **nerve deafness** (or **deafness**) is cumulative and permanent damage to the hair cells of the organ of Corti.

15. A hawk's vision is about eight times more sensitive than that of humans because it has about six times more sensors per square mm of **fovea** than humans.

16. A receptor potential that causes action potentials by causing voltage-gated sodium channels to open is called a **generator** potential.

17. Hair cells have a set of projections called **stereocilia** that look like organ pipes.

18. Many animals can focus sounds by moving their ear **pinnae** toward the sound.

Multiple Choice

1. A sensor converts a stimulus
   a. from one form of physical stimulus to another.
   b. **into some type of membrane potential.**
   c. from an action potential to a synaptic signal.
   d. by summing incoming action potentials.
   e. into different forms to be sent to the brain.

2. The magnitude of a receptor potential
   a. depends on the strength of the incoming action potential.
   b. remains high even after a long period of stimulation.
   c. is the same no matter what the type of stimulus.
   d. depends on the amount of neurotransmitter released.
   e. **affects the frequency of resulting action potentials.**

3. Which of the following is an example of adaptation of sensors?
   a. Going into deep sleep
   b. Discriminating different colors
   c. **Ignoring your shoes as you walk**
   d. Ignoring a boring lecture
   e. Detecting sound and light simultaneously

4. Pheromones are chemical signals that can signal, for example, from
   a. one neuron to another.
   b. the peripheral nervous to the central nervous system.
   c. prey to predator.
   d. parasite to host.
   e. **female to male.**

5. The chemosensory hairs that flies use to detect and taste food are located
   a. near the mouthparts.
   b. at the base of the wings.
   c. on the tip of the proboscis.
   d. **on the feet.**
   e. on the antennae.

6. How does a male silkworm moth locate a female at a distance?
   a. **He flies toward a chemical signal.**
   b. He flies toward a sound signal.
   c. He flies toward a shape like a female moth.

d. He emits a sound and she approaches.

e. He emits a chemical and she approaches.

7. Sensitivity of the sense of smell is proportional to
   a. the amount of mucus in the nose.
   b. the surface area of nasal epithelium.
   c. **the density of olfactory nerve endings.**
   d. the number of capillaries in the nose.
   e. the typical body temperature.

8. The greatest intensity of perceived smell comes from the
   a. enzyme that binds with the most odorant molecules.
   b. **odorant that binds to the most receptors.**
   c. most different types of odorant molecules.
   d. greatest threshold of depolarization.
   e. greatest number of odorant molecules entering the cell.

9. Which of the following are *not* mechanoreceptors?
   a. Stretch receptors
   b. Hair cells
   c. Pressure receptors
   d. **Olfactory receptors**
   e. Airflow receptors

10. Action potentials are generated in a mechanoreceptor when
    a. ion channels close in response to membrane distortion.
    b. **ion channels open in response to membrane distortion.**
    c. receptors bind chemicals in response to pressure.
    d. sensitivity of the membrane to neurotransmitters increases.
    e. signals from other mechanoreceptors are summated.

11. Which of the following is most descriptive of the Pacinian corpuscle?
    a. It has high two-point discrimination ability.
    b. It responds to steady pressure for a long time.
    c. **It has concentric layers of connective tissue.**
    d. It has extensive, long dendritic processes.
    e. It is extremely sensitive to light touch.

12. When the stereocilia of hair cells are bent, those hair cells then
    a. trigger muscle contraction.
    b. **release neurotransmitter.**
    c. undergo action potentials.
    d. contract their stereocilia.
    e. become less sensitive.

13. Stereocilia in hair cells in the canals of the fish lateral line
    a. **are embedded in gelatinous material.**
    b. are moved individually by pressure.
    c. are immobilized within a cupula.
    d. bend only after electrical stimulation.
    e. bend under the influence of gravity.

14. In invertebrates, the functional equivalent of the vertebrate vestibular apparatus is the
    a. hair cell.
    b. lateral line.
    c. Meissner's corpuscle.

d. **statocyst.**

e. stretch receptor.

15. Vertebrates rely on information from which sensory structure to keep their balance?
    a. Eustachian tube
    b. Otoliths
    c. **Semicircular canal**
    d. Statocyst
    e. Tympanic membrane

16. The middle ear serves which auditory function?
    a. **It converts air pressure waves into fluid pressure waves.**
    b. It converts fluid pressure waves into air pressure waves.
    c. It converts air pressure waves into nerve impulses.
    d. It converts fluid pressure waves into nerve impulses.
    e. It converts pressure waves into hair cell movements.

17. Hair cells in the ear that give auditory information are concentrated in the
    a. oval window.
    b. tympanic membrane.
    c. **basilar membrane.**
    d. semicircular canals.
    e. Reissner's membrane.

18. What is the basis for auditory response distinguishing different sound frequencies?
    a. The three bones of the middle ear respond differentially.
    b. The loops of the semicircular canals respond differentially.
    c. The oval window and round window respond differentially.
    d. **Different sections of the basilar membrane respond differentially.**
    e. Individual hair cells have different peak frequency responses.

19. The molecular mechanism for absorbing light into visual systems is
    a. shape change in the protein opsin.
    b. depolarization of the rhodopsin molecule.
    c. **isomerization of the molecule retinal.**
    d. a combination of opsin and retinal.
    e. oxidation of the rhodopsin molecule.

20. When an individual rod cell is stimulated with light, its membrane potential
    a. **becomes more negative than before stimulation.**
    b. becomes more positive than before stimulation.
    c. becomes more positive than other neurons.
    d. begins to generate action potentials.
    e. begins to reduce membrane polarization.

21. The activation of a rhodopsin molecule sets off a chain reaction that leads to
    a. the opening of a sodium channel.
    b. **the closing of a million sodium channels.**
    c. the formation of an activated phosphodiesterase molecule.
    d. the activation of a million transducin molecules.
    e. the activation of other rhodopsin molecules.

22. Arthropods have evolved compound eyes; these consist of large numbers of
    a. retinas.
    b. cones.
    c. eye cups.
    d. ommatidia.
    e. pupils.

23. The amount of light entering the eye can be decreased when
    a. an optician puts in atropine drops.
    b. the shape of the lens changes.
    c. the autonomic nervous system opens the pupil.
    d. the light is focused.
    e. the iris constricts.

24. Which of the following does *not* affect the focus of an image on the retina?
    a. The shape of the lens
    b. The shape of the retina
    c. The ligaments suspending the lens
    d. The ciliary muscles
    e. The elasticity of the lens

25. What causes the blind spot in the eye?
    a. An unusually high density of rod cells
    b. An unusually high density of cone cells
    c. The location where incoming light is focused
    d. The location where neural axons exit
    e. The location where photoreceptors are saturated

26. Which of the following is a difference between rods and cones?
    a. Cones are more sensitive at low light intensity.
    b. Rods are responsible for color vision.
    c. There are more cones than rods in the human retina.
    d. Strictly nocturnal animals have more cones than rods.
    e. Cones provide greatest acuity of vision.

27. Which of the following best describes the visual processing that occurs within the retina?
    a. After processing within individual photoreceptors, impulses travel to the brain.
    b. Light signals are processed in retinal cells just in front of the photoreceptors.
    c. Signals pass from photoreceptor cells directly to ganglion cells.
    d. Action potentials in photoreceptor cells affect bipolar cells before reaching ganglion cells.
    e. Membrane potential changes in a network of retinal cells activate ganglion cells.

28. The tapetum, a reflective layer at the back of the eye, functions so that
    a. nocturnal animals can see in the dark.
    b. backscattered light does not reduce acuity.
    c. light rays do not strike ganglion cells.
    d. specific wavelengths of light can enter.
    e. eyes can glow in the dark.

29. Which of the following describes the receptive field of a retinal ganglion cell?
    a. There is no overlap with the receptive fields of neighboring ganglion cells.
    b. The ganglion cell receives input from one rod or cone cell.

    c. The receptive field depends on connections of horizontal and bipolar cells.
    d. A signal at the edge of the receptive field is more significant than one from the center.
    e. The receptive field is defined by signals from neighboring ganglion cells.

30. Which of these statements about animal sensitivity is true?
    a. Humans and snakes can see ultraviolet rays, but insects cannot.
    b. Humans and snakes can see infrared rays, but insects cannot.
    c. Humans and insects can see infrared rays, but snakes cannot.
    d. Humans cannot detect ultraviolet rays, but insects can.
    e. Humans cannot detect ultraviolet rays, but snakes can.

31. In general, _____ are cells of the nervous system that transduce physical or chemical stimuli into signals that are transmitted to other parts of the nervous system for processing and interpretation.
    a. sensors
    b. effectors
    c. glial cells of the blood–brain barrier
    d. "nuclei" within the midbrain
    e. None of the above

32. Which of the following statements is true regarding the stimulation of receptor proteins within plasma membranes?
    a. The receptor proteins of mechanosensors, thermosensors, and electrosensors are themselves the ion channels.
    b. The receptor proteins of chemosensors and photosensors initiate biochemical cascades that eventually open and close ion channels.
    c. Receptor proteins are integral to the sensory process.
    d. a, b, and c
    e. a and b

33. Receptor potentials produce action potentials in two ways by generating action potentials within the sensors or by causing the release of _____ that induces an associated neuron to generate action potentials.
    a. hormone
    b. ATP
    c. interleukin
    d. neurotransmitter
    e. glucagon

34. That human beings tend to join bird-watching societies more often than mammal-smelling societies suggests
    a. humans are descended from the same reptilian ancestors that gave rise to birds.
    b. humans have a limited sense of smell.
    c. humans are unusual among mammals in that we depend more on vision than on olfaction.
    d. a and b
    e. None of the above

35. Attaching tiny magnets to the heads of homing pigeons may prevent them from finding their home roost. The hypothesis is that
    a. the magnets may simply be a physical impediment, and they cause homing failure by nonelectromagnetic factors.
    b. pigeons may home via some kind of magnetic mechanism.
    c. magnetic lines of force around Earth may serve as navigational cues to certain animals.
    d. *a*, *b*, and *c*
    e. **b and c**

36. Which of the following statements about sensors is *not* true?
    a. Most sensors are modified neurons.
    b. **Specific sensors can respond to all types of stimuli.**
    c. Changes in the stimulus strength lead to changes in the sensor's receptor potential.
    d. Sensors show the phenomenon called adaptation.
    e. How information from sensors is interpreted depends on where the information is sent.

37. Which of the following is *not* a necessary part of the definition of a sensor?
    a. A cell
    b. Transduces energy into action potentials
    c. **Causes the opening of sodium channels in the membrane**
    d. Has a receptor potential
    e. Can become insensitive to a source of continuous stimulation

38. As a male silkworm moth nears a female that is releasing bombykol,
    a. the female starts to release more bombykol.
    b. **the action potentials in the antennal nerve increase.**
    c. 200 bombykol-sensitive hairs are being stimulated per second.
    d. a larger number of the bombykol-sensitive hairs undergo adaptation.
    e. the receptor potential in bombykol-sensitive hairs is reduced.

39. Which of the following molecular events triggered by the binding of an odorant molecule to a receptor protein occurs third?
    a. Adenylate cyclase is activated
    b. **cAMP is produced from ATP**
    c. G protein is activated
    d. Second messenger binds sodium channel
    e. Membrane is depolarized

40. Which of the following statements about gustation is *not* true?
    a. **All taste buds are specialized for one of four distinct tastes: sweet, sour, salty, and bitter.**
    b. Stimulated taste sensors respond by releasing neurotransmitter.
    c. Microvilli increase the surface area of taste sensors.
    d. Individual taste buds are replaced every few days.
    e. Taste sensors form synapses with dendrites of neurons.

41. Which of the following statements about mechano-sensors is *not* true?
    a. Mechanosensors transduce mechanical forces into changes in receptor potential.
    b. If a mechanosensor is subject to increased distortion, more ion channels within its membrane open.
    c. **Usually, sensory nerves are not involved in circuits returning from mechanoreceptors.**
    d. If the receptor potential of a mechanosensor rises above a threshold, an action potential is propagated.
    e. Stimulus strength determines the rate of generated action potentials.

42. Which of the following skin receptors is specifically adapted for sensing pressure?
    a. Meissner's corpuscle
    b. **Pacinian corpuscle**
    c. Expanded-tip tactile receptors
    d. Neuron-wrapped hair follicles
    e. Bare nerve endings

43. Hair cells are associated with all but one of the following sensory systems. Select the exception.
    a. **Golgi tendon organ**
    b. Lateral line sensory systems
    c. Statocysts
    d. Vestibular apparatus
    e. Semicircular canals

44. Which of the following statements about the lateral line system is *not* true?
    a. The cupulae of the lateral line system each contain several hair cells.
    b. **Movement of mucus within the canal of the lateral line system displaces the cupulae.**
    c. The lateral line system allows the fish to sense the presence of other fish even in total darkness.
    d. The lateral line system responds to pressure waves in the surrounding water.
    e. The canal of the lateral system has numerous openings to the external environment.

45. Which of the following structures of the mammalian auditory system is responsible for signal amplification?
    a. The tympanic membrane
    b. **The ear ossicles**
    c. The oval window
    d. The cochlea
    e. The round window

46. Which of the following statements about the auditory functioning of the cochlea is *not* true?
    a. The flexing of the round window follows the flexing of the oval window in a delayed fashion.
    b. The hair cells on the organ of Corti are moved against the rigid tectorial membrane.
    c. The intensity of the sound determines how many hair cells will be stimulated.
    d. The frequency of the sound determines which hair cells will be stimulated.
    e. **Lower frequency sounds result in the stimulation of hair cells closer to the round window.**

47. Which of the following structures of the mammalian auditory system is involved in transduction of pressure changes into changes in receptor potential?
    a. The tympanic membrane
    b. The ear ossicles
    c. The oval window
    d. The organ of Corti
    e. The basilar membrane

48. Which of the following statements about rhodopsin is *not* true?
    a. Rhodopsin consists of a protein and a light-absorbing molecule.
    b. When 11-*cis*-retinal absorbs light, it becomes all-*trans*-retinal.
    c. Energy is required to convert all-*trans*-retinal into 11-*cis*-retinal.
    d. Rhodopsin is a transmembrane protein.
    e. All animals that can sense light do so using rhodopsin.

49. Which of the following statements about the functioning of a rod cell is *not* true?
    a. Rhodopsin is located in the stack of disks in the end of the rod cell that is farthest from the light source.
    b. A rod cell is a modified neuron.
    c. The resting potential of a rod cell in the dark is less negative than a typical neuron.
    d. A membrane potential of a rod cell exposed to light becomes more positive.
    e. The plasma membrane of a rod cell is fairly permeable to Na^+ ions.

50. Which of the following statements about the molecular events of photoreception is *not* true?
    a. A single photon of light can excite a rhodopsin molecule.
    b. A rod cell in the light would have most of its sodium channels open.
    c. cGMP keeps the sodium channels open.
    d. Activated phosphodiesterase (PDE) catalyzes the reaction hydrolyzing cGMP into GMP.
    e. Activated transducin activates PDE.

51. If a planarian is positioned relative to a stationary light source so that more light-sensitive cells in its right eye cup are stimulated than those in its left eye cup, the planarian will
    a. turn to the left.
    b. turn to the right.
    c. make a complete clockwise circle.
    d. make a complete counterclockwise circle.
    e. stop moving.

52. Which of the following statements about the compound eyes of arthropods is *not* true?
    a. Ommatidia are the optical units of compound eyes.
    b. The number of ommatidia per eye can vary greatly in different species.
    c. The light-sensitive cell of the ommatidium is called a rhabdom.
    d. A single sensory neuron leaves each ommatidium.
    e. Ommatidia have a lens system that focus the light on the rhabdom.

53. Which of the following statements about the functioning of the vertebrate eye is *not* true?
    a. The cornea is transparent so it can transmit light.
    b. The size of the pupil varies with light levels.
    c. The iris is under control of the autonomic nervous system.
    d. The lens focuses the image on the retina.
    e. The fovea is the part of the retina with the lowest density of photoreceptors.

54. Which of the following statements about accommodation is *not* true?
    a. Accommodation is the process whereby objects from different portions of the visual field are focused on the retina.
    b. Vertebrates like fish and reptiles accommodate by moving the lens relative to the retina.
    c. The suspensory ligaments keep the lens flattened.
    d. The ciliary muscles contract when you attempt to focus on a distant object.
    e. The ciliary muscles change the shape of the eye by counteracting the action of the suspensory ligaments.

55. Which of the following statements about vertebrate vision is *not* true?
    a. Visual acuity varies with the density of photoreceptors in the retina.
    b. Some vertebrates have two foveas per eye.
    c. A blind spot is always located where the optic nerve leaves the eye.
    d. Unlike a camera, the vertebrate eye does *not* project inverted images on the retina.
    e. Two major types of photoreceptor cells are found in the retina.

56. Which of the following statements about the vertebrate retina is *not* true?
    a. There are more rods than cones in the retinas of humans.
    b. There is a higher proportion of cones than rods in the foveas of humans.
    c. Cones give us our highest visual acuity.
    d. Our peripheral vision involves more rod than cone cells.
    e. Some animals that are entirely nocturnal have only cones in their retinas.

57. Which of the following statements about color vision is *not* true?
    a. There are three different types of cone cells in the retina.
    b. The absorption spectra of cone cells differ because of molecular differences in the retinal molecules.
    c. Rods do not contribute to color vision.
    d. There are several different genes that code for opsin molecules.
    e. The proportion of the several different cone cells that respond to light determines the color that is perceived.

58. Which of the following cell layers of the retina is responsible for the prevention of backscattering of light?
    a. Horizontal cell layer
    b. Pigmented cell layer

   c. Bipolar cell layer
   d. Amacrine cell layer
   e. Ganglion cell layer

59. Which of the following cell layers of the retina is responsible for producing action potentials that are sent to the brain?
   a. Horizontal cell layer
   b. Pigmented cell layer
   c. Bipolar cell layer
   d. Amacrine cell layer
   e. **Ganglion cell layer**

60. Which of these is the correct order of flow of information?
   a. Stimulus, ion channel, action potential, receptor protein, neurotransmitter release
   b. Stimulus, neurotransmitter release, action potential, ion channel, receptor protein
   c. **Stimulus, receptor protein, ion channel, neurotransmitter release, action potential**
   d. Stimulus, action potential, neurotransmitter release, receptor protein, ion channel
   e. Stimulus, receptor protein, action potential, neurotransmitter release, ion channel

61. Chemosensors
   a. are universal among animals.
   b. can cause strong behavioral responses.
   c. do not undergo adaptation.
   d. **a and b**
   e. a, b, and c

62. Meissner's corpuscles
   a. adapt very slowly.
   b. are present uniformly on skin surfaces.
   c. sense pressure.
   d. have concentric layers of connective tissue.
   e. **sense light touch.**

63. The Golgi tendon organ
   a. **causes muscles to relax and protects against tearing.**
   b. senses light touch.
   c. increases muscle contraction.
   d. is found in high densities on lips and fingertips.
   e. provides steady-state information about pressure.

64. Which of the following statements is *not* true of receptor potentials?
   a. They can cause the release of a neurotransmitter.
   b. They can cause an action potential.
   c. They are a change in membrane potential of the sensor cell.
   d. **They can spread over long distances.**
   e. They can be amplified.

65. Conduction deafness is caused by loss of function of
   a. the inner ear.
   b. the eustachian tube.
   c. **the tympanic membrane.**
   d. hair cells in the organ of Corti.
   e. the cochlea.

43 The Mammalian Nervous System: Structure and Higher Functions

Fill in the Blank

1. The basic functional unit of the brain is the **neuron**.

2. The brain and spinal cord together are called the **central nervous system**.

3. The **peripheral nervous system** is made up of a network of nerves throughout the body.

4. Vision, hearing, touch, and balance make up **afferent** information.

5. **Efferent** information is sent from the brain to the muscles and glands.

6. Efferent pathways that are involuntary are also called the **autonomic** division.

7. The nervous system can engage in many tasks at the same time. This is called **parallel** processing.

8. A spinal **reflex** occurs when afferent information is converted to efferent information without involvement of the brain.

9. A group of neurons that is anatomically or neurochemically distinct is called a **nucleus**.

10. The transfer of short-term memory to long-term memory is the function of the **hippocampus**.

11. The cortex is folded into ridges called **gyri** and valleys called **sulci**.

12. If a person is blind in one eye, he has difficulty discriminating **distances**.

13. Sleep researchers use an **electroencephalograph** to measure electric potential differences between neurons.

14. When experiences modify behavior, it is called **learning;** the ability of the brain to retain this is **memory**.

15. When two unrelated stimuli are linked to the same response, it is called a **conditioned** reflex.

16. **Procedural** memory cannot be consciously recalled. It is remembering how to perform a motor skill, such as using a computer keyboard.

17. A deficit in the ability to use or understand words is called an **aphasia**.

18. The autonomic nervous system is crucial to the maintenance of **homeostasis** in the body.

19. The **neocortex** is an evolutionarily more recent part of the telencephalon found in birds and mammals.

20. In the visual cortex, cells that receive information from both eyes are called **binocular** cells.

Multiple Choice

1. What is the brain mostly made of?
   a. neurons
   b. axons
   c. water
   d. gray matter
   e. white matter

2. Afferent information flows _____ the CNS, and efferent information flows _____ the CNS.
   a. to; to
   b. to; from
   c. from; to
   d. from; from
   e. from and to; to

3. The hindbrain develops into which structure?
   a. Medulla
   b. Pons
   c. Cerebellum
   d. All of the above
   e. None of the above

4. The forebrain develops into which structure?
   a. Telencephalon
   b. Diencephalon
   c. Cerebellum
   d. a and b
   e. All of the above

5. The thalamus and hypothalamus develop from
   a. the telencephalon.
   b. the diencephalon.
   c. the cerebrum.
   d. the cerebellum.
   e. the hindbrain.

6. Select the correct direction of flow of afferent information.
   a. Medulla, pons, midbrain, thalamus
   b. Medulla, pons, thalamus, midbrain
   c. Telencephalon, thalamus, pons, medulla
   d. Telencephalon, midbrain, pons, medulla
   e. Pons, medulla, thalamus, midbrain

7. In general, the more autonomic functions are found in the _____, and the more complex functions are found in the _____.
   a. forebrain; hindbrain
   b. telencephalon; diencephalon
   c. thalamus; hypothalamus
   d. midbrain; hindbrain
   e. hindbrain; forebrain

8. The biggest difference between the brains of humans and fish is the size of the
   a. medulla.
   b. cerebellum.
   c. cerebrum.
   d. diencephalon.
   e. thalamus.

9. In the spinal cord, the gray matter contains the _____, and the white matter contains the _____.
   a. axons; cell bodies
   b. cell bodies; axons
   c. dorsal horn; ventral horn
   d. ventral horn; dorsal horn
   e. afferent information; efferent information

10. Efferent nerves leave the spinal cord through the
    a. ventral roots.
    b. dorsal roots.
    c. gray matter.
    d. interneurons.
    e. ventral and dorsal horns.

11. Interneurons are found in the
    a. midbrain.
    b. thalamus.
    c. white matter of the spinal cord.
    d. gray matter of the spinal cord.
    e. telencephalon.

12. The function of the reticular system is to
    a. conduct impulses through the spinal cord.
    b. distribute information to its proper location in the forebrain.
    c. regulate the level of arousal of the nervous system.
    d. regulate physiological drives and emotion.
    e. transfer short-term memory to long-term memory.

13. The reticular system is located in the
    a. spinal cord.
    b. midbrain.
    c. hindbrain.
    d. b and c.
    e. a, b, and c.

14. The function of the limbic system is
    a. regulation of instincts, emotions, and physiological drives.
    b. to transfer short-term memory to long-term memory.
    c. to regulate levels of arousal.
    d. a and b
    e. a, b, and c

15. What structure constitutes the largest part of the human brain?
    a. Telencephalon
    b. Diencephalon
    c. Medulla
    d. Pons
    e. Cerebellum

16. The telencephalon is divided into two hemispheres covered by a sheet of gray matter called the
    a. hippocampus.
    b. reticular system.
    c. sulci.
    d. gyri.
    e. cerebral cortex.

17. The part of the brain that is involved in high-order information processing is called the
    a. association cortex.
    b. thalamus.
    c. limbic system.
    d. central sulcus.
    e. hippocampus.

18. A patient can see and hear, but cannot recognize a familiar face. Which lobe of the cerebrum is most likely damaged?
    a. Temporal
    b. Occipital
    c. Parietal
    d. Frontal
    e. Cerebellum

19. A patient cannot feel pressure applied to her hand even though nothing is wrong with her hand. Which lobe of the cerebrum is most likely damaged?
    a. Temporal
    b. Occipital
    c. Parietal
    d. Frontal
    e. Cerebellum

20. A patient cannot see motion, even though nothing is wrong with his eyes. Which lobe of the cerebrum is most likely damaged?
    a. Temporal
    b. Occipital
    c. Parietal
    d. Frontal
    e. Cerebellum

21. A patient suffers from a personality defect and cannot plan for future events. Which lobe of the cerebrum is most likely damaged?
    a. Temporal
    b. Occipital
    c. Parietal
    d. Frontal
    e. Cerebellum

22. A patient suffers from contralateral neglect syndrome, ignoring stimuli from the left side of the body. Which part of the brain has most likely been damaged?
    a. The left parietal lobe
    b. The right parietal lobe
    c. The left frontal lobe
    d. The right frontal lobe
    e. The left temporal lobe

23. The primary motor cortex is found in the _____ lobe and controls _____.
    a. parietal; feeling sensation
    b. parietal; movement
    c. temporal; movement
    d. frontal; movement
    e. frontal; feeling sensation

24. The primary somatosensory cortex is located in the _____ lobe and controls _____.
    a. parietal; feeling sensation
    b. parietal; movement
    c. temporal; feeling sensation
    d. temporal; movement
    e. frontal; feeling sensation

25. _____ have the largest brains in the animal kingdom, but _____ have the largest brain to body size ratio.
    a. Dolphins; humans
    b. Elephants; whales
    c. Elephants; humans
    d. Dolphins; humans
    e. Humans; rodents

26. The fight-or-flight mechanisms are the function of the _____ branch of the autonomic nervous system.
    a. sympathetic
    b. parasympathetic
    c. contralateral
    d. efferent
    e. afferent

27. The parasympathetic division controls
    a. fight-or-flight response.
    b. increased heart rate and blood pressure.
    c. increased digestion and decreased heart rate.
    d. increased release of epinephrine and glucose production.
    e. memory.

28. Preganglionic neurons use _____ as the neurotransmitter, while postganglionic neurons use _____ as the neurotransmitter.
    a. norepinephrine; acetylcholine
    b. acetylcholine; norepinephrine
    c. norepinephrine or acetylcholine; norepinephrine
    d. acetylcholine; norepinephrine or acetylcholine
    e. norepinephrine or acetylcholine; norepinephrine or acetylcholine

29. The anatomy of the sympathetic and parasympathetic divisions differs. For instance, the preganglionic neurons of the sympathetic division are located
    a. mostly in the brain stem.
    b. in the upper regions of the spinal cord.
    c. lined up like chains along the spinal cord.
    d. near the target organs.
    e. in the midbrain.

30. Visual information follows the pathway:
    a. Eye, thalamus, occipital lobe, optic chiasm
    b. Eye, thalamus, optic chiasm, occipital lobe
    c. Eye, optic chiasm, occipital lobe, thalamus
    d. Eye, thalamus, occipital lobe, optic chiasm
    e. Eye, optic chiasm, thalamus, occipital lobe

31. Complex cells in the visual cortex are stimulated by
    a. specific colors.
    b. bars of light with specific orientations and locations on the retina.
    c. bars of light with any orientation at a specific location on the retina.
    d. bars of light with specific orientations at any location on the retina.
    e. any type of light flashed on the retina.

32. Each retina sends _____ axons to the brain, received by about _____ neurons in the visual cortex.
    a. 1 million; 200 million
    b. 1 million; 2 million
    c. 100 million, 2 million

d. 100 million; 200 million
e. 200 million; 1 million

33. David Hubel and Torsten Wiesel, in their studies on vision, found
    a. many areas of the retina can stimulate a single cell in the visual cortex.
    b. cells in the visual cortex respond to a receptive field on the retina.
    c. cats can see bars of light at specific orientations only.
    d. simple cells make connections to complex cells in the visual cortex.
    e. visual information crosses over the optic chiasm.

34. During REM sleep _____ occurs.
    a. dreaming
    b. loss of motor output by the brain
    c. sleepwalking
    d. a and b
    e. a, b, and c

35. On the cellular level, sleep occurs because of
    a. hyperpolarization of the cells of the thalamus and cortex.
    b. depolarization of the cells of the thalamus and cortex.
    c. increased synaptic input between axons and neurons in the thalamus and cortex.
    d. desynchronization of electrical impulses in the cortex.
    e. decreased opening of potassium and calcium channels in the membrane of cortical cells.

36. Increased levels of adenosine in the brain probably contribute to
    a. REM sleep.
    b. muscle twitching during sleep.
    c. deep, slow-wave sleep.
    d. depolarization of cells in the thalamus and cortex.
    e. waking after sleep.

37. Long-term potentiation is
    a. increased sensitivity to an electrical stimulation.
    b. decreased sensitivity to an electrical stimulation.
    c. habituation to a stimulus.
    d. the application of a high frequency electrical stimulation.
    e. a decreased entry of calcium ions into the postsynaptic cell.

38. Long-term depression is
    a. increased sensitivity to an electrical stimulation.
    b. decreased sensitivity to an electrical stimulation.
    c. the application of a continuous low-level stimulus.
    d. the inability of neurons to fire an electrical impulse.
    e. the inability to retain long-term memory.

39. The Russian physiologist Ivan Pavlov and his dog became famous for demonstrating
    a. how short-term memory converts to long-term memory.
    b. associative learning.
    c. long-term potentiation.
    d. eye blink reflex.
    e. muscle twitches during REM sleep.

40. In eye blink reflex studies, the conditioned reflex was localized to a region in the
    a. medulla.
    b. spinal cord.
    c. **cerebellum.**
    d. thalamus.
    e. frontal lobe.

41. In order to try to cure a patient's severe epilepsy, both sides of his hippocampus were removed. An unfortunate side effect was
    a. he could no longer feel emotions.
    b. he lost his short-term memory.
    c. **he couldn't convert short-term memory into long-term memory.**
    d. he lost his immediate memory.
    e. he could no longer recognize faces.

42. _____ memory is almost perfectly photographic.
    a. **Immediate**
    b. Short-term
    c. Long-term
    d. Declarative
    e. Procedural

43. Short-term memory lasts about
    a. a few seconds.
    b. **5 or 10 minutes.**
    c. 20 or 30 minutes.
    d. about one hour.
    e. a few days.

44. Roger Sperry won the Nobel prize for his work on
    a. the hippocampus and memory loss.
    b. the conditioned reflex.
    c. eye puffs on rabbits.
    d. **lateralization of language to the left hemisphere.**
    e. REM sleep.

45. When the corpus callosum is cut in a young child's brain, a likely result would be
    a. language functions are learned by the right hemisphere.
    b. each hemisphere would take on a separate personality.
    c. loss of long-term memory.
    d. **a and b**
    e. a, b, and c

46. Damage to Broca's area results in
    a. loss of understanding language.
    b. **loss of, or poor, speech.**
    c. loss of ability to recognize faces.
    d. loss of ability to read.
    e. loss of hearing.

47. Damage to Wernicke's area results in
    a. **loss of understanding language.**
    b. loss of, or poor, speech.
    c. loss of ability to recognize faces.
    d. loss of hearing.
    e. loss of long-term memory.

48–51. From the following list, match the structure with its function in the following questions.
    a. Medulla
    b. Cerebellum
    c. Diencephalon
    d. Telencephalon

48. Contains the final relay station for sensory information going to the telencephalon **(c)**

49. Controls physiological functions such as breathing **(a)**

50. Orchestrates and refines motor commands **(b)**

51. Plays major roles in conscious behavior, learning, and memory **(d)**

52. Which of the following is *not* a function of the spinal cord?
    a. Generation of repetitive motor patterns
    b. Reflexes
    c. Conduction of motor impulses from the brain
    d. **Refinement of motor and behavioral processes**
    e. Conversion of afferent to efferent information

53. An agnosia is a disorder in which
    a. **the individual is aware of a stimulus but cannot identify it.**
    b. the individual can understand language but has poorly articulated speech.
    c. the individual ignores stimuli from the right side of the body.
    d. the individual cannot understand written or verbal language.
    e. the individual cannot convert short-term memory to long-term memory.

54. Norepinephrine _____ the heart rate, and acetylcholine _____ the heart rate.
    a. decreases; increases
    b. **increases; decreases**
    c. increases; increases
    d. decreases; decreases
    e. increases; neither increases nor decreases

55. The human brain has about
    a. 1 million neurons.
    b. 10 million neurons.
    c. 1 billion neurons.
    d. **10 billion neurons.**
    e. 100 billion neurons.

44 *Effectors*

Fill in the Blank

1. Microtubules are comprised of subunits of **tubulin** protein.

2. Intercalated discs join together cells in **cardiac** muscle tissue.

3. Huxley and Huxley proposed the **sliding-filament** theory of muscle contraction.

4. An action potential spreading through the muscle fiber causes a minimal contraction known as a **twitch**.

5. **Tetanus** is the maximum tension of a muscle contraction.

6. Simple **hydrostatic** skeletons consist of a fluid-filled cavity controlled by muscle movement.

7. Cartilage tissue consists of **collagen** fibers in a rubbery matrix of proteins and polysaccharides.

8. Cells that break down or reabsorb bone are the **osteoclasts**.

9. Muscles at joints are often arranged in antagonistic pairs of extensors and **flexors**.

10. A type of cellular effector, **nematocysts** are fired like miniature missiles by jellyfish to capture prey.

11. **Effectors** are adaptations that animals use to respond to information that is sensed, integrated, and transmitted by their neural and endocrine systems.

12. **Smooth** muscle provides the contractile forces for most of our internal organs, which are under control of the autonomic nervous system.

13. The **sarcoplasmic reticulum** is a highly specialized network of intracellular membranes of striated muscle cells that enables them to sequester and release calcium for muscle contraction.

14. The living cells of bone that are responsible for the remodeling of bone structure are called **osteoblasts**.

Multiple Choice

1. What mechanism is responsible for movements of cilia and flagella?
   a. Muscle contraction
   b. The microfilament system
   c. **The microtubule system**
   d. Crystal formation
   e. The skeletal system

2. Which of the following does *not* rely on ciliary movement?
   a. Movement of an egg through an oviduct
   b. Movement of particles through the airways of lungs
   c. Feeding movements of a paramecium
   d. Movement of particles over the gills of a clam
   e. **Amoeboid movement**

3. The movement of eukaryote flagella resembles
   a. turning a corkscrew.
   b. a swimmer's stroke.
   c. **cracking a whip.**
   d. an airplane propeller.
   e. rowing a boat.

4. The beating of cilia and flagella occurs by
   a. contraction of individual tubules.
   b. contraction of individual tubulin molecules.
   c. attraction of sliding filaments to each other.
   d. **moving dynein attachments between filaments.**
   e. contraction of dynein in individual tubules.

5. Which of the following substances is required for any movement of cilia or flagella?
   a. Actin
   b. **ATP**
   c. Auxin
   d. Proteolytic enzymes
   e. *a* and *b*

6. Which observation would suggest that axons grow by polymerization of tubulin subunits?
   a. Energy is expended in the process of chromosome movement.
   b. **Chemicals that block tubulin interactions block chromosome movement.**
   c. Chemicals that block dynein interactions block chromosome movement.
   d. Chromosomes move by sliding past one another.
   e. Chromosomes are individually attached to microtubules.

7. The extension of a pseudopod in amoeboid movement occurs by
   a. extension of actin microfilaments next to the membrane.
   b. contraction of the plasmagel next to the membrane.
   c. the changing of plasmagel to plasmasol, which expands and bulges outward.
   d. **microfilament-generated cytoplasmic streaming of the plasmasol.**
   e. rapid expansion of cell membrane and filling with plasmagel.

8. The resting potential of smooth muscle cells is
   a. not subject to forming action potentials.
   b. affected by stretching the cells.
   c. more negative than in most cells.
   d. unaffected by nearby potential change.
   e. nearly zero.

9. The striated appearance of skeletal muscle is due to the
   a. dark color of myosin.
   b. multiple nuclei per fiber.
   c. regular arrangement of filaments.
   d. dense array of microtubules.
   e. dense packing of ATP molecules.

10. An individual sarcomere unit consists of
    a. a stack of actin fibers.
    b. a stack of myosin units.
    c. overlapping actin and membrane.
    d. overlapping myosin and membrane.
    e. overlapping actin and myosin.

11. How do muscle fibers shorten during contraction?
    a. Individual protein filaments contract.
    b. More cross-bridges between filaments are formed.
    c. Arrays of filaments overlap each other.
    d. Protein filaments coil up more tightly.
    e. Subunits of protein polymers detach.

12. How do actin and myosin molecules interact?
    a. Globular myosin heads bind to actin filaments.
    b. Globular actin heads bind to myosin filaments.
    c. Other proteins connect between the two.
    d. Myosin filaments bend to connect to actin.
    e. Actin filaments bend to connect to myosin.

13. Why do muscles stiffen in rigor mortis when animals die?
    a. Without ATP, muscles cannot contract.
    b. Without ATP, actin and myosin cannot bind.
    c. Without ATP, actin and myosin cannot separate.
    d. ATP is required for synthesis of protein filaments.
    e. ATP forms cross-bridges between filaments.

14. Vertebrate skeletal muscles are excitable cells because
    a. they can be stimulated by ATP.
    b. they can be stimulated by an electric charge.
    c. they can secrete neurotransmitter.
    d. they possess voltage-gated sodium channels.
    e. they can attain a high level of activity.

15. What is the role of the sarcoplasmic reticulum in muscle contraction?
    a. It stores calcium ions for release during contraction.
    b. It surrounds and protects the muscle filaments.
    c. It provides sites of ATP synthesis.
    d. It depolarizes when stimulated by an impulse.
    e. It synthesizes actin and myosin filaments.

16. How does tropomyosin control the muscle contraction?
    a. It provides a bridge between actin and myosin.
    b. It provides a site where ATP can be utilized.
    c. Changing its position exposes myosin binding sites.
    d. It transmits electric charge to the filaments.
    e. Changing its shape opens membrane channels.

17. How can muscle fibers show a range of responses to different levels of stimulation?
    a. Each muscle fiber contraction is all or none.

b. Calcium ion availability sets an upper limit.
   c. A new contraction can occur only after resting condition is reached.
   d. Following a stimulation, the fiber stays contracted.
   e. Individual twitches in the same fiber can summate.

18. The legs of cross-country skiers and long-distance runners are likely to have
    a. almost all slow-twitch fibers.
    b. almost all fast-twitch fibers.
    c. about the same number of slow-twitch as fast-twitch fibers.
    d. more slow-twitch fibers.
    e. more fast-twitch fibers.

19. Fast-twitch skeletal muscle fibers, called white muscle, are characterized by
    a. high concentration of myoglobin.
    b. abundant mitochondria.
    c. rapid development of high tension.
    d. sustaining activity for a long time.
    e. higher oxygen requirements.

20. What is a feature of cardiac muscle that helps the heart withstand high pressures without tearing?
    a. The fibers branch and intertwine.
    b. The fibers are arranged in parallel.
    c. Each fiber has a single nucleus.
    d. The fibers have gap junctions between them.
    e. Some fibers have pacemaking functions.

21. The exoskeleton of clams contains
    a. calcium carbonate.
    b. cartilage.
    c. chitin.
    d. collagen.
    e. cuticle.

22. A disadvantage of an arthropod's exoskeleton is that
    a. it protects against abrasion.
    b. it prevents water loss.
    c. it provides attachment sites for muscles.
    d. it bends at the animal's joints.
    e. it is soft just after molting.

23. An advantage of an endoskeleton over an exoskeleton is that
    a. it provides muscle attachment sites.
    b. it provides protection.
    c. it grows as the animal grows.
    d. it supports the animal's weight.
    e. it gives structure to the animal.

24. What is the role of cartilage in the skeleton?
    a. To bear a heavy load
    b. To add flexibility
    c. To be lightweight
    d. To sustain vibrations
    e. To grow rapidly

25. Bone tissue consists of _____ cells in a matrix of _____.
    a. osteocyte; polysaccharide
    b. osteocyte; collagen and calcium phosphate
    c. osteocyte; myosin and actin
    d. collagen; calcium phosphate
    e. collagen; polysaccharide

26. How do osteoblasts cause bone to grow?
    a. They lay down new matrix until surrounded.
    b. They increase the number of cells within the matrix.
    c. **They lay down new matrix on bone surface.**
    d. They tear down old bone and deposit new bone.
    e. They tear down old cartilage and deposit bone.

27. Haversian systems consist of
    a. osteocytes that connect different cavities.
    b. osteoblasts that will give rise to osteoclasts.
    c. a meshwork of cancellous bone tissue.
    d. **cylindrical units surrounding a canal of blood vessels.**
    e. a reinforced system of bone that resists fracturing.

28. The role of tendons is to
    a. join two bones together.
    b. join two ligaments together.
    c. join bone and ligament together.
    d. join muscle and ligament together.
    e. **join muscle and bone together.**

29. Which of the following does *not* act as an effector?
    a. The poisonous secretion of pufferfish
    b. The mercaptan spray of skunks
    c. Pheromone secretion by butterflies
    d. **Sensory neuron stimulation**
    e. The secretion of saliva in the mouth

30. Which of the following functions is *not* governed by microfilaments?
    a. Formation of daughter cells following mitosis
    b. Support of intestinal cell microvilli
    c. **Skeletal muscle contraction**
    d. Changes in cell shape
    e. Phagocytosis

31. Whether a muscle contraction is strong or weak depends both on how many motor neurons to that muscle are firing, and the rate at which those neurons are firing. These two factors can be thought of as spatial _____ and temporal _____, respectively.
    a. transmission; transduction
    b. transduction; transmission
    c. organization; summation
    d. **summation; summation**
    e. coordination; summation

32. What function does the hydrostatic skeleton of invertebrates perform?
    a. Skeletal muscle contraction
    b. Skeletal muscle expansion
    c. Locomotion
    d. Propulsion
    e. **b, c, and d**

33. Which of the following joints is *not* found in humans?
    a. Ball and socket joint
    b. Hinge joint
    c. Pivot joint
    d. Saddle and ellipsoid joints
    e. **All of the above are found in humans.**

34. One group of invertebrate effectors that enables animals to change color is called
    a. nematocysts.
    b. **chromatophores.**

c. electroplates.
d. poison glands.
e. retractable stingers.

35. Which of the following general statements about effectors is *not* true?
    a. Any mechanism that an animal uses to respond is an effector.
    b. All effector action that involves movement is dependent on the interaction of microfilaments or microtubules.
    c. **Muscle action depends on the movement of microtubules.**
    d. Movement due to microfilaments and microtubules depends on the sliding action of long protein molecules.
    e. Effectors rely on energy made available by ATP.

36. Which of the following statements about cilia and flagella is *not* true?
    a. The cilium encounters the greatest resistance to movement during the power stroke.
    b. Flagella are longer and less numerous than cilia.
    c. The same molecular mechanism powers both the cilium and the flagellum.
    d. **The pattern of movement of cilium and flagellum is the same.**
    e. Flagella are used to create feeding currents in sponges and swimming in sperm of most animals.

37. In which of the following situations would you *not* expect to find cilia?
    a. Within the air passages of the lungs in humans
    b. Within the female reproductive tract in humans
    c. **On the surface of intestinal cells in the human digestive tract**
    d. On the surface of a clam gill
    e. On a filter feeding animal

38. Which of the following statements about the molecular mechanism powering both the cilium and the flagellum is *not* true?
    a. Nine pairs of microtubules are found in the axonemes of both cilia and flagella.
    b. The globular protein tubulin is polymerized to form the hollow, tubular microtubules.
    c. Radial cross-arms of microtubules consist of a molecule of dynein.
    d. A mechanoenzyme converts chemical energy of ATP into movement.
    e. **The microtubules in a cilium cannot slide past each other because at any given time some cross-arms link the microtubules together.**

39. What happens to axonemes that are isolated from the cell when all components are removed except for microtubules and the dynein cross-arms?
    a. The microtubules will flex normally if ATP and Ca^{2+} are available.
    b. **The microtubules will telescope past each other if ATP and Ca^{2+} are available.**
    c. The microtubules lose all motility even if ATP and Ca^{2+} are available.
    d. The microtubules will flex normally even without ATP and Ca^{2+}.
    e. The microtubules will telescope past each other even without ATP and Ca^{2+}.

40. The polymerization and depolymerization of tubulin is involved in helping to cause all of the following phenomena except one. Select the exception.
    a. Movement of chromosomes during mitosis and meiosis
    b. Growth of neurons during development
    c. Changing cell shape
    d. **Amoeboid movement**
    e. Movement of organelles

41. The dominant microfilament component in cells is
    a. **actin.**
    b. myosin.
    c. tubulin.
    d. dynein.
    e. troponin.

42. Microfilaments are involved in helping to cause all of the following phenomena except one. Select the exception.
    a. **Movement of chromosomes during mitosis and meiosis**
    b. The division of cytoplasm during animal cell mitosis
    c. Changing cell shape
    d. Amoeboid movement
    e. Phagocytosis and pinocytosis

43. Which of the following statements about amoeboid movement is *not* true?
    a. Plasmasol is located in the interior of the cell.
    b. Plasmagel contains a network of actin microfilaments that interact with myosin.
    c. The cell moves by flowing into its pseudopodia.
    d. **During amoeboid movement, the plasmasol contracts.**
    e. The pseudopod stops forming when the cytoplasm at the leading edge converts to gel.

44. Muscle cells are involved in helping to cause all of the following phenomena except one. Select the exception.
    a. Thrashing of a nematode
    b. Swimming of a jellyfish
    c. **Color change in chameleons**
    d. Peristalsis
    e. Flight of a bumblebee

45. Which of the following statements about smooth muscle is *not* true?
    a. Smooth muscle is under the control of the autonomic nervous system.
    b. **Smooth muscle cells are multinucleate.**
    c. Smooth muscle does not have a striated appearance when viewed with the microscope.
    d. Gap junctions are common in smooth muscle.
    e. Stretched smooth muscle will contract.

46. Which of the following events does *not* occur during muscle contraction?
    a. **The distance between Z lines increases.**
    b. The sarcomere shortens.
    c. The H zone is reduced.
    d. The I band is reduced.
    e. The area with both actin and myosin increases.

47. Which of the following statements about the molecular arrangement of actin and myosin in myofibrils is *not* true?
    a. A thin filament consists of actin and tropomyosin.
    b. Two chains of actin monomers are twisted into a helix.
    c. Two strands of tropomyosin lie in the grooves of the actin.
    d. **Troponin forms the head of the myosin molecule.**
    e. The myosin heads have ATPase activity and interact with the actin.

48. Starting with the arrival of an action potential at the neuromuscular junction, which of the following is the correct order of events?
    a. Calcium is released from the sarcoplasmic reticulum, an action potential travels down the T tubules, depolarization spreads through the T tubule, myosin binds actin
    b. **An action potential travels down the T tubules, depolarization spreads through the T tubule, calcium is released from the sarcoplasmic reticulum, myosin binds actin**
    c. An action potential travels down the T tubules, depolarization spreads through the T tubule, calcium is taken up by the sarcoplasmic reticulum, myosin binds actin
    d. An action potential travels down the T tubules, depolarization spreads through the T tubule, ATP binds to myosin, myosin binds actin
    e. A T tubule is depolarized, calcium is released from the sarcoplasmic reticulum, an action potential is created in the muscle cell, myosin binds actin

49. Which of the following statements about a muscle twitch is *not* true?
    a. A twitch is usually caused by a single action potential.
    b. A twitch is an all or none response.
    c. **During tetanus, the twitching of many muscle cells keeps the whole muscle in a state of tension.**
    d. During a twitch, the actin and myosin filaments return to their resting positions before the next stimulus.
    e. During a twitch, the Ca^{2+} are pumped back into the sarcoplasmic reticulum before the next stimulus.

50. During tonus
    a. different muscles are taking responsibility for maintaining posture.
    b. the muscle has generated maximum tension.
    c. the muscle will eventually exhaust its supply of ATP.
    d. **a small but changing number of motor units are active.**
    e. the maximum number of action potentials is being received by the muscle.

51. Fast- and slow-twitch fibers differ in all of the following ways except one. Select the exception.
    a. Number of mitochondria
    b. Amount of myoglobin
    c. Amount of glycogen and fat
    d. **Number of neuromuscular junctions**
    e. Strength of twitch

52. Which of the following phenomena does *not* involve a hydrostatic skeleton?
    a. Burrowing in earthworms
    b. Swimming in scallops
    c. Retraction of tentacles and head in the sea anemone
    d. Jet propulsion in squid
    e. Burrowing in clams

53. Which of the following statements about exoskeletons is *not* true?
    a. During molting, the old exoskeleton is reabsorbed and a new exoskeleton forms.
    b. A clam shell is a exoskeleton.
    c. The outer layer of the arthropod cuticle prevents water loss in terrestrial species.
    d. The inner layer of the arthropod exoskeleton is used for muscle attachment.
    e. The arthropod exoskeleton is a continuous covering that is thinned at the joints.

54. Which of the following statements about the endoskeletons of vertebrates is *not* true?
    a. The muscles are attached to a living support structure.
    b. The rib bones are part of the appendicular skeleton.
    c. Molting is unnecessary in animals with endoskeletons.
    d. Endoskeletons consist of bone and cartilage.
    e. At least two bones are required to form a joint.

55. Which of the following statements about cartilage and bone is *not* true?
    a. Some vertebrates are composed entirely of cartilage and no bone.
    b. Some vertebrates are composed entirely of bone and no cartilage.
    c. Only bone has a mineral salt added to it.
    d. The endoskeleton in many vertebrates serves as a reservoir of calcium.
    e. The principal protein in cartilage is collagen.

56. Long bones begin as _____ bones. At maturity the shaft of a long bone consists of a _____ of _____ bone.
    a. dermal; cylinder; compact
    b. dermal; solid rod; cancellous
    c. cartilage; cylinder; compact
    d. cartilage; cylinder; cancellous
    e. cartilage; solid rod; compact

57. Which of the following statements about skeletal systems as a series of joints and levers is *not* true?
    a. An extensor muscle moves a bone closer to the body.
    b. Antagonistic pairs include an extensor muscle and a flexor muscle associated with the same joint.
    c. Ligaments hold joints together.
    d. Tendons attach muscles to bones.
    e. Ligaments sometimes hold tendons in position.

58. If you were designing a skeletal lever system for maximum power, you would
    a. insert both the muscle and the load force close to the joint.
    b. insert both the muscle and the load force far from the joint.
    c. insert the muscle close to the joint and place the load far from the joint.
    d. insert the muscle far from the joint and place the load close to the joint.
    e. use a large muscle.

59. Which of the following mechanisms is *not* known to occur in chromatophore-mediated color adaptation?
    a. Rapid synthesis or destruction of pigment within chromatophores
    b. Dispersal or aggregation of pigment granules within chromatophores
    c. Shape change of chromatophores due to amoeboid movement
    d. Shape change of chromatophores due to interaction with muscle tissue
    e. All of the above

60. Glands are involved in all of the following, except
    a. intracellular communication.
    b. intercellular communication.
    c. communication between different individuals.
    d. defense.
    e. temperature regulation.

61. Which of the following statements about electric organs is *not* true?
    a. In some species production of electric fields is used for defense or prey capture.
    b. In some species production of electric fields is used for orientation.
    c. Electric organs are derived from nervous tissue.
    d. Nerves, muscles, and electric organs use the same mechanisms to produce electric potentials.
    e. Large electric currents are achieved by coordinating the discharge of the cells in the electric organ.

62. Which of these muscle types has gap junctions?
    a. Smooth
    b. Cardiac
    c. Skeletal
    d. a and b
    e. a, b, and c

63. Which of these muscle types is multinucleated?
    a. Smooth
    b. Cardiac
    c. Skeletal
    d. b and c
    e. a, b, and c

64. Which of these muscle types has regular arrangements of actin and myosin?
    a. Smooth
    b. Cardiac
    c. Skeletal
    d. a and b
    e. a, b, and c

65. Which of these muscle types uses calcium to trigger actin–myosin interactions for movement?
    a. Smooth
    b. Cardiac
    c. Skeletal
    d. a and b
    e. a, b, and c

66. Which of these muscle types has pacemaking function?
    a. Smooth
    b. Cardiac
    c. Skeletal
    d. a and b
    e. a, b, and c

67. _____ complexes with calcium and acts as a second messenger in muscle cells.
    a. Myosin
    b. Actin
    c. Troponin
    d. cAMP
    e. Calmodulin

68. Which type of joint is found in the elbow?
    a. Ball and socket
    b. Pivotal
    c. Hinge
    d. Saddle
    e. Plane

69. Which type of joint allows almost complete rotational movement?
    a. Ball and socket
    b. Pivotal

c. Hinge
d. Saddle
e. Plane

70. An annelid moves by
    a. stretching its longitudinal muscles and pushing the fluid-filled body cavity forward.
    b. forming pseudopods.
    c. ciliary movement.
    d. alternating contractions of longitudinal and circular muscles.
    e. contraction of circular muscles which puts pressure on the hydrostatic skeleton causing it to extend.

71. When troponin binds calcium,
    a. it allows tropomyosin to bind actin.
    b. ion channels open and sodium rushes into the muscle cells.
    c. it changes conformation, twisting tropomyosin and exposing the actin–myosin bind site.
    d. it changes conformation, exposing the ATP binding site and allowing the actin–myosin bond to break.
    e. calmodulin binds to calcium and starts a cascade by activating myosin kinase.

45 *Gas Exchange in Animals*

Fill in the Blank

1. External **gills** are highly branched folds of the body surface for gas exchange.

2. The surface openings of an insect's breathing tubes are on its **abdomen**.

3. The lungs of **birds** are the most efficient lungs of vertebrates.

4. In fish gas exchange systems, blood flows through gill lamellae in a direction opposite to water flow, a phenomenon known as **countercurrent exchange**.

5. In the mammalian lung, the amount of air being moved in normal breathing is called the **tidal volume**.

6. After extreme exhalation, the lungs and airways still contain a **residual volume** of air.

7. The oxygen-binding molecule abundant in muscle is **myoglobin**.

8. Hemoglobin consists of **four** protein chains linked together.

9. The notation P_{O_2} stands for the **partial pressure** of oxygen.

10. The shape of the oxygen dissociation curve of hemoglobin is said to be **sigmoid**.

11. An unusual feature of **birds'** lungs is that they expand and contract relatively little during a breathing cycle. They also contract during inhalation and expand during exhalation.

12. Energy to carry out essential functions comes in the form of ATP, which can be sustained only where **oxygen** gas is present.

13. There are no active transport mechanisms for respiratory gases. Therefore, **diffusion** is the only means by which respiratory gases are exchanged.

14. In Fick's law of diffusion, the term $(C_1 - C_2)/L$ is a **concentration gradient**.

15. Respiration is unique in birds. In addition to lungs, birds have **air sacs** at several locations in their bodies.

16. Mammalian lungs have two adaptive features: the production of mucus, and **surfactant**, a substance that reduces the surface tension of the liquid lining the insides of the alveoli.

17. **Ventilate** is what fish do when they move water over their gills, and **perfuse** when blood is moved across the internal side of gas exchange membranes.

Multiple Choice

1. Breathing provides the body with oxygen required to support the energy metabolism of all cells. Breathing also eliminates _____, one of the waste products of cell metabolism.
   a. carbon monoxide
   b. carbon dioxide
   c. carbon tetrachloride
   d. calcium carbonate
   e. carbonic acid

2. Which of the following explains why oxygen can be exchanged more easily in air than in water?
   a. The oxygen content of air is higher than that of water.
   b. Oxygen diffuses more slowly in water than in air.
   c. More energy is required to move water than air because water is denser.
   d. *a* and *b*
   e. *a*, *b*, and *c*

3. Aquatic _____ are in a double bind, because as the temperature of their environment increases, so does their demand for oxygen, but the oxygen content of water declines with increasing water temperatures.
   a. insects
   b. ectotherms
   c. endotherms
   d. plants
   e. None of the above

4. Which factor accounts for the efficiency of gas exchange in fish gills?
   a. Maximizing surface area
   b. Minimizing path length for diffusion
   c. Countercurrent flow of blood and water over opposite sides of the gas exchange surfaces
   d. *a* and *b*
   e. *a*, *b*, and *c*

5. Gas exchange in animals always involves
   a. cellular respiration.
   b. breathing movements.
   c. neural control of exchange.
   d. diffusion across membranes.
   e. active transport of gases.

6. The amount of gas exchange required to support an animal's metabolism increases with
   a. decreased oxygen in the air.
   b. decreased temperature.
   c. decreased body size.
   d. decreased water breathing.
   e. decreased body movement.

7. Rapid gas exchange is easier in air than in water because
   a. the oxygen content of water is higher than that of air.
   b. the carbon dioxide content of water is higher than that of air.
   c. oxygen diffuses more rapidly in water.
   d. water is more dense and viscous than air.
   e. more energy is required to move air than water.

8. Why do humans have difficulty breathing at high elevations?
   a. Oxygen makes up a lower percentage of the air there.
   b. The temperature is lower there.
   c. The barometric pressure is higher there.
   d. The partial pressure of oxygen is lower there.
   e. The air is dryer there.

9. Small insects may take a bubble of air underwater when they dive. The bubble can serve as an air tank for some time because
   a. as carbon dioxide increases, the bubble inflates.
   b. as oxygen is consumed, air pressure in the bubble decreases.
   c. as oxygen is consumed, oxygen diffuses into the bubble.
   d. the partial pressure of nitrogen remains constant.
   e. for each oxygen molecule used, a carbon dioxide molecule replaces it.

10. The respiratory system of insects consists of
    a. branched air tubes called spiracles that supply capillaries.
    b. branched air tubes called tracheae that supply capillaries.
    c. branching gill systems that end in openings called tracheae.
    d. branching gill systems that end in openings called spiracles.
    e. extensive layers of gas exchange tissue just under the exoskeleton.

11. The delicate gills of fish are supported by
    a. opercular flaps and gill arches.
    b. opercular flaps and gill filaments.
    c. gill filaments and gill arches.
    d. opercular flaps and a diaphragm.
    e. gill arches and a diaphragm.

12. Which of the following is *not* true about the air sacs of birds?
    a. They connect with each other.
    b. They occur in anterior and posterior pairs.
    c. They make the bird's respiratory system more efficient than a mammal's.
    d. They allow for one-way air flow through the lungs.
    e. They provide extra gas exchange surface.

13. Which of the following characterizes the lungs of mammals?
    a. Joined with air sacs
    b. Tidal ventilation system
    c. Complete emptying in exhalation
    d. Crosscurrent air flow
    e. Countercurrent air flow

14. Which of the following structures is the site of gas exchange in the lungs?
    a. Alveoli
    b. Bronchi
    c. Bronchioles
    d. Trachea
    e. *a* and *b*

15. Which factor maximizes the rate of gas exchange in mammals?
    a. Exceedingly high partial pressures of oxygen in the blood
    b. The larynx (voice box) opening into the trachea
    c. Enormous surface area for gas exchange
    d. The small increase in size during inhalation
    e. Production of mucus and surfactants

16. Surfactant produced by cells lining the alveoli serves to
    a. increase surface tension.
    b. increase stretching of the alveolar walls.
    c. assist muscular movements of breathing.
    d. reduce cohesion of surface molecules.
    e. reduce ciliary movement.

17. One effect of cigarette smoking on the lungs is to
    a. remove mucus from the bronchi.
    b. immobilize cilia lining the bronchi.
    c. speed air flow through the bronchi.
    d. increase surfactant action.
    e. increase surface area in alveoli.

18. The lower side of the lung cavity is formed by the
    a. diaphragm.
    b. esophagus.
    c. stomach.
    d. ribs.
    e. intercostal muscles.

19. The lungs expand in inhalation because
    a. the diaphragm contracts upward.
    b. the shoulder girdle moves upward.
    c. the volume of the thoracic cavity increases.
    d. lung tissue actively stretches.
    e. the lung tissue rebounds from exhalation.

20. The process of exhalation is begun mainly due to
    a. the contraction of intercostal muscles.
    b. the relaxation of muscles.
    c. the contraction of the diaphragm.
    d. complete collapse of the lung tissue.
    e. low pressure in the thoracic cavity.

21. If the thoracic wall is punctured, air leaking in will cause the lung to collapse. Thus in the normal, intact thoracic cavity
    a. punctured lungs will collapse.
    b. air pressure is the same as outside.
    c. a slight suction keeps lungs inflated.
    d. breathing movements keep lungs inflated.
    e. lung expansion is passive.

22. What part of the blood is most efficient at carrying oxygen?
    a. Blood plasma solution
    b. Blood plasma proteins
    c. Blood platelets
    d. Membrane molecules of red blood cells
    e. Hemoglobin molecules of red blood cells

23. Which of the following would increase the amount of oxygen diffusing from the lungs into the blood?
    a. **Increasing the binding rate of oxygen to hemoglobin**
    b. Decreasing the partial pressure of oxygen in the lung
    c. Increasing the partial pressure of oxygen in the blood
    d. Decreasing the red blood cell count of the blood
    e. Increasing the water vapor of air in the lungs

24. Each molecule of hemoglobin when fully saturated carries how many molecules of oxygen?
    a. One
    b. Two
    c. **Four**
    d. Twenty
    e. Nearly one hundred

25. Hemoglobin delivers oxygen to body cells from the red blood cells
    a. until the hemoglobin is depleted of oxygen.
    b. **until the partial pressures of oxygen in the cells are equivalent.**
    c. although the hemoglobin is never completely saturated.
    d. by releasing oxygen to cells with higher partial pressure of oxygen.
    e. until the fluid pressure in the red blood cells is lower.

26. In which of the following P_{O_2} environments will hemoglobin release its oxygen most easily?
    a. **30 mm Hg**
    b. 50 mm Hg
    c. 70 mm Hg
    d. 90 mm Hg
    e. 110 mm Hg

27. What is the advantage of having myoglobin in muscle cells?
    a. It binds with oxygen just as easily as hemoglobin does.
    b. It contributes to the dark color of flight muscle in birds.
    c. It increases the effectiveness of muscles used in short bursts.
    d. **It releases bound oxygen at lower P_{O_2} conditions than hemoglobin does.**
    e. It uses hemoglobin as a reserve source of oxygen.

28. What is the advantage of having fetal hemoglobin?
    a. Fetal hemoglobin pumps oxygen from adult hemoglobin to the fetus.
    b. Fetal hemoglobin releases oxygen at a lower partial pressure than adult hemoglobin does.
    c. **In the placenta, oxygen diffuses from adult hemoglobin to fetal hemoglobin.**
    d. At low oxygen pressures, adult hemoglobin is more likely to pick up oxygen.
    e. Fetal hemoglobin occurs in higher density in red blood cells than adult hemoglobin does.

29. In rapidly metabolizing tissues where conditions are more acidic, how does hemoglobin behave in comparison to its action in less acidic environments?
    a. **It releases more oxygen.**
    b. It releases less oxygen.

    c. It releases more carbon dioxide.
    d. It releases less carbon dioxide.
    e. There is no difference in behavior.

30. After humans acclimate to high altitudes, their hemoglobin
    a. is more often saturated with oxygen.
    b. **delivers more oxygen to the tissues.**
    c. is more concentrated in red blood cells.
    d. shifts its oxygen-binding curve to the left.
    e. resembles fetal hemoglobin in oxygen binding.

31. Most of the carbon dioxide in the blood is transported
    a. bound with hemoglobin.
    b. as dissolved gas.
    c. **as bicarbonate ion.**
    d. as carbonic acid.
    e. as calcium carbonate.

32. Carbonic anhydrase is an enzyme in red blood cells that catalyzes a reaction between carbon dioxide and
    a. bicarbonate.
    b. carbonic acid.
    c. hemoglobin.
    d. oxygen.
    e. **water.**

33. Neural control of breathing is in the
    a. cerebrum.
    b. diaphragm.
    c. **medulla.**
    d. olfactory lobe.
    e. spinal cord.

34. The breathing center initiates ventilation in response to
    a. a decrease in air pressure.
    b. a decrease in oxygen.
    c. **an increase in carbon dioxide.**
    d. the time since the last breath.
    e. the rate of gas exchange in the alveoli.

35. Nodes that are sensitive to oxygen and blood pressure levels are located in the
    a. **aorta and carotid arteries.**
    b. brain stem capillaries.
    c. hypothalamus.
    d. pulmonary artery.
    e. pulmonary vein.

36. Select the choice from the following list that makes an *incorrect* statement. It is easier to achieve a high level of gas exchange in air than water because
    a. water has a higher density and viscosity than air.
    b. **cold water can hold less oxygen than warm water, yet oxygen demand increases with temperature.**
    c. it takes more energy to breath water than air.
    d. gases diffuse more slowly in water than air.
    e. the O_2 content of a volume of water is less than an equal volume of air.

37. Which of the following is *not* an expected response from a fish to a drop in water temperature?
    a. **Metabolism increases**
    b. Amount of oxygen available increases
    c. Blood flow decreases
    d. Oxygen consumption decreases
    e. All of the above

38. A planet is discovered where the barometric pressure is 2,000 mm of mercury and the air is 15% oxygen. What is the partial pressure of oxygen at sea level on this planet?
    a. 15%
    b. 15 mm Hg
    c. 133 mm Hg
    d. 300 mm Hg
    e. 30%

39. Which of the following statements does not explain why diffusion of CO_2 from an animal is not as great a problem as the diffusion of O_2 into the animal?
    a. CO_2 is more soluble in water than O_2.
    b. Cellular respiration produces less CO_2 than the O_2 that is consumed.
    c. The CO_2 content of air is less than the O_2 content.
    d. The atmospheric partial pressure of O_2 is greater than the atmospheric partial pressure of CO_2.
    e. There is a greater CO_2 concentration gradient from the cell to the atmosphere than is true for oxygen.

40. Which one of the following organisms does *not* require both external ventilation of its respiratory surfaces with the medium containing O_2 and internal ventilation of its respiratory surfaces with blood?
    a. A crayfish
    b. A rabbit
    c. An insect
    d. A squid
    e. A bird

41. Which of the following statements about respiratory adaptations is *not* true?
    a. Internalization of respiratory surfaces leads to the need for ventilation.
    b. External respiratory surfaces are more subject to environmental damage.
    c. Internal gills and lungs are both inpocketings of the body surface.
    d. To function properly, gills need to be supported by the environment.
    e. Desiccation of the respiratory surface is more likely to occur in lungs than gills.

42. Which of the following statements about insect respiration is *not* true?
    a. Spiracles are the air tubes that carry the respiratory gases.
    b. Because oxygen diffuses much faster in air than water, the insect respiratory system is highly efficient.
    c. Aquatic insects that carry air bubbles with them underwater are using the air bubble as a source of oxygen.
    d. When oxygen reaches the ends of the insect respiratory tubes, it must dissolve in water before it can enter a cell.
    e. None of the above

43. Which of the following statements about the structure of the fish gill is *not* true?
    a. Blood in the efferent and afferent vessels flows in opposite directions.

b. Exchange of respiratory gases occurs within the lamellae of the gill filaments.
    c. The efferent and afferent vessels are the countercurrent flow system of the gill.
    d. The lamellae greatly increase the surface area for gas exchange.
    e. The opercular flaps enclose the gill chambers.

44. Which of the following adaptations is *not* seen in fish gills?
    a. A countercurrent exchange system
    b. Bidirectional ventilation of the gills
    c. Morphological features to increase the surface area available for gas exchange
    d. Morphological features to decrease the path length for diffusion of the respiratory gases
    e. Morphological features to maximize the efficiency of oxygen extraction

45. Including the trachea, during the breathing cycle of a bird, which of the following structures would be third to receive the inspired air? (Some structures can be involved twice.)
    a. Parabronchi of lungs
    b. Trachea
    c. Air capillaries
    d. Posterior air sacs
    e. Anterior air sacs

46. Bird and fish respiratory systems are similar because
    a. both employ a countercurrent exchanger.
    b. both have air sacs.
    c. both have unidirectional flow of the environmental medium over the gas exchange membranes.
    d. both are infoldings of the body.
    e. All of the above

47. In the following equation, which quantity represents the volume of your normal breath? Total lung capacity = residual volume + expiratory reserve volume + inspiratory reserve volume + tidal volume.
    a. Total lung capacity
    b. Residual volume
    c. Expiratory reserve volume
    d. Inspiratory reserve volume
    e. Tidal volume

48. Which of the following is *not* an adverse consequence of tidal breathing as seen in mammals?
    a. Precludes countercurrent gas exchange
    b. Dead space
    c. Residual volume
    d. Short diffusion path length
    e. Limits the O_2 concentration gradient between air and blood

49. How many of the following respiratory system structures do *not* contain cartilage? Pharynx, bronchi, alveoli, larynx, trachea, bronchioles
    a. 0
    b. 1
    c. 2
    d. 3
    e. 4

50. Which of the following statements about surfactants in the lungs is *not* true?
    a. Water coating the respiratory surfaces of the lungs creates surface tension.
    b. **Surfactants increase the cohesive forces between water molecules.**
    c. Certain cells in the alveoli produce surfactants.
    d. Surface tension influences the amount of effort required to inflate the lungs.
    e. Premature infants suffer respiratory distress syndrome if natural surfactants are not present within the lungs.

51. Which of the following statements about the mechanics of ventilation is *not* true?
    a. If the pleural cavity is punctured, the lung may collapse.
    b. As the diaphragm relaxes, air is expelled from the respiratory system.
    c. As the diaphragm relaxes, the pressure within the pleural cavities increases.
    d. Less energy is expended during the exhalation phase.
    e. **There is a slight positive pressure within the pleural cavities between breaths.**

52. Which of the following statements about the transport of respiratory gases by the blood is *not* true?
    a. **The amount of oxygen that can dissolve directly in plasma is sufficient to support the resting metabolism.**
    b. By binding oxygen, hemoglobin helps to maintain a steeper oxygen concentration gradient
    c. Internal ventilation of the gas exchange surfaces is dependent on blood flow.
    d. Each hemoglobin molecule can carry a maximum of four O_2 molecules.
    e. Molecules of O_2 are associated with the heme groups of hemoglobin.

53. Which of the following statements about the binding of oxygen by hemoglobin is *not* true?
    a. The percent of completely oxygen-saturated hemoglobin increases as P_{O_2} increases.
    b. The oxygen-binding curve is S-shaped.
    c. There is a narrow range of the oxygen-binding curve where oxygen-dissociation-association is rapid.
    d. **It requires a very low P_{O_2} to cause all four hemoglobin subunits to bind to O_2.**
    e. If one subunit of a completely oxygen-saturated hemoglobin loses its O_2, the affinity of the remaining subunits for their O_2 increases.

54. Which of the following statements about oxygen-binding curves is *not* true?
    a. **It requires a smaller decrease in P_{O_2} to go from 100% to 75% saturated hemoglobin than it does to go from 75% to 50%.**
    b. During normal metabolism, the body is only using the upper one-fourth of its oxygen-binding curve.
    c. The blood normally carries an enormous reserve of O_2.
    d. The steepest part of the oxygen-binding curve results in a great release of O_2 for only a modest drop in P_{O_2}.
    e. The P_{O_2} of blood only goes below about 40 mm Hg during extreme exertion.

55. Select the *incorrect* statement about myoglobin.
    a. Iron in the myoglobin molecule can bind to oxygen.
    b. Myoglobin is found in muscle cells.
    c. **Myoglobin has a lesser affinity for oxygen than hemoglobin does.**
    d. Diving mammals have high concentrations of myoglobin in their muscles.
    e. "White" and "dark" muscle differ in their myoglobin content.

56. If you plotted the oxygen-binding curves of the following molecules, which curve would be to the left of all the others?
    a. Fetal human hemoglobin
    b. Adult llama hemoglobin
    c. Adult human hemoglobin
    d. **Myoglobin**
    e. Fetal llama hemoglobin

57. If you plotted the oxygen-binding curves of the following hemoglobins subject to the pH levels shown, which curve would be to the right of all the others?
    a. Fetal human hemoglobin at pH 7.6
    b. Adult human hemoglobin at pH 7.6
    c. **Adult human hemoglobin at pH 7.2**
    d. Fetal human hemoglobin at pH 7.2
    e. Adult llama hemoglobin at pH 7.2

58. If you plotted the oxygen-binding curves of adult hemoglobins subject to the following conditions, which curve would be to the right of all the others?
    a. pH 7.6, little diphosphoglyceric acid
    b. pH 7.6, much diphosphoglyceric acid
    c. pH 7.2, little diphosphoglyceric acid
    d. **pH 7.2, much diphosphoglyceric acid**
    e. pH 7.4, little diphosphoglyceric acid

59. Most of the CO_2 is transported in the blood as _____, and it is located mainly in the _____.
    a. CO_2; plasma
    b. H_2CO_3; plasma
    c. **HCO_3^-; plasma**
    d. carboxyhemoglobin; erythrocytes
    e. HCO_3^-; erythrocytes

60. Due to the activity of carbonic anhydrase, within the lung there is a concentration gradient of CO_2 from the erythrocyte to the _____, and a concentration gradient of _____ from the plasma to the _____.
    a. plasma; HCO_3^-; plasma
    b. plasma; HCO_3^-; lung
    c. lung, CO_2; lung
    d. **lung; HCO_3^-; erythrocyte**
    e. lung; HCO_3^-; lung

61. The breathing rhythm is generated in the _____ and is influenced by variation in levels of _____ in the blood.
    a. **medulla; CO_2 and O_2**

b. medulla; CO_2
c. medulla; O_2
d. frontal lobe; CO_2 and O_2
e. frontal lobe; CO_2

62. Which of the following statements does *not* support the notion that the CO_2 level of the blood is a better feedback stimulus for ventilation than the O_2 level?
a. The P_{O_2} level of the blood fluctuates little.

b. Normally, hemoglobin maintains a reserve supply of oxygen.
c. **CO_2 is more highly soluble in plasma than O_2.**
d. Small changes in metabolism influence the P_{CO_2} levels directly.
e. The P_{CO_2} level of the blood fluctuates more than the P_{O_2} level.

46 *Circulatory Systems*

Fill in the Blank

1. From fishes to amphibians to reptiles to mammals and birds, the complexity and number of chambers in the heart increases. An important consequence of this increased complexity is the gradual separation of the circulatory system into two independent circuits–**pulmonary** and **systemic**.

2. The "lub-dub" sounds of the cardiac cycle are caused by **the shutting of the heart valves.**

3. A blood clot that becomes established within the lumen of an artery is termed a thrombus. A piece of a thrombus breaking loose can travel through the arteries and eventually become lodged in a vessel of small diameter. If this occurs in the brain, the resulting condition is called a **stroke**.

4. The interstitial fluid that accumulates outside the capillaries is returned to the heart by a separate system of vessels called the **lymphatic** system.

5. In a human being under normal conditions, the bone marrow can produce about two million **red blood cells, or erythrocytes,** per minute.

6. In open circulatory systems, fluid enters the **heart** through holes called ostia.

7. In humans, blood is pumped from heart to lungs and back through the **pulmonary circuit.**

8. The walls of large **arteries** have elastic fibers and can withstand high pressures.

9. The condition of **edema** results from accumulation of interstitial fluids and tissue swelling.

10. If too little blood is pumped to the brain, **fainting** will result.

11. Blood cells form from stem cells in the **bone marrow**.

12. **Platelets** are cell fragments responsible for blood clotting.

13. The brain's cardiovascular control center is in the **medulla**.

14. The **diving reflex** slows the heart when the individual is plunged into cold water.

15. A blockage in a coronary artery is called a coronary **thrombosis**.

16. The lymphatic system empties intercellular fluid into the **thoracic duct**, which is emptied into the superior vena cava.

17. Stem cells in the bone marrow that can produce all other types of blood cells are called **totipotent**.

Multiple Choice

1. A cardiovascular system is *not* necessary in the hydra because
   a. it is an aquatic animal.
   b. is has no skeleton.
   c. **it is only two cells thick.**
   d. it has tentacles to move water.
   e. it does not move rapidly.

2. In an open circulatory system,
   a. there is no heart.
   b. there is no blood.
   c. there are no blood vessels.
   d. blood flows out of the body.
   e. **blood flows between cells.**

3. An example of an animal with an open circulatory system is the
   a. bat.
   b. earthworm.
   c. hydra.
   d. **snail.**
   e. lizard.

4. How can insects be successful with open circulatory systems?
   a. They have multiple hearts.
   b. They have low activity levels.
   c. **Their gas exchange is not via the circulatory system.**
   d. Their nervous system is extensive and rapid.
   e. They have numerous valves in their vessels.

5. In vertebrates, exchange of substances between the blood and the interstitial fluids occurs in the
   a. arteries.
   b. arterioles.
   c. **capillaries.**
   d. veins.
   e. venules.

6. In the fish circulatory system, blood
   a. **moves from the muscular ventricle to the gills.**
   b. entering the aorta is under high pressure.
   c. leaving the heart moves to body tissues.
   d. leaving the gills is under high pressure.
   e. is received from the body into a muscular atrium.

7. Circulatory systems of adult amphibians demonstrate which of the following advantages over those of fishes?
   a. A three-chambered heart
   b. Heart metamorphosis
   c. **Partial separation of pulmonary and systemic circulation**

    d. A pocket of the gut that serves as an air bladder
    e. Separation of oxygenated from unoxygenated blood

8. The advantage of having a heart with two atria is that
    a. oxygenated and unoxygenated blood are separated.
    b. blood is pumped directly from heart to tissues.
    c. blood can be slowed down before going to tissues.
    d. the body can support a higher blood pressure.
    e. there are two muscular regions for pumping.

9. In which way are crocodilians different from other reptiles?
    a. They have a separate pulmonary circulation.
    b. They have lungs and no gills.
    c. They have an open circulatory system.
    d. They have more muscular atria.
    e. They have a four-chambered heart.

10. In the human heart, blood is pumped from the left ventricle into the
    a. right ventricle.
    b. left atrium.
    c. right atrium.
    d. pulmonary circuit.
    e. systemic circuit.

11. What vessel transports oxygenated blood from the lung into the heart?
    a. Pulmonary artery
    b. Pulmonary vein
    c. Superior vena cava
    d. Inferior vena cava
    e. Coronary artery

12. In the cardiac cycle, blood pressure is at a maximum when
    a. the atria are contracting during systole.
    b. the atria are contracting during diastole.
    c. the ventricles are contracting during systole.
    d. the ventricles are relaxing during systole.
    e. the ventricles are relaxing during diastole.

13. The heart beats because of cardiac muscle's unique properties, such as
    a. high resistance to outside stimulation.
    b. cell communication via gap junctions.
    c. low level of electric activity.
    d. external pacemaking system.
    e. rapid chemical communication.

14. The specific location of the heart pacemaker is the
    a. sinoatrial node.
    b. atrioventricular node.
    c. Purkinje fibers.
    d. bundle of His.
    e. ventricular mass.

15. The timing of the spread of the action potential from atrium to ventricle is controlled by the
    a. sinoatrial node.
    b. atrioventricular node.
    c. Purkinje fibers.
    d. bundle of His.
    e. ventricular mass.

16. The diameter of a capillary is about that of
    a. an arteriole.
    b. a nerve.
    c. a red blood cell.
    d. a valve.
    e. a venule.

17. As blood enters the capillaries from the arterioles,
    a. the blood pressure decreases and the osmotic potential decreases.
    b. the blood pressure decreases and the osmotic potential increases.
    c. the blood pressure increases and the osmotic potential decreases.
    d. the blood pressure decreases and the osmotic potential remains steady.
    e. the blood pressure increases and the osmotic potential remains steady.

18. Histamine causes swelling by
    a. making blood vessels expand and decreasing the permeability of capillaries.
    b. making blood vessels contract and decreasing the permeability of capillaries.
    c. making blood vessels expand and increasing the pressure in capillaries.
    d. decreasing the permeability of and increasing the pressure in capillaries.
    e. increasing the permeability of and increasing the pressure in capillaries.

19. What is the function of lymph nodes?
    a. They are sites where bacteria can grow.
    b. They are sites where inflammation occurs.
    c. They are sites regulating blood fluid.
    d. They are sites where new cells enter.
    e. They are sites of mechanical filtering.

20. What causes blood to move in the veins toward the heart?
    a. Gravity
    b. The contraction of venous walls
    c. Pulsing movement from the heart
    d. The contraction of nearby muscles
    e. Venous capacitance

21. The Frank–Starling law suggests that during exercise
    a. the flow of blood through the heart speeds up.
    b. the heart is stretched and contracts more forcefully.
    c. the veins expand to handle increased blood flow.
    d. the heart fills to capacity with blood at each beat.
    e. the valves in veins can no longer prevent backflow.

22. The hematocrit, or percentage of the blood made up of cells, in normal humans is about
    a. 10%.
    b. 25%.
    c. 40%.
    d. 70%.
    e. 90%.

23. The most abundant cells in the blood are
    a. erythrocytes.
    b. leukocytes.
    c. phagocytes.
    d. platelets.
    e. None of the above

24. The hormone erythropoietin is released by the kidney
    a. to remove old red blood cells from circulation.
    b. in response to high levels of oxygen in circulation.
    c. in response to low levels of hemoglobin.
    d. to stimulate production of red blood cells.
    e. to stimulate platelet formation.

25. The lifetime of an individual red blood cell is approximately
    a. 4 hours.
    b. 4 days.
    c. 4 weeks.
    d. 40 days.
    e. 4 months.

26. The mature red blood cells of humans
    a. consist of over 75% hemoglobin.
    b. contain no nuclei or endoplasmic reticulum.
    c. have a nearly spherical shape.
    d. have a relatively low surface area-to-volume ratio.
    e. have a rigid shape and low flexibility.

27. When is a platelet activated to initiate clotting?
    a. During a histamine reaction
    b. When leukocytes are activated
    c. When collagen fibers are encountered
    d. When blood flow rates drop
    e. In the presence of prothrombin

28. How does the circulating protein fibrinogen contribute to blood clotting?
    a. It polymerizes to form fibrin threads.
    b. It acts as a catalyst for thrombin activation.
    c. It binds to red blood cells.
    d. It activates the platelets.
    e. It triggers phagocytosis by the leukocytes.

29. Transferrin molecules in blood plasma function in
    a. carrying oxygen.
    b. carrying carbon dioxide.
    c. acting as a buffer.
    d. energy storage.
    e. carrying iron.

30. Which of the following will lead to autoregulatory increase of blood flow to a tissue?
    a. High blood pressure
    b. High carbon dioxide concentration
    c. High oxygen concentration
    d. High ATP concentration
    e. High glucose concentration

31. Which of the following statements about gastrovascular cavities is *not* true?
    a. Organisms with gastrovascular cavities tend to be only a few cells thick.
    b. Gastrovascular cavities are always highly branched.
    c. Gastrovascular cavities are filled with interstitial fluid.
    d. The diffusion path length in organisms with gastrovascular cavities is usually short.
    e. Large organisms do not usually have gastrovascular cavities.

32. Which of the following statements about open circulatory systems is *not* true?
    a. A pump is usually present.
    b. Blood is pumped out of openings in the heart called ostia.
    c. The blood is the interstitial fluid.
    d. Hemolymph bathes the tissues directly.
    e. Many mollusks and all arthropods have open circulatory systems.

33. Which of the following structures is *not* part of the circulatory system of an earthworm?
    a. Contractile vessels
    b. A major dorsal vessel
    c. Capillary beds in the lungs
    d. Capillary beds in each segment
    e. A major ventral vessel

34. Which of the following is *not* a function of the circulatory system of an insect?
    a. Transport of nutrients
    b. Removal of waste products
    c. Transport of hormones
    d. Internal ventilation of the gas exchange surfaces
    e. All of the above

35. Which of the following vertebrates have a three-chambered heart with both pulmonary and systemic circuits?
    a. Fish
    b. Lungfish
    c. Amphibians
    d. Crocodiles
    e. b and c

36. Which of the following structures in a vertebrate with a four-chambered heart would have blood with the lowest oxygen concentration?
    a. Pulmonary vein
    b. Pulmonary artery
    c. Left atrium
    d. Aorta
    e. Arteriole end of a capillary

37. When submerged, frogs receive most of their O_2 from capillaries within the skin. Which of the following structures would contain blood with the highest O_2 concentration in a submerged frog that is not able to breath using its lungs?
    a. Right atrium
    b. Left atrium
    c. Aorta
    d. Ventricle
    e. Pulmonary vein

38. Beginning in a systemic vein, which structure listed below would be encountered fourth during passage through the human circulatory system?
    a. Vena cava
    b. Aorta
    c. Left atrium
    d. Pulmonary vein
    e. Right ventricle

39. A blood pressure represented as 140/100 means that
    a. the pressure during ventricular contraction is 140, while the pressure during ventricular relaxation is 100.
    b. the pressure during ventricular contraction is 100, while the pressure during ventricular relaxation is 140.

   c. the pressure during ventricular contraction is 140, while the pressure during atrial contraction is 100.

   d. the pressure during atrial contraction is 140, while the pressure during ventricular contraction is 100.

   e. the diastolic pressure is 140 and the systolic pressure is 100.

40. During the cardiac cycle, the blood pressure in the aorta is at a minimum
   a. at the end of systole.
   b. **at the beginning of systole.**
   c. at the beginning of diastole.
   d. when the physician hears the "dub" sound with the stethoscope.
   e. None of the above

41. Which of the following heart regions is third in sequence of action potential propagation during a normal heartbeat?
   a. **Bundle of His**
   b. Atrioventricular node
   c. Purkinje fibers
   d. Sinoatrial node
   e. Ventricular muscle cells

42. Which of the following statements concerning the heartbeat is true?
   a. Only the ventricles contract together.
   b. Only the atria contract together.
   c. **The atria contract together and the ventricles contract together.**
   d. Only the atrium and ventricle on the right side of the heart contract together.
   e. Only the atrium and ventricle on the left side of the heart contract together.

43. Gap junctions
   a. electrically isolate cardiac muscle cells.
   b. **allow large numbers of cardiac muscle cells to contract in unison.**
   c. provide enough strength to withstand the large pressure generated by the ventricles.
   d. are only found in the cardiac muscle cells of the ventricles.
   e. are especially abundant in the muscles between the atria and ventricles.

44. The blood vessels with the greatest total cross-sectional area are the
   a. arteries.
   b. arterioles.
   c. **capillaries.**
   d. venules.
   e. veins.

45. The _____ have blood with the lowest pressure.
   a. arteries
   b. arterioles
   c. **capillaries**
   d. venules
   e. veins

46. In the electrocardiogram (EKG) of a normal cardiac cycle, the wave called T corresponds to
   a. depolarization of the atria.
   b. depolarization of the ventricles.
   c. repolarization of the atria.
   d. **repolarization of the ventricles.**
   e. More than one of the above

47. Blood pressure is _____ than osmotic pressure at the arterial end of a capillary bed, and the process of _____ occurs there.
   a. greater; reabsorption
   b. **greater; filtration**
   c. lesser; reabsorption
   d. lesser; filtration
   e. lesser; edema

48. Edema can be caused by an increase in _____ pressure in capillaries, as in _____ or a decrease in _____ pressure, as in _____.
   a. **blood; inflammation; osmotic; kwashiorkor**
   b. blood; kwashiorkor; osmotic; inflammation
   c. osmotic; inflammation; blood; kwashiorkor
   d. osmotic; kwashiorkor; blood; inflammation
   e. osmotic; thrombosis; blood; kwashiorkor

49. Which of the following statements about the lymphatic system is *not* true?
   a. The lymphatic system conducts lymph from the lymph capillaries to the thoracic duct, where it enters the circulatory system.
   b. **Lymph is identical to blood except it does not contain erythrocytes.**
   c. Lymphoid capillaries are distributed throughout the body.
   d. Contraction of skeletal muscles propels the lymph.
   e. Lymph nodes are an important part of the defense system that combats infections.

50. All of the following except one are mechanisms that facilitate return of venous blood to the heart. Select the exception.
   a. Fainting
   b. Contraction of skeletal muscles
   c. Breathing
   d. **Contraction of smooth muscle in most of the veins of the body**
   e. Valves in veins

51. The blood vessels with the greatest capacity to store blood are the
   a. arteries.
   b. arterioles.
   c. capillaries.
   d. venules.
   e. **veins.**

52. Which of the following statements about blood cells is *not* true?
   a. There are at least 500 times as many red as white blood cells.
   b. Erythrocytes are anucleate in mammals.
   c. The biconcave shape of the erythrocyte creates a large surface area-to-volume ratio.
   d. **Leukocytes and erythrocytes are restricted to the circulatory and lymphatic vessels.**
   e. Leukocytes flow with the blood but can also move by amoeboid motion.

53. Which of the following circulatory components is *not* normally found circulating in the plasma?
   a. Platelets
   b. **Thrombin**
   c. Albumin
   d. Fibrinogen
   e. Transferrin

54. The purpose of precapillary sphincters is
    a. to increase blood pressure to the arteries.
    b. to decrease blood pressure to the veins.
    c. **to shut off the supply of blood to the capillary bed.**
    d. to increase blood pressure to the capillaries.
    e. to provide blood to smooth muscles surrounding capillary beds.

55. Which of the following neurotransmitters or hormones does *not* cause contraction of smooth muscle?
    a. Norepinephrine
    b. Epinephrine
    c. Angiotensin
    d. Vasopressin
    e. **Acetylcholine**

56. Which one of the following responses characterizes the diving reflex?
    a. **Reduction in the heartbeat rate**
    b. Increase in systolic pressure
    c. Increase in diastolic pressure
    d. Tachycardia
    e. Dilation of arteries to most organs

57. Advantages of closed circulatory systems over open circulatory systems include which of the following?
    a. Exchange occurs more rapidly.
    b. Closed systems can direct blood to specific tissues.
    c. Cells and large molecules can be kept separate from the animal's intercellular material.
    d. Closed circulatory systems can support higher levels of metabolic activity.
    e. **All of the above**

58. In the evolution of the vertebrate circulatory system, the _____ is the organism that reveals the transition step leading to separate pulmonary and systemic circuits.
    a. **lungfish**
    b. ancient ostracoderm
    c. bird
    d. amphibian
    e. reptile

59. _____ and _____ are terms that describe the contraction and relaxation, respectively, of the ventricles in mammals.
    a. Diastole; systole
    b. **Systole; diastole**
    c. Systole; disystole
    d. Diastole; disystole
    e. None of the above

60. Which of the following is *not* a risk factor for developing atherosclerosis?
    a. A high fat and high cholesterol diet
    b. Smoking
    c. **Exercise**
    d. Sedentary living
    e. Hypertension

61. The condition kwashiorkor is caused by
    a. severe carbohydrate starvation.
    b. **severe protein starvation.**
    c. endoparasites.
    d. growth hormone deficiency.
    e. None of the above

62. When a piece of thrombus breaks loose and lodges in a smaller vessel, blocking the flow of blood, it is called a(n)
    a. thrombosis.
    b. infarction.
    c. **embolism.**
    d. atherosclerosis.
    e. hematocrit.

63. When whole blood is centrifuged, it is separated into a liquid component called _____ and a bottom layer of _____.
    a. hematocrit; cells
    b. **plasma; cells**
    c. plaque; cells
    d. hematocrit; plasma
    e. packed cell volume; plasma

64. The function of leukocytes is
    a. **defense against infection.**
    b. transport of respiratory gases.
    c. blood clotting.
    d. to distribute nutrients to tissues.
    e. to produce all the different types of blood cells.

65. Which of these hormones causes constriction of the blood vessels?
    a. Epinephrine
    b. Vasopressin
    c. Angiotensin
    d. b and c
    e. **a, b, and c**

66. The purpose of stretch sensors is
    a. to detect atherosclerosis.
    b. **to detect changes in blood pressure and relay to the brain.**
    c. to constrict arteries when available oxygen is low.
    d. to detect when a flexor muscle is stretched too far and pump more blood to the muscle.
    e. None of the above

67. Which of these factors contributes to increased blood flow to tissues?
    a. Low oxygen concentration
    b. Increased concentration of lactate
    c. High concentration of carbon dioxide
    d. a and b
    e. **a, b, and c**

47 Animal Nutrition

Fill in the Blank

1. The teeth of mammals are covered with a hard material called **enamel**.

2. **Hydrolytic** enzymes break down macromolecules into their monomeric units by adding water.

3. Many digestive enzymes are produced in an inactive **zymogen** form.

4. A wave of smooth muscle contraction called **peristalsis** pushes food along the digestive tract.

5. In addition to serving as an endocrine gland, the pancreas secretes the digestive enzyme precursor **trypsinogen**.

6. Many species of **bacteria** live in the human large intestine.

7. The initial digestion of protein in the vertebrate digestive tract begins in the **stomach**.

8. The **liver** is the organ that manages most of the balance between circulating and storing nutrients.

9. The liver converts pyruvic acid and other molecules to glucose by the process of **gluconeogenesis**.

10. The regulation of blood glucose is accomplished mainly by hormones produced by the **pancreas**.

11. Most plants, some monerans, and some protists are autotrophs. Animals, however, are **heterotrophs**, because they derive both their energy and their structural molecules from their food.

12. Most animals process their food through **extracellular** digestion in a digestive cavity called a gut.

13. If 1 kcal = 4.184 joules, then the number of kcal found in a diet consisting of 13,398 joules is **3,200** kcal.

14. From acetyl units obtained from carbohydrates or fats, we can synthesize almost all the lipids required by the body, but we must have a dietary source of essential **fatty acids**, notably linoleic acid.

15. Bile prevents fat droplets from **aggregating, or clumping together,** so that the maximum surface area is exposed to lipase action.

16. Crabs and earthworms that actively feed on dead organic matter are called **detritivores**.

17. An unusual feature of the vertebrate digestive tract is that it has its own intrinsic **nervous system**.

Multiple Choice

1. The order in which stored fuels are utilized during starvation is
   a. fats, then glycogen, then proteins.
   b. glycogen, then proteins, then fats.
   c. proteins, then fats, then glycogen.
   d. fats, then proteins, then glycogen.
   e. **glycogen, then fats, then proteins.**

2. Why are certain amino acids called essential amino acids?
   a. They are required for making protein.
   b. **They cannot be made from other amino acids.**
   c. They are universally needed by all animals.
   d. They are essential as an energy source.
   e. They are required for making nucleic acids.

3. Why are vitamins essential nutrients for cells?
   a. Vitamins are used as an energy source.
   b. Vitamins are used to digest foods.
   c. Animals cannot synthesize any vitamins.
   d. Some vitamins are fat soluble.
   e. **Vitamins act as cofactors with enzymes.**

4. Organisms that derive both their energy and molecular nutrients from other organisms are called
   a. autotrophs.
   b. herbivores.
   c. **heterotrophs.**
   d. photosynthetic.
   e. protists.

5. The energy content of food is described in terms of calories because
   a. the amount of energy in food depends upon the temperature.
   b. food heats up as it is being digested.
   c. **the energy in food ultimately becomes heat.**
   d. heat is the main product of digestion.
   e. heat is the main product of respiration.

6. The major form of stored energy in animal bodies is
   a. protein, because it is a long-term energy storage form.
   b. glycogen, because it breaks down into readily usable carbohydrates.
   c. glycogen, because it is lightweight.
   d. **fat, because it has the highest energy content per gram.**
   e. fat, because it is readily stored with water.

7. Vitamin C has all except which of the following properties?
   a. Functions in collagen production
   b. Important for healthy skin
   c. An essential vitamin for humans
   d. A fat-soluble vitamin
   e. Excess will be excreted

8. Vitamin D is obtained by different people in various ways, including
   a. from high-latitude sunlight by dark-skinned people.
   b. from tropical sunlight by dark-skinned people.
   c. from sunlight by Eskimos.
   d. by absorption of sunlight by protein molecules.
   e. from water-soluble components of foods.

9. The nutritional disease kwashiorkor results from
   a. protein deficiency.
   b. vitamin deficiency.
   c. calorie deficiency.
   d. overdose of fat-soluble vitamins.
   e. overdose of thyroxine.

10. Which of the following diseases is *not* due to a vitamin deficiency?
    a. Pellagra
    b. Rickets
    c. Goiter
    d. Pernicious anemia
    e. Beriberi

11. All of the following represent feeding adaptations of carnivores, except
    a. spider webs.
    b. rattlesnake venom.
    c. bat echolocation.
    d. elephant trunks.
    e. jellyfish tentacles.

12. How does a filter feeder acquire food items?
    a. By using poison to restrain prey
    b. By using claws and jaws to restrain prey
    c. By ingesting mud and extracting particles
    d. By filtering food substances from blood
    e. By extracting particles suspended in water

13. A mammal with a diet of grain and leaves would be expected to have what kind of teeth?
    a. Prominent canine teeth and small molars
    b. Prominent molars and small incisors
    c. Prominent molars and canine teeth
    d. A balanced set of incisors, molars, and canines
    e. Prominent canine teeth and small incisors

14. For most animals, digestion of food occurs
    a. intracellularly.
    b. in the gastrovascular cavity.
    c. in the coelomic body cavity.
    d. in the midgut.
    e. in the crop.

15. Which of the following animals need no digestive system?
    a. Jellyfish and hydras
    b. Intestinal tapeworms
    c. Clams and mussels
    d. Spiders and sea stars
    e. Baleen whales

16. Which of the following is true about digestive systems in all animals?
    a. Food in the gut is outside the animal.
    b. There are two openings to the tract.
    c. Animals break food into smaller pieces before ingestion.
    d. All digestion occurs inside the gut.
    e. The digestive tract has specialized segments.

17. Which of the following structures does *not* serve to increase surface area for nutrient absorption?
    a. Intestinal villi
    b. Earthworm typhlosole
    c. Shark spiral valve
    d. Bird gizzard
    e. Cell microvilli

18. An endolipase is an enzyme that breaks down
    a. carbohydrates.
    b. nucleic acids.
    c. proteins.
    d. molecules by cutting in the middle.
    e. molecules by snipping at the ends.

19. Why is the lining of the intestine *not* usually attacked by digestive enzymes?
    a. It is covered with mucus.
    b. Enzymes there are in their inactive form.
    c. The pH there is close to neutral.
    d. A layer of dead cells lines the intestine.
    e. Digestive enzymes are in low concentration.

20. In the small intestine, the blood vessels that carry away absorbed nutrients lie in what layer?
    a. Microvilli
    b. Mucosa
    c. Submucosa
    d. Circular muscle layer
    e. Longitudinal muscle layer

21. Movement of food from the stomach into the esophagus is normally prevented by
    a. peristalsis.
    b. reverse peristalsis.
    c. the pyloric sphincter.
    d. a sphincter.
    e. the pharynx.

22. The major enzyme produced by the stomach is
    a. amylase.
    b. chyme.
    c. mucus.
    d. pepsin.
    e. trypsin.

23. What activates the inactive form of stomach enzymes?
    a. Activating enzymes
    b. ATP
    c. Low pH
    d. The appropriate substrate molecule
    e. The presence of water

24. Most absorption of nutrients in the digestive tract is in the
    a. stomach.
    b. small intestine.
    c. large intestine.
    d. liver.
    e. pancreas.

25. Bile aids in the breakdown of lipids by
    a. hydrolyzing lipids.
    b. activating hydrolytic enzymes.
    c. aggregating droplets of lipids.
    d. emulsifying lipids.
    e. making lipids water-soluble.

26. What neutralizes the acidic chyme in the small intestine?
    a. Bicarbonate from the pancreas
    b. Buffers from the jejunum
    c. Bile from the liver
    d. Trypsin activation
    e. A variety of zymogens

27. How does sodium cotransport facilitate glucose absorption?
    a. Active transport of sodium aids in salt uptake.
    b. When sodium diffuses into cells, the carrier also binds a glucose.
    c. When sodium is pumped into cells, glucose moves out.
    d. Sodium and glucose both diffuse into cells from the gut.
    e. Glucose is actively transported into the gut.

28. The major function of the colon or large intestine is
    a. digestive breakdown of foods.
    b. nutrient absorption of foods.
    c. housing parasitic bacteria.
    d. secretion of bile and enzymes.
    e. reabsorption of water.

29. The hormone secretin is a chemical message
    a. secreted by the pancreas.
    b. secreted by the stomach.
    c. whose release is stimulated by the nervous system.
    d. that triggers the intestine to release enzymes.
    e. that triggers pancreatic secretion.

30. Insulin is released by the pancreas when
    a. blood glucose falls.
    b. blood glucagon falls.
    c. blood glucose rises.
    d. blood glucagon rises.
    e. blood insulin falls.

31. The effects of glucagon include
    a. stimulating glucose uptake into cells.
    b. stimulating liver cells to break down glycogen.
    c. controlling blood glucose in the absence of insulin.
    d. stimulating cells to store energy as fat.
    e. None of the above

32. Which of the following is true?
    a. Carbohydrates are stored in the liver and in muscle as glycogen. The total glycogen stores are usually not more than the equivalent of a day's energy requirements.
    b. Fat is an important form of stored energy.
    c. Fat has the highest energy content per gram.
    d. Protein is the most important energy storage component.
    e. a, b, and c

33. Digestive enzymes are classified according to the substances they hydrolyze and where the enzyme cleaves the given molecule. An endopeptidase hydrolyzes a _____ at _____ site along the length of the molecule.
    a. carbohydrate; an external
    b. protein; an internal
    c. fat; an internal
    d. carbohydrate; an internal
    e. peptide; an external

34. The gut of an animal is often described as an elongated tube consisting of four layers of different cell types. These layers are
    a. submucosa, cartilage, mucosa, and endoplasmic reticulum.
    b. smooth muscle layers, submucosa, mucosa, and cartilage.
    c. cartilage, smooth muscle layers, mucosa, and circular muscle.
    d. mucosa, submucosa, circular muscle, and longitudinal muscle.
    e. cartilage, mucosa, endoplasmic reticulum, and submucosa.

35. _____ ingest both plant and animal food and process their food through _____ digestion.
    a. Predators; intracellular
    b. Omnivores; intracellular
    c. Carnivores; intracellular
    d. Herbivores; extracellular
    e. Omnivores; extracellular

36. Which of the following statements about undernourishment is *not* true?
    a. One-fifth of the world's population is undernourished.
    b. Several weeks of fasting are required to deplete glycogen reserves.
    c. Self-imposed starvation is called anorexia nervosa.
    d. The loss of blood proteins during starvation leads to edema.
    e. During starvation, fat reserves are metabolized before body protein.

37. Herbivores have _____ essential amino acids than carnivores and consequently are less likely to suffer from _____ than carnivores.
    a. fewer; undernourishment
    b. fewer; malnourishment
    c. more; undernourishment
    d. more; malnourishment
    e. more; overnourishment

38. Which of the following statements about the essential amino acids is *not* true?
    a. Most animals have some essential amino acids.
    b. Ingesting a surplus of one essential amino acid cannot compensate for a shortage of another.
    c. Some plant foods such as legumes supply all eight of the essential amino acids in humans.
    d. Acetyl groups can be combined with amino groups to produce many of the nonessential amino acids.
    e. As with amino acids, some fatty acids are also essential.

39. Which of the following macronutrients is especially difficult for herbivores to obtain?
    a. Calcium
    b. Magnesium
    c. Sodium
    d. Potassium
    e. Phosphorus

40. A strictly vegetarian diet with no vitamin B_{12} supplements can lead to
    a. beriberi.
    b. pellagra.
    c. pernicious anemia.
    d. scurvy.
    e. night-blindness.

41. The nutritional deficiency disease beriberi is caused by an inadequate supply of the water-soluble vitamin
    a. B_1, thiamin.
    b. B_2, riboflavin.
    c. B_{12}, cobalamin.
    d. folic acid.
    e. C, ascorbic acid.

42. The nutritional deficiency disease rickets is caused by an inadequate supply of the fat-soluble vitamin
    a. A, retinol.
    b. D, calciferol.
    c. E, tocopherol.
    d. K, menadione.
    e. biotin.

43. The nutritional deficiency disease simple goiter is caused by inadequate supply of the micronutrient
    a. fluoride.
    b. iodine.
    c. chromium.
    d. zinc.
    e. copper.

44. Which of the following statements about vitamins is *not* true?
    a. Vitamins, like essential amino acids and fatty acids, are organic molecules, but they are needed in micro quantities.
    b. Most vertebrates require the very same vitamins.
    c. Vitamins function mostly as, or as parts of, coenzymes.
    d. Vitamins are required only in very small amounts.
    e. Humans require more water-soluble vitamins than fat-soluble vitamins.

45. Which one of the following organisms does *not* have adaptations for filter feeding?
    a. A sponge
    b. A toothed whale
    c. A barnacle
    d. An oyster
    e. A clam

46. The root of a typical tooth contains _____ and _____, but *not* _____.
    a. enamel; dentine; a pulp cavity
    b. enamel; a pulp cavity; dentine
    c. dentine; a pulp cavity; enamel
    d. dentine; enamel; a pulp cavity
    e. cement; dentine; a pulp cavity

47. Which of the following teeth are least likely to be found in animals with a diet consisting mainly of plants?
    a. Incisors
    b. Canines
    c. Premolars
    d. Molars
    e. Cheek teeth

48. Which one of the following structures of a tubular digestive tract is mainly involved with water and ion recovery?
    a. Gizzard
    b. Buccal cavity
    c. Stomach
    d. Hindgut
    e. Midgut

49. All of the following structures found in a gut tube except one facilitate the absorption of nutrients. Select the exception.
    a. Villi
    b. Spiral valve
    c. Crop
    d. Microvilli
    e. Typhlosole

50. Which layer of the vertebrate gut shows adaptations for increasing absorptive surface area?
    a. Lumen
    b. Mucosa
    c. Submucosa
    d. Serosa
    e. Peritoneum

51. Which layer of the vertebrate gut is responsible for peristalsis?
    a. Smooth muscle layer
    b. Mucosa
    c. Submucosa
    d. Serosa
    e. Peritoneum

52. Which structure is *not* encountered by a food bolus when being swallowed?
    a. Epiglottis
    b. Larynx
    c. Soft palate
    d. Pharynx
    e. Esophagus

53. Which of the following statements about movement of food in the gut is *not* true?
    a. Three sphincters are found in the vertebrate gut.
    b. Peristalsis can move food in both directions.
    c. Stretched smooth muscle contracts.
    d. Peristalsis begins when food enters the glottis.
    e. The muscle in a sphincter is normally contracted.

54. The sweetness that develops if you continually chew a piece of bread is due to the enzyme
    a. maltase.
    b. sucrase.
    c. lactase.
    d. amylase.
    e. pepsin.

55. Pepsinogen is converted into pepsin by
    a. **low pH.**
    b. chyme.
    c. enterokinase.
    d. trypsinogen.
    e. amylase from the salivary glands.

56. Which of the following is *not* brought about by the HCl secreted in the stomach?
    a. Activation of the principal zymogen of the stomach
    b. Proper pH for the digestive enzyme of the stomach
    c. Breakdown of ingested tissues
    d. **Formation of chylomicrons**
    e. Death of ingested bacteria

57. Which of the following statements about digestion in the small intestine is *not* true?
    a. Most digestion occurs in the duodenum of the small intestine.
    b. Bile is produced by the liver, stored in the gall bladder, and emulsifies fat in the small intestine.
    c. **Bile molecules have one end that is lipophobic and one end that is hydrophilic.**
    d. The pancreas secretes zymogens and bicarbonate ions into the duodenum.
    e. Enterokinase is secreted by intestinal mucosal cells.

58. Which one of the following proteases is produced by the small intestine?
    a. **Dipeptidase**
    b. Chymotrypsin
    c. Trypsin
    d. Carboxypeptidase
    e. Pepsin

59. Which of the following statements about the role of the large intestine in digestion is *not* true?
    a. If too little water is reabsorbed from the feces, diarrhea results.
    b. **The last section of the large intestine is called the colon.**
    c. *Escherichia coli* is a normal inhabitant of the large intestine.
    d. Flatulence results from the metabolism of intestinal bacteria.
    e. The appendix is a vestigial caecum.

60. Which of the following activities is *not* carried out by the liver in its involvement in the regulation of fuel metabolism?
    a. Conversion of nutrients into glycogen and fat
    b. Gluconeogenesis
    c. Synthesis of plasma proteins from amino acids
    d. **Production of high density lipoproteins (HDL) for deposition in adipose tissue**
    e. Processing of chylomicrons from the small intestine

61. Which of the following statements about the hormonal control of fuel metabolism is *not* true?
    a. **During the absorptive period, the liver converts glycogen into glucose.**
    b. During the postabsorptive period, the body cells preferentially use fatty acids for metabolic fuel.
    c. Cells of the nervous system depend almost exclusively on glucose for metabolic fuel.

    d. Glucagon plays a major role during the postabsorptive period by causing the liver to convert glycogen into glucose and stimulating gluconeogenesis.
    e. Adrenaline and glucocorticoid cortisol have effects similar to glucagon.

62. People living in impoverished regions and chronic alcoholics frequently have a niacin deficiency called
    a. beriberi.
    b. kwashiorkor.
    c. **pellagra.**
    d. scurvy.
    e. rickets.

63. Primates should eat citrus fruit to prevent scurvy, a deficiency of _____.
    a. niacin
    b. thiamin
    c. calciferol
    d. **ascorbic acid**
    e. biotin

64. Ruminant animals, such as goats and cows,
    a. produce enzymes in their guts that break down cellulose.
    b. practice coprophagy.
    c. have a four-chambered stomach for increased surface area for better absorption of nutrients.
    d. produce hydrogen sulfide, an important greenhouse gas.
    e. **get much of their protein from the digestion of microorganisms.**

65. These molecules accept cholesterol and probably remove it from the tissues to the liver. Smoking lowers their levels.
    a. Chylomicrons
    b. Very-low-density lipoproteins
    c. Low-density lipoproteins
    d. **High-density lipoproteins**
    e. Cholecystokinin

66. Which of the following molecules contain mostly triglycerides and transport them to fat cells in tissues.
    a. Chylomicrons
    b. **Very-low-density lipoproteins**
    c. Low-density lipoproteins
    d. High-density lipoproteins
    e. Cholecystokinin

67. Which of the following is true of lipophilic components?
    a. They dissolve in water.
    b. **They are often stored for a long time.**
    c. They metabolize quickly.
    d. They are easily filtered by the kidney.
    e. They are hydrophilic.

68. Which of the following is an example of a bioaccumulation?
    a. **A bird that eats fish, which eat invertebrates, which eat algae, which pick up a pesticide in a stream**
    b. Lead building up in a child's liver
    c. A mouse eating poison in a mouse trap
    d. A factory constantly dumping toxins in a river
    e. A bird accidentally getting sprayed with pesticide and infecting its offspring

69. The purpose of cytochrome P450 is to
    a. make energy in the electron transport chain.
    b. break down cytochrome into simple diglycerides.
    c. **detoxify synthetic chemicals by adding an ——OH or an ——SO$_3$ group.**
    d. break down toxins into smaller, less harmful molecules.
    e. break down glycogen into glucose in the liver.

70. Polychlorinated biphenyls are dangerous because
    a. they cause hormones to shut off.
    b. **they mimic hormones, causing loss of control of hormonal functioning.**
    c. they bind vitamin B$_1$ and cause a deficiency.
    d. they cause a thinning of bird eggshells.
    e. they replace iron in blood and calcium in bones.

48 Salt and Water Balance and Nitrogen Excretion

Fill in the Blank

1. Organisms that allow their body fluids to have the same osmotic potential as the environment's are called **osmoconformers**.

2. A cell is **hypertonic** to its environment if its fluids are more concentrated than the environment.

3. Marine bony fishes actively excrete salt from their kidneys and gills, and nitrogenous wastes are lost as ammonia from the **gills**.

4. Tadpoles living in fresh water excrete nitrogenous wastes in the form of **ammonia.**

5. The Malpighian tubules of insects conserve water and excrete **uric acid.**

6. In the nephron of the vertebrate kidney, the glomerulus is a dense knot of thin-walled **capillaries**.

7. In humans, urine from the collecting ducts passes through the **ureter** to the urinary bladder.

8. Reabsorption in the renal tubules is facilitated by **microvilli** providing extra surface area in the cells lining the tubule.

9. A normal kidney has several **autoregulatory** mechanisms that are adaptations to monitor and maintain kidney functions.

10. High levels of antidiuretic hormone cause the collecting ducts to become **more** permeable to water, and highly concentrated urine is produced.

11. The human kidneys filter about 180 liters of blood per day, but produce about 2 to 3 liters of urine per day. Therefore, the percentage of fluid volume that ends up in urine is about **1 to 2% (2 to 3 L/180 L)**.

12. In an artificial kidney, the blood of the patient comes into very close contact with **a dialyzing fluid**, which is separated from the blood only by a semipermeable membrane.

13. Renin is a regulatory enzyme released by the kidney. It acts on a circulating protein to begin converting that protein into an active hormone called **angiotensin**.

14. In addition to water and carbon dioxide, the metabolism of proteins and nucleic acids also produces nitrogenous waste found most frequently in the form of **ammonia or NH_3**.

15. Birds that eat marine animals ingest an excess of salts, which they excrete through their **nasal salt** glands.

Multiple Choice

1. Which of the following is true of the extracellular fluids in our bodies?
   a. They have the same composition as seawater.
   b. They have much higher osmotic concentration than fluids in cells.
   c. They contain no proteins.
   d. They have a fixed water concentration.
   e. They can supply water and nutrients to the cells.

2. Water moves into tissues as a result of all of the following, except
   a. osmotic gradient.
   b. fluid pressure.
   c. active transport.
   d. principles of diffusion.
   e. salinity gradient.

3. Ammonia and urea are waste products derived from the metabolic breakdown of
   a. carbohydrates.
   b. lipids.
   c. sugars
   d. proteins.
   e. salts.

4. Marine invertebrates in which the salinity of body fluids changes with the osmotic potential of their environments are known as
   a. osmoconformers.
   b. osmoregulators.
   c. osmoexcretors.
   d. hypotonic.
   e. hypertonic.

5. Which of the following molecules is most toxic to cells?
   a. Water
   b. Sodium chloride
   c. Ammonia
   d. Urea
   e. Uric acid

6. Which of the following animals is most likely to excrete mostly urea or uric acid instead of ammonia?
   a. Freshwater fishes
   b. Saltwater fishes
   c. Tadpoles
   d. Shrimp
   e. Seagulls

7. Which of the following substances is the least soluble in water?
   a. Ammonia
   b. Uric acid
   c. Urea
   d. Sodium chloride
   e. Amino acids

8. Organisms that are ionic regulators maintain
   a. an osmotic concentration that may differ from the environment.
   b. an osmotic concentration that is the same as the environment.
   c. **an ion concentration that may differ from the environment.**
   d. an ion concentration that is the same as the environment.
   e. an ion concentration by diffusion only.

9. Which of the following operates by filtering the body fluids into a tube, then secreting or reabsorbing specific substances?
   a. Flame cells of flatworms
   b. Metanephridia of annelid worms
   c. Malpighian tubules of insects
   d. Vertebrate nephrons
   e. **All of the above**

10. In which of the following are nitrogenous waste solutes eliminated via the gut?
    a. Flame cells of flatworms
    b. Metanephridia of annelid worms
    c. **Malpighian tubules of insects**
    d. Vertebrate nephrons
    e. All of the above

11. The functional unit of the kidney is the
    a. Bowman's capsule.
    b. capillary.
    c. glomerulus.
    d. **nephron.**
    e. renal tubule.

12. What drives the process of filtration from the capillaries into the glomerulus?
    a. Active transport
    b. **Arterial blood pressure**
    c. Venous blood pressure
    d. Osmotic pressure
    e. Secretion

13. During filtration, which does *not* enter the Bowman's capsule from the bloodstream?
    a. Water
    b. Glucose
    c. Ions
    d. Amino acids
    e. **Plasma proteins**

14. Analysis of kidney function in simple vertebrates suggests that the earliest vertebrate nephron functioned mainly
    a. **to remove excess fluid.**
    b. to remove excess salts.
    c. to remove excess nutrients.
    d. to filter nitrogenous wastes.
    e. to conserve water.

15. Marine bony fishes acquire excess salt when they drink seawater. This salt load is handled in all of the following ways, except
    a. **producing very dilute urine.**
    b. producing very little urine.
    c. excreting ions from gills.
    d. secreting salt from the renal tubules.
    e. a reduced number of glomeruli.

16. Cartilaginous fishes maintain an osmotic concentration in their blood that matches that of seawater by
    a. excreting large amounts of water.
    b. drinking large amounts of seawater.
    c. concentrating seawater salts in their blood.
    d. **concentrating urea in their blood.**
    e. excreting salts at their gills.

17. Water conservation adaptations of reptiles include all of the following, except
    a. scaly skin.
    b. excretion of uric acid.
    c. shelled eggs.
    d. internal fertilization of gametes.
    e. **excretion of salts.**

18. Water conservation adaptations of amphibians living in dry environments include all of the following, except
    a. waxy secretions on the skin.
    b. estivation during dry periods.
    c. **production of concentrated urine.**
    d. burrowing into the ground.
    e. large urinary bladders.

19. Which part of nephrons makes up the inner medulla of the kidney?
    a. Bowman's capsule
    b. Convoluted tubule
    c. Glomerulus
    d. **Loop of Henle**
    e. *a* and *d*

20. Of the fluid that is filtered into the Bowman's capsule, approximately how much of it is reabsorbed back into the blood within the kidney?
    a. Less than 5%
    b. About 25%
    c. About 50%
    d. About 75%
    e. **Over 95%**

21. The cells lining the proximal convoluted tubule have numerous microvilli and mitochondria. Such structures indicate that which of the following activities are conducted in these cells?
    a. Rapid diffusion of water
    b. **Active transport**
    c. Conservation of water
    d. Storage of salts
    e. Production of urea

22. Valuable molecules like glucose, amino acids, and vitamins are reabsorbed into the blood at which location in the nephron?
    a. Bowman's capsule
    b. Collecting duct
    c. Glomerulus
    d. Loop of Henle
    e. **Convoluted tubule**

23. The loops of Henle are considered to function as a countercurrent multiplier because
    a. ascending and descending limbs have different permeabilities.
    b. the amount of water conserved is related to the length of the loop.
    c. **as urine flows in the loop, a concentration gradient is established in the medulla.**

d. as urine flows in the loop, it becomes more concentrated in the nephron.

e. active transport in the ascending and descending limbs multiplies the concentration gradient.

24. Which of the following responses would *not* correct for a drop in glomerular blood pressure?
    a. Angiotensin elevates the body's blood pressure.
    b. Angiotensin stimulates thirst in the brain.
    c. **Angiotensin constricts arterioles entering the kidney.**
    d. Stretch receptors trigger antidiuretic hormone release.
    e. Aldosterone stimulates sodium reabsorption.

25. The effect of antidiuretic hormone is to
    a. reduce permeability in the loop of Henle.
    b. reduce permeability of the collecting ducts.
    c. reduce blood volume.
    d. increase the volume of urine.
    e. **increase the concentration of urine.**

26. Antidiuretic hormone secretion increases when the hypothalamus is stimulated by
    a. angiotensin sensors.
    b. glucose receptors.
    c. **osmosensors.**
    d. renin receptors.
    e. stretch receptors.

27. Urine that is more concentrated than their body fluids is produced by
    a. **birds.**
    b. catfish.
    c. crocodiles.
    d. frogs.
    e. sharks.

28. Kidney dialysis units involve
    a. active pumping of urea from the patient's blood into dialysis fluid.
    b. **diffusion of ions from the patient's blood into dialysis fluid.**
    c. transfusion of blood from a dialysis unit into the patient's blood.
    d. transport of large molecules between the patient's blood and dialysis fluid.
    e. transport of cells between the patient's blood and dialysis fluid.

29. Which of the following is a function of excretory systems?
    a. Help regulate osmotic potential and the volume of extracellular fluids
    b. Excrete molecules that are present in excess
    c. Conserve molecules that are valuable or in short supply
    d. Eliminate toxic waste products of phosphorus metabolism
    e. **a, b, and c**

30. Excretory systems control
    a. filtration.
    b. excretion.
    c. reabsorption.
    d. homeostasis.
    e. **All of the above**

31. Which of the following is true of the human kidney?
    a. Functional units are called hepatocytes.
    b. Nephrons consist of proximal and distal convoluted tubules as well as collecting ducts and loops of Henle.
    c. Most of the substances and water filtered from the blood in the glomerulus return to the venous blood draining the kidney.
    d. Anything coming into the kidney has to come from the renal artery.
    e. **b, c, and d**

32. Which of the following is *not* true of the evolution of the vertebrate nephron?
    a. The function of the earliest vertebrate nephron was to eliminate excess coelomic fluid while conserving important molecules.
    b. The embryology of vertebrates offers some information about the sequence of the evolutionary stages of the vertebrate nephron.
    c. An examination of ancestral living vertebrates may yield information on the evolution of the vertebrate nephron.
    d. **Over the course of vertebrate evolution, nephrons evolved into a complex system of nephrostomes.**
    e. *a* and *c*

33. The terms "ammonotelic," "ureotelic," and "urecotelic" are used to describe
    a. the actions of hormones on the excretory system.
    b. **the types of nitrogenous waste produced by various classes of vertebrates.**
    c. pathways of kidney evolution.
    d. modifications of kidney tubules to enhance excretion.
    e. modes of excretory system development.

34. Organisms living in a freshwater environment normally
    a. **excrete copious dilute urine and retain salts.**
    b. excrete a small volume of dilute urine and retain salts.
    c. excrete copious concentrated urine.
    d. excrete small amounts of concentrated urine.
    e. conserve water and conserve salts.

35. A ureotelic organism excretes most of its nitrogenous waste as
    a. ammonia.
    b. water.
    c. carbon dioxide.
    d. uric acid.
    e. **urea.**

36. Terrestrial organisms must conserve water. The least amount of water is lost with the excretion of which nitrogenous waste product?
    a. Carbon dioxide
    b. **Uric acid**
    c. Ammonia
    d. Water
    e. Urea

37. Gout results from an accumulation of _____ in the joints.
    a. carbon dioxide
    b. water
    c. ammonia
    d. **uric acid**
    e. urea

38–42. Choose the appropriate excretory system or component from the list of answer choices below.
   a. Protonephridia
   b. Nephron
   c. Malpighian tubule
   d. Green gland
   e. Metanephridia

38. Annelid worms are associated with which type of excretory system? (*e*)

39. The vertebrate system is associated with which type of functional unit? (*b*)

40. Flatworms are distinguished by which type of excretory unit? (*a*)

41. Insects and most terrestrial arthropods have an excretory system composed of which type of functional unit? (*c*)

42. The primary excretory system of humans utilizes what type of functional unit? (*b*)

43. Which of the following is *not* part of a nephron?
   a. Podocyte
   b. Flame cell
   c. Renal corpuscle
   d. Glomerulus
   e. Bowman's capsule

44. Blood enters a nephron's vascular component by way of the
   a. peritubular capillaries.
   b. glomerulus.
   c. efferent arteriole.
   d. afferent arteriole.
   e. renal vein.

45–54. Choose the appropriate anatomical section from the list below to respond to Questions 45–54. There may or may not be more than one correct answer.
   a. Renal pyramids
   b. Proximal convoluted tubule
   c. Distal convoluted tubule
   d. Loop of Henle
   e. Collecting duct

45. Resides anatomically closest to Bowman's capsule (*b*)

46. Functions as a countercurrent multiplier system (*d*)

47. Comprise(s) the internal core of the medulla (*a*)

48. Site of glucose and amino acid reabsorption (*b*)

49. Site of major water reabsorption (*b*)

50. Permeability is under the control of antidiuretic hormone (*e*)

51. Length of this particular section relates to the potential maximum urine concentration possible (*d*)

52. Connects the proximal and distal convoluted tubules (*d*)

53. Segment where microvilli are located (*b*)

54. Located principally within the medulla (*a, d,* and *e*)

55. The proper sequence of urine passage from the kidney is
   a. renal pyramid, ureter, bladder, rectal gland, urethra.
   b. ureter, renal pyramid, rectal gland, bladder urethra.
   c. renal pyramid, urethra, bladder, ureter.
   d. renal pyramid, ureter, bladder, urethra.
   e. rectal gland, renal pyramid, ureter, bladder, urethra.

56. A person with diabetes insipidus fails to respond to ADH. What is a symptom of a person with this condition?
   a. Urine contains glucose
   b. Copious hyperosmotic urine
   c. Copious dilute urine
   d. Small volume of concentrated urine
   e. Failure to urinate

57. Beer contains ethyl alcohol, which inhibits ADH production. After consuming beer, the kidney
   a. produces copious concentrated urine.
   b. produces a small volume of concentrated urine.
   c. produces a small volume of dilute urine.
   d. produces copious dilute urine.
   e. does nothing.

58. A freshwater fish is continuously faced with regulating its internal environment. The fish tends to _____ water because it is _____ osmotic to fresh water.
   a. gain; hypo-
   b. gain; hyper-
   c. lose; hypo-
   d. lose; hyper-
   e. lose; iso-

59. To regulate to its environment, a freshwater fish must _____ salts and produce _____ urine.
   a. conserve; copious dilute
   b. excrete; copious dilute
   c. conserve; small volumes of dilute
   d. excrete; small volumes of dilute
   e. absorb; large volumes of concentrated

60. The brine shrimp, *Artemia*, can live in extremely high environmental salinities because it
   a. has active transport of salt across its gill membranes.
   b. has active transport of water across its gill membranes.
   c. excretes salts through its nasal salt glands.
   d. osmoconforms with its surroundings.
   e. excretes concentrated uric acid, so it conserves water.

61. Why don't terrestrial animals excrete nitrogen as ammonia?
   a. It takes more energy to convert nitrogen to ammonia than to urea.
   b. They lose more water by excreting ammonia.
   c. Ammonia is very toxic to terrestrial animals.
   d. It is not stable, and quickly converts to urea.
   e. It is insoluble in water, and so cannot be excreted as urine.

62. Which molecule contains the most nitrogen?
    a. Ammonia
    b. Uric acid
    c. Urea
    d. b and c
    e. All have the same number of nitrogen atoms.

63. This excretory system contains beating cilia and is found in flatworms.
    a. Protonephridia
    b. Metanephridia
    c. Malpighian tubules
    d. Nephron
    e. Aquaporin

64. This excretory system actively transports uric acid and ions into blind tubules and is found in insects.
    a. Protonephridia
    b. Metanephridia
    c. Malpighian tubules
    d. Nephron
    e. Aquaporin

65. This excretory system has capillary beds that come into close contact with podocytes where there is a transfer of water and small molecules. It is found in amphibians.
    a. Protonephridia
    b. Metanephridia
    c. Malpighian tubules
    d. Nephron
    e. Aquaporin

66. This excretory system has ciliated openings in the coelom, called nephrostomes, which lead to a tubule that opens to the outside of the animal. It is found in annelids.
    a. Protonephridia
    b. Metanephridia
    c. Malpighian tubules
    d. Nephron
    e. Aquaporin

49 *Animal Behavior*

Fill in the Blank

1. A species-specific behavior performed the same way every time is called a **fixed action pattern**.

2. The scientific study of the ecology and evolution of animal behavior is **ethology**.

3. Male songbirds learn, and later express, their songs under the influence of the hormone **testosterone**.

4. Changes in **motivation** can explain why animals are aggressive to others in one season and cooperative in another.

5. Learning a particular releaser during a critical period is the phenomenon of **imprinting**.

6. A bird's song is said to be **crystallized** when it has been learned in the form in which it will be sung thereafter.

7. Konrad Lorenz investigated courtship behavior in hybrid **ducks**.

8. If behavioral patterns can be artificially selected for in laboratory populations, then behavior in that species must be at least partially under **genetic** control.

9. Many observations indicate that organisms have an internal or **endogenous** clock.

10. A circadian rhythm can be reset with an environmental cue in a process called **entrainment**.

11. The simplest form of navigation, using landmarks, is **piloting**.

12. If the **pineal** gland of a bird is removed, the bird will no longer have circadian rhythms.

13. The regular departure and return of organisms seasonally is called **migration**.

14. Morphological structures, physiological apparatus, and behavior of organisms are all shaped by **natural selection (or evolution)**.

15. Molecules used for chemical communication between animals are called **pheromones**.

16. **Culture**, the transmission of learned behavior through generations, is a hallmark of humans.

Multiple Choice

1. Imprinting
   a. **must occur during a critical period in development.**
   b. always occurs between parents and offspring.
   c. is genetically determined without a component of learning.
   d. is an encoding of simple information.
   e. occurs equally well with a variety of stimuli.

2. What outcome is expected when adult rats have their ovaries or testes removed and are subsequently treated with female sex steroids?
   a. Both sexes will exhibit lordosis (female mating behavior).
   b. Neither sex will exhibit lordosis.
   c. **Females will exhibit lordosis, and males will exhibit no response.**
   d. Females will exhibit lordosis, and males will exhibit mounting behavior (male mating behavior).
   e. Females will exhibit no response, and males will exhibit lordosis.

3. What outcome, in terms of lordosis (female mating behavior) or mounting (male mating behavior), is expected when newborn rats have their ovaries or testes removed, are treated with testosterone as newborns, and are treated again with testosterone as adults?
   a. **Both exhibit mounting behaviors.**
   b. Neither sex exhibits mounting or lordosis behaviors.
   c. Males exhibit mounting, and females have no response.
   d. Males exhibit mounting, and females exhibit lordosis.
   e. Both sexes exhibit lordosis.

4. Experiments on songbirds determined that the reason female birds normally do *not* sing is due to
   a. lack of the proper muscles.
   b. lack of specific parts of the nervous system.
   c. **the absence of testosterone.**
   d. the inability to learn the song.
   e. learning sex-specific behavior from a nonsinging mother.

5. Anatomical comparisons of male and female songbirds reveal that _____ of male birds increase(s) in volume during periods of active singing.
   a. the throat muscles
   b. **specific regions of the brain**
   c. the pineal gland
   d. the suprachiasmatic nuclei
   e. the auditory nerves

243

6. When purebred nonhygienic bees are crossed with purebred hygienic bees, they produce dihybrid non-hygienic offspring. If these offspring are backcrossed with the purebred hygienic bees, which type of behavior is *not* exhibited in the next set of offspring?
   a. Nonhygienic
   b. Nonhygienic, but will remove pupae from uncapped cells
   c. Nonhygienic, but will uncap cells of dead pupae
   d. Hygienic, but only when neighboring bees are hygienic
   e. Hygienic

7. Male gypsy moths travel thousands of meters to reach female gypsy moths. Their behavior is a response to
   a. visual signals.
   b. auditory signals.
   c. pheromones.
   d. random search patterns.
   e. electric signals.

8. The oldest form of animal communication is probably
   a. visual.
   b. auditory.
   c. tactile.
   d. chemical.
   e. electric.

9. Chemical territory marking can convey all of the following information, except
   a. height.
   b. direction of travel.
   c. individual identity.
   d. reproductive status
   e. elapsed time.

10. Which of the following is *not* a feature of visual communication?
    a. Easy to produce
    b. Comes in an endless variety
    c. Effective over long distances
    d. Can be changed rapidly
    e. Indicates the position of the signaler

11. When a food source is located toward the sun, the direction of a honeybee's waggle dance on a vertical surface will be
    a. straight up.
    b. straight down.
    c. 90° to the right.
    d. 90° to the left.
    e. 45° to the right.

12. The feature of a honeybee's waggle dance that indicates the distance to a food source is the _____ of the waggles.
    a. frequency
    b. speed
    c. direction
    d. size
    e. shape

13. Insects use their antennae in which two forms of communication?
    a. Tactile and auditory
    b. Tactile and chemical
    c. Chemical and auditory

d. Visual and tactile
e. Visual and chemical

14. Which of the following are most likely to communicate by sound?
    a. Fish
    b. Amphibians
    c. Reptiles
    d. Insects
    e. Mammals

15. Visual communication is better than auditory communication at
    a. being used in dark environments.
    b. conveying complex information.
    c. going around objects.
    d. getting the attention of the receiver.
    e. communicating over long distances.

16. Which of the following would suggest that learning plays a role in a spider's spinning a web?
    a. A newly hatched spider spins a perfect web.
    b. The web structure is the same every time.
    c. Web structure varies from species to species.
    d. The web size changes as the individual grows.
    e. The web is changed after contact with particular prey.

17. If behavior is under partial genetic control,
    a. no variability in the behavior will be seen.
    b. it is not subject to natural selection.
    c. it may be modifiable by experience.
    d. it is acquired only through learning.
    e. it will be species-specific.

18. What is the objective of a deprivation experiment?
    a. To determine the effect of isolation on an animal
    b. To determine if behavior is a fixed action pattern
    c. To determine how an animal learns
    d. To determine the influence of parents on behavior
    e. To determine the influence of mates on behavior

19. Which of the following is an example of a deprivation experiment?
    a. Removing a bird's mate in the middle of courtship
    b. Removing half of a litter of puppies
    c. Transferring a newborn rat to a different mother
    d. Hand-rearing a bird from the egg
    e. Taking all nest material away from nesting birds

20. A sign stimulus or releaser
    a. is required in order for learning to occur.
    b. is required in order for a fixed action pattern to occur.
    c. triggers the onset of aggressive behavior.
    d. facilitates recognition of appropriate mates.
    e. causes the animal to remember the full stimulus.

21. Experiments on interactions between herring gulls and their chicks showed that
    a. simple models of releasers were effective.
    b. the more lifelike the releaser, the more likely the behavior.
    c. chicks preferred parents with contrasting marks on their bills.
    d. appropriate sounds as well as visual stimuli were required.
    e. parents gave more food to visually discriminating chicks.

22. The presence of supernormal releasers indicates that
    a. natural selection operates against unusual behavior.
    b. animals accept a narrow range of releaser parameters.
    c. animals always prefer bigger and better stimuli.
    d. animals can prefer unnatural stimuli over natural stimuli.
    e. animals are not discriminating about those stimuli.

23. What determines the motivational state of an animal?
    a. Its previous experience
    b. Its physiological state
    c. Its social environment
    d. Its physical environment
    e. The stimuli available

24. Which of the following behaviors would be expected to be genetically programmed?
    a. Behavior that can be learned by trial and error
    b. Behavior that leads to successful mating
    c. Behavior that does not endanger the animal
    d. Behavior that involves social cues
    e. Behavior that is important in variable circumstances

25. Chaffinches must learn their song during a critical period by being exposed to
    a. a normal chaffinch song only.
    b. any male songbird's song.
    c. any bird vocalizations.
    d. a normal chaffinch song along with other bird vocalizations.
    e. normal song with rearranged elements.

26. What can behaviorists learn from hybridization experiments?
    a. A behavior is often coded by a single gene.
    b. Behavioral traits are usually all inherited from a single parent.
    c. A hybrid may show behavior intermediate between that of its parents.
    d. Hybrids show behavior patterns completely different than either parent.
    e. Only simple animals have genetically programmed behavior.

27. One behavior that is inherited in a simple Mendelian manner is
    a. fruit fly courtship.
    b. lovebird nest building.
    c. honeybee brood cleaning.
    d. dog retrieving behavior.
    e. rattlesnake prey capture.

28. Circadian rhythms are characterized by
    a. seasonal changes.
    b. lack of synchrony with the environment.
    c. tendency to lengthen with time.
    d. persistence in the absence of cues.
    e. changes with the animal's age.

29. When a person changes time zones suddenly, as in jet travel, the circadian rhythm
    a. is destroyed temporarily.
    b. needs entrainment.
    c. adjusts within a few hours.
    d. becomes random.
    e. takes on a different periodicity.

30. The circadian clock in mammals appears to be controlled by neurons located near the
    a. hypothalamus.
    b. optic nerve.
    c. pituitary.
    d. reticular activating system.
    e. visual cortex.

31. A bird will be unable to entrain its circadian rhythms if light is obscured from its
    a. eyes.
    b. optic nerve.
    c. pineal gland.
    d. pituitary.
    e. visual cortex.

32. The adaptive advantage of circannual rhythms is
    a. in timing breeding to coincide with peak resources.
    b. maximum in tropical species.
    c. that it entrains circadian rhythms.
    d. minimal in migratory birds.
    e. maximal in photoperiodic species.

33. Homing pigeons released at a remote, unfamiliar site find their way home by
    a. following visual landmarks.
    b. retracing the route along which they were carried.
    c. searching randomly until they locate home.
    d. following other birds.
    e. navigating without visual images.

34. Which of the following is an experimental method to test how migrating animals navigate?
    a. Deprive them of foodstores
    b. Blindfold them
    c. Capture and displace them
    d. Keep them in captivity
    e. Get them to migrate at the wrong season

35. Biological rhythms play a role in migratory behavior, as displayed by the
    a. tendency to migrate over great distances.
    b. presence of migratory restlessness.
    c. tendency to migrate at night.
    d. presence of orientation toward the sun.
    e. tendency to form flocks.

36. Clock-shifting experiments revealed that birds tell direction from
    a. the sun.
    b. the stars.
    c. the wind direction.
    d. sound cues.
    e. air pressure cues.

37. Experiments with young songbirds in a planetarium suggested that birds can use star clues if
    a. they are tested in a rotating star pattern.
    b. they learned in a stationary star pattern.
    c. they learned in a rotating star pattern.
    d. they learned in the Northern Hemisphere.
    e. they do not see the stars until adulthood.

38. What would be the advantage of orienting with respect to a fixed point of light in the night sky?
    a. It is easier to fly to a steady coordinate.
    b. In the Northern Hemisphere it would always be in the north.

c. It would give information about the sun compass.

d. It would give information if the night were overcast.

e. It would be the brightest spot in the sky.

39. Which of the following patterns is seen in animal orientation systems?
    a. Each species has one sensory mechanism for orientation.
    b. All seem to rely on light systems.
    c. They seem to rely on bicoordinate navigation.
    d. They are based on redundant cues.
    e. All seem to need an internal clock.

40. What information suggests that pigeons use magnetic clues for orientation?
    a. A neurophysiological magnetic transducer has been described.
    b. They can orient even on cloudy days.
    c. Attachment of small magnets disrupts their homing.
    d. Magnetic field lines could give latitude information.
    e. Magnetite particles have been found in their cells.

41. Which of the following suggests that humans have some innate, unlearned motor patterns?
    a. All humans feel anger and fear.
    b. All cultures teach their young.
    c. Humans have a large capacity to learn.
    d. Blind infants do not smile normally.
    e. All cultures have similar facial expressions.

42. Fixed action patterns are expected to be common in species in all of the following circumstances, except
    a. when there is no opportunity to learn.
    b. when it is possible to learn the wrong behavior.
    c. when organisms are long-lived.
    d. when generations do not overlap.
    e. when dealing with dangerous prey.

43. Human behavior is largely influenced by
    a. fixed action patterns.
    b. genetic traits.
    c. instinct.
    d. learning.
    e. orienting.

44. Which of the following statements about spiders' web building is *not* true?
    a. Webs are built perfectly the first time.
    b. Instinctive behavior, such as web building, is not usually included within the province of the field of ethology.
    c. The building of webs in spiders is genetically determined.
    d. Young spiders build their first webs without prior experience.
    e. Web design and construction is species-specific.

45. Which of the following is *not* a part of the definition of instinctive behavior?
    a. Stereotypic
    b. Genetically determined
    c. Not modified by learning
    d. Inherited in a Mendelian fashion
    e. Species-specific

46. Which of the following is *not* an example of a fixed action pattern?
    a. Web construction in orb-weaving spiders
    b. Egg cocoon construction in spiders
    c. Behaviors that are expressed during a deprivation experiment
    d. Song acquisition in the white-crowned sparrow
    e. Bill pecking in herring gull chicks

47. Which one of the following is *not* a valid reason why an animal did *not* show a specific behavior during a deprivation experiment?
    a. The behavior is not instinctive.
    b. Isolation began after the behavior's critical period had passed.
    c. The correct sign stimulus was not present.
    d. The normal releasers were not observed.
    e. The animal was in an inappropriate developmental or physiological state.

48. Which of the following is *not* a releaser?
    a. Shape of the adult head in herring gull chicks
    b. The bill spot of the adult in herring gull chicks
    c. A tuft of red feathers for a male European robin
    d. Egg shape in the oystercatcher
    e. The presence of a nut for a tree squirrel

49. Which of the following statements about supernormal releasers is *not* true?
    a. A supernormal releaser includes novel stimuli not found in the normal releaser.
    b. A supernormal releaser triggers a greater response in an animal than does a normal releaser.
    c. Supernormal releasers are exaggerations of normal stimuli.
    d. A supernormal releaser would not trigger an appropriate response in an animal subjected to a deprivation experiment.
    e. Supernormal releasers are possible because of the simplicity of the properties of normal releasers.

50. Which of the following statements about imprinting is *not* true?
    a. Imprinting is rapid learning that occurs during a critical period.
    b. Imprinting is restricted to the early life of an animal.
    c. Imprinting makes it possible to encode complex information in the nervous system.
    d. Imprinting occurs during song acquisition of some species of birds.
    e. Parent–offspring recognition is often dependent on imprinting.

51. Which of the following statements about hygienic behavior in honeybees is *not* true?
    a. Hygienic behavior in honeybees can be explained in a simple Mendelian manner.
    b. The strain showing hygienic behavior is resistant to a disease that kills larvae.
    c. In a cross of the hygienic and nonhygienic strains, all of the F_1 offspring were of one phenotype.
    d. The offspring from a backcross showed a 3:1 ratio.
    e. Behavior of the hybrids suggested that there are not separate genes for uncapping cells and for removing infected larvae.

52. Which of the following statements about studies on egg laying in *Aplysia* is *not* true?
    a. In *Aplysia*, egg laying involves a series of fixed action patterns.
    b. Egg-laying hormone is a peptide produced by certain cells within the nervous system of *Aplysia*.
    c. **Egg-laying hormone elicits all of the normal behaviors associated with egg laying.**
    d. The gene that codes for egg-laying hormone actually codes for a much larger protein.
    e. Egg-laying hormone only works within the context of a highly organized nervous system.

53. All of the following except one must be true before a behavior will be favored by natural selection. Select the exception.
    a. The behavior must be genetically determined.
    b. **The behavior must not expose the individual to increased predation.**
    c. The behavior must confer a net gain to the reproductive success of the individual, relative to other population members.
    d. The benefits of the behavior must be greater than the costs.
    e. The behavior cannot be culturally based.

54. Which of the following statements about circadian rhythms is *not* true?
    a. Circadian rhythms are seldom exactly 24 hours.
    b. **Entrainment is the process of bringing two rhythms into phase.**
    c. A rhythm that is less than 24 hours must be phase-advanced to remain in phase with the 24-hour cycle of light and dark.
    d. The effect of jet lag on your circadian rhythms is the same if you fly 1,000 miles east or 1,000 miles west.
    e. The period of a rhythm is the length of one cycle.

55. Which of the following statements about endogenous clocks is *not* true?
    a. There are limits to the entrainment ability of endogenous clocks.
    b. The endogenous clock of mammals is located in the suprachiasmatic nuclei.
    c. Endogenous clocks are found in virtually every animal group.
    d. **Cells in the two eyes of birds function as endogenous clocks.**
    e. Many plants have endogenous clocks.

56. A blind bird with a circadian rhythm of 22 hours is fitted with a black cap that covers the top of its head. It is placed into an environmental chamber with a light-dark period of exactly 12 hours. After several days, the bird will
    a. become phase-advanced.
    b. become phase-delayed.
    c. lose all circadian rhythms.
    d. **develop a free-running rhythm.**
    e. be unaffected.

57. Which of the following statements about piloting is *not* true?
    a. Piloting need not rely on specific landmarks.
    b. **Piloting is only useful for short-distance migrations.**

    c. Gray whales use piloting while moving from their wintering grounds to their summering grounds.
    d. Water currents can be used as piloting cues.
    e. Piloting is not an entirely learned behavior.

58. Which of the following statements about homing is *not* true?
    a. Simple piloting can explain many cases of homing.
    b. Homing pigeons fitted with frosted contact lenses were still able to home.
    c. Piloting is not sufficient to explain homing in the homing pigeon.
    d. Marine birds show many examples of homing over great distances.
    e. **In most migratory bird species, the young learn to home by following the adults.**

59. Studies done on European starlings transported and released from different areas during their migratory period suggest that the starlings use the type of navigation called
    a. piloting.
    b. **distance-and-direction navigation.**
    c. bicoordinate navigation.
    d. random search navigation.
    e. homing.

60. Research has shown that there is a correlation between the duration of _____ and the required _____ to the goal.
    a. **migratory restlessness; distance**
    b. migratory restlessness; direction
    c. homing; distance
    d. homing; direction
    e. circadian rhythms; distance

61. In bicoordinate navigation, the animal knows all of the following, except
    a. its initial latitude and longitude.
    b. the latitude of its goal.
    c. the distance to the goal.
    d. **the sun's position at its current location and at its goal.**
    e. its initial longitude.

62. A sun compass requires that the animal know
    a. its current location.
    b. its latitude and longitude.
    c. **the current time.**
    d. the sun's direction at the goal.
    e. the latitude and longitude of its goal.

63. A bird is trained to feed from the southeast end of a circular cage. At sunrise, a mirror is used to shift the sun 45° to the right of it real position. Where will the bird attempt to feed?
    a. North
    b. East
    c. **South**
    d. West
    e. Southwest

64. A bird is trained to feed from the west end of a circular cage. At sunrise, a fixed light is set up at the south end of the cage. If the bird treats the light as the real sun, where will the bird attempt to feed at noon?
    a. North
    b. East

c. South
d. West
e. Northwest

65. A bird is trained to feed from the south end of a circular cage. The bird is then placed into a light-controlled room and phase-advanced by 6 hours. If the bird is returned to the circular cage at noon, where will it search for food?
    a. North
    b. East
    c. South
    d. West
    e. Northwest

66. An animal is displaced such that the sun rises earlier and is lower in the sky than it would be at its home position. What direction should the animal move in order to home?
    a. Southwest
    b. Southeast
    c. Northwest
    d. Northeast
    e. South

67. A king snake must clamp down on the mouth of its prey, the rattlesnake, to avoid being bitten and killed by the rattlesnake. This type of behavior is
    a. learned.
    b. genetically programmed.
    c. imprinting.
    d. dependent upon hormone release.
    e. entrainment.

68. The fruit fly, *Drosophila*, makes a good subject for hybridization studies because
    a. it has a short life span.
    b. it produces large numbers of offspring.

c. its behavior traits can be traced to simple Mendelian segregation.
d. a and b
e. a, b, and c

69. Slime molds use a chemical signal to cause aggregation of individual cells in times of stress. The signal they use is
    a. cAMP.
    b. testosterone.
    c. pheromone.
    d. bombykol.
    e. gyplure.

70. Glass knife fish communicate by _____ since they live in murky waters and cannot see well.
    a. sound waves
    b. chemical signals
    c. tactile signals
    d. electric signals
    e. water vibrations

71. Which of the following is *not* a class of behavior that Darwin identified in his book, *The Expression of the Emotions in Man and Animals*?
    a. Autonomic responses
    b. Intention movements
    c. Displacement behavior
    d. Irrelevant behavior
    e. Fixed action patterns

72. When an animal grooms while being threatened by a predator, it is an example of a(n)
    a. intention movement.
    b. autonomic response.
    c. learned behavior.
    d. fixed action pattern.
    e. displacement behavior.

50 *Behavioral Ecology*

Fill in the Blank

1. Many **fish** protect their young, but do not feed them.

2. Species with social systems having sterile classes are termed **eusocial**.

3. Members of social groups often have **increased** competition for food compared to solitary animals.

4. Members of social groups usually have **increased** protection from predators compared to solitary animals.

5. **Sexual selection** is the spread of traits that confer advantages to their bearers in competition for mates or for resources needed for courtship.

6. **Territorial behavior** involves the defense of an area that contains food, nesting sites, or other resources.

7. In white-fronted bee-eaters, about half the nonbreeding individuals help at the nests of other individuals. Helping relatives **increases** fitness, and thus white-fronted bee-eaters usually help relatives rather than nonrelatives.

8. Ecologists often analyze their observations of animal behavior in terms of the costs and **benefits** to the performer.

9. An important variable in determining the costs and benefits of a behavior is the **relatedness** between the performer and recipient.

10. The **energetic cost** of a behavior is the difference between the energy the animal would have expended had it rested and the energy expended in performing the behavior.

11. The **risk cost** of a behavior is the increased chance of being injured or killed as a result of performing it, compared with resting.

12. The **opportunity cost** of a behavior is the sum of the benefits the animal forfeits by not being able to perform other behaviors during the same time interval. For example, an animal forfeits the opportunity to feed when it defends its territory.

13. **Reciprocal altruism** evolves if the performer is in turn the recipient of beneficial acts from the individuals it has helped.

14. In order for behaviors to evolve, they must have a **genetic** basis.

15. **Leks** are display grounds for some species of males to court females.

16. In several species of weaverbirds the males are **polygynous**, having more than one mate.

Multiple Choice

1. An elephant seal's environment is defined as
   a. all other elephant seals in the same area.
   b. all other organisms that influence it.
   c. the physical factors of its area.
   d. all other organisms of its food web.
   e. **all the physical and biological factors that influence it.**

2. An appropriate study in the field of ecology would be
   a. a comparison of vertebrate skeletal structures.
   b. the effect of adrenaline on human heart rate.
   c. an investigation of the roles of hormones in plant growth.
   d. **the effect of fire on forest animals' populations.**
   e. a classification of fungi into various groups.

3. Which of the following dogs has the greatest level of fitness relative to the others?
   a. A young female dog that has agility and cleverness
   b. An old male dog with great strength and muscle tone
   c. **An old female dog that had once given birth to three puppies**
   d. A young female dog currently raising six adopted puppies
   e. A young male dog that is maturing sexually

4. The term for the environment in which an organism lives is its
   a. community.
   b. ecology.
   c. **habitat.**
   d. niche.
   e. population.

5. Red abalone larvae will settle on coralline algae substrates only after recognizing a peptide molecule that the algae produce. As an example of a habitat selection cue, one would expect that this peptide is
   a. required for abalone reproductive success.
   b. produced by the algae in order to attract the abalone.
   c. **a good predictor of conditions suitable for abalone survival.**
   d. a food source for the abalone.
   e. a waste by-product of the algae.

6. Bluegill sunfish are put in two tanks with their prey, water fleas. Each tank contains an equal proportion of small, medium, and large water fleas. One tank has a low density of water fleas, and the other has a

high flea density. According to foraging theory, the bluegill sunfish should eat
a. an equal proportion of the three flea sizes at both densities.
b. a greater proportion of large fleas at both densities.
c. equal proportions of the three flea sizes at low density and mostly large fleas at high density.
d. mostly large fleas at low density and equal proportions of the three flea sizes at high density.
e. mostly medium and large fleas at low density and mostly large fleas at high density.

7. Which of the following is *not* a necessary assumption of foraging theory?
a. Efficient predators spend less time on predation than nonefficient predators.
b. Superior predators can produce more offspring.
c. Superior foraging ability can be genetically based.
d. Predators choose prey in ways that maximize their energy intake rate.
e. Efficient predators always choose the most abundant prey.

8. A hyena that takes food away from another hyena is exhibiting
a. altruistic behavior.
b. cooperative behavior.
c. selfish behavior.
d. spiteful behavior.
e. territorial behavior.

9. An example of reciprocal altruism is
a. two chimpanzees picking parasitic fleas off one another.
b. a parent bird feeding its young.
c. one penguin guarding its territory against other penguins.
d. wild dogs hunting together to capture prey.
e. two male deer competing for the same female.

10. When wood pigeons draw together in a tight flock in response to the presence of a hawk, they are exhibiting
a. altruistic behavior.
b. cooperative behavior.
c. selfish behavior.
d. spiteful behavior.
e. territorial behavior.

11. When one dog defends its food by attacking a second dog, and meanwhile a third dog takes the food, the first dog has exhibited
a. altruistic behavior.
b. cooperative behavior.
c. selfish behavior.
d. spiteful behavior.
e. territorial behavior.

12. Kin selection is
a. the mating of relatives.
b. the recognition of relatives among societal animals.
c. the adoption of young by generally unrelated adults.
d. a process of forcing young males out of a society while keeping the females.
e. a behavior that increases the survivorship of an individual's relatives.

13. Altruistic acts are most likely to evolve into behavior patterns when the participants
a. are capable of learning.
b. have individual fitness.
c. are genetically related.
d. are largely nonsocial.
e. are mating partners.

14. Two individuals have a coefficient of relatedness of $r = 0.125$. Most likely, they are
a. parent and offspring.
b. grandparent and grandchild.
c. siblings.
d. full cousins.
e. half cousins.

15. The coefficient of relatedness for a typical grandparent and grandchild is
a. 1.00.
b. 0.75.
c. 0.50.
d. 0.25.
e. 0.125.

16. The coefficient of relatedness for typical full siblings is
a. 1.00.
b. 0.75.
c. 0.50.
d. 0.25.
e. 0.125.

17. Male courtship behavior accomplishes all of the following, except to
a. induce a female to mate with the male.
b. display that the male is in good health.
c. show that the male is a good provider.
d. convey that the male has successfully mated in the past.
e. signal that the male has a good genotype.

18. After copulation, a male may remain with a female for a time and prevent her from copulating with other males. Although this behavior has its benefits, it also carries a high
a. energetic cost.
b. opportunity cost.
c. risk cost.
d. competitive cost.
e. altruistic cost.

19. After copulation, many male insects are genetically programmed to stay near the female for a time. This behavior is most adaptive because it
a. guards the female against injury.
b. prevents the female from copulating again until after fertilization.
c. allows the male to recover from the energetic cost of mating.
d. defends a nesting territory.
e. creates a social bond between the male and female.

20. Within display grounds called leks,
a. each male defends a small space and females choose mates.
b. each female defends a small space and males choose mates.

c. one male prevents other males from entering the lek.

d. one female prevents other females from entering the lek.

e. one male prevents other males and all but one female from entering the lek.

21. Females may choose a mate based on all of the following, except
    a. the quality of the resources in his territory.
    b. his display of food items.
    c. his apparent good health.
    d. his physical features.
    e. **the increased probability that his sperm will fertilize the eggs.**

22. Female reproductive success is improved by choosing a male with all of the following features, except his
    a. high genetic quality.
    b. good health.
    c. **interaction with other female mates.**
    d. high parental care.
    e. control of abundant resources.

23. Parental investment
    a. increases the chances of survival of the parent.
    b. increases the parent's ability to produce additional offspring.
    c. **increases the chances of survival of the offspring.**
    d. decreases as the number of offspring increases.
    e. decreases when costs to parents are lowered.

24. Which of the following characteristics is most likely to occur in a bird species having brightly colored females and dull-colored males?
    a. **The females pursue the males during courtship.**
    b. The species has a monogamous mating system.
    c. Offspring are raised by relatives of the parents.
    d. The species forms large nesting colonies.
    e. The relative size of the females is small compared to the males.

25. Which of the following statements does *not* support the theory that the large size of males in many vertebrate species is a result of sexual selection?
    a. Larger males tend to defend larger territories.
    b. Larger males win fights with smaller males.
    c. Larger males attract more females than smaller males.
    d. Polygynous species have larger males than similar monogamous species.
    e. **Parents feed larger male offspring more food than smaller offspring.**

26. The most widespread social system among animals is
    a. **parents and offspring.**
    b. peer grouping.
    c. several females and their offspring.
    d. several pairs and their offspring.
    e. a dominant pair, their offspring, and unrelated subordinates.

27. Among mammals, which offspring, if any, tend to leave their parent's group?
    a. Both male and female offspring leave.
    b. Females leave, but males leave only if a new territory becomes available.
    c. **Males leave but females do not leave.**
    d. Neither males nor females leave.
    e. Males leave, but females leave only if a new set of offspring are born.

28. Large numbers of sterile individuals occur in all of the following, except
    a. termites.
    b. ants.
    c. bees.
    d. **beetles.**
    e. wasps.

29. In species with diploid females and haploid males, the average coefficient of relatedness of full sisters is $r =$
    a. 1.00.
    b. **0.75.**
    c. 0.50.
    d. 0.25.
    e. 0.125.

30. In species with diploid females and haploid males, which pair of relatives is most similar genetically?
    a. Mother and daughter
    b. Mother and son
    c. Father and son
    d. **Full sisters**
    e. Full brothers

31. Which of the following groups exhibits eusocial behavior and has diploid females and haploid males?
    a. Penguins
    b. **Bees**
    c. Jackals
    d. Termites
    e. Naked mole rats

32. W. D. Hamilton hypothesized that eusociality evolved because sterile worker females are genetically more related to their sisters than to any other relatives. According to this hypothesis, which one of the following would *not* be expected?
    a. Queens should produce equal numbers of sons and daughters.
    b. Workers should feed their sisters better than their brothers.
    c. If the original queen is lost, the workers would not favor the new queen's daughters.
    d. Worker females would have the same father and mother.
    e. **Both males and females are diploid.**

33. Which of the following is *not* true of eusociality?
    a. Eusocial species form elaborate nests or burrows.
    b. Helping behavior may be a necessary prerequisite for the evolution of eusociality.
    c. Some eusocial species contain only diploid members.
    d. Eusociality may occur in one species and be lacking in a closely related species.
    e. **Eusocial individuals typically live part of their lives away from the society.**

34. Large, hoofed, plant-eating mammals tend to
    a. feed preferentially on high-protein foods.
    b. **live in herds.**
    c. feed in forests.
    d. have monogamous mating patterns.
    e. have higher metabolic rates than smaller mammals.

35. Which of the following features is *not* common among forest-dwelling birds?
    a. Feed alone
    b. Build well-hidden nests
    c. Monogamy
    d. Sexes have identical plumage
    e. Nest in colonies

36. Which is *not* true of primate social systems?
    a. Nocturnal species tend to feed alone.
    b. Day-active species tend to eat animal food when available.
    c. Arboreal forest-dwelling species live in large groups.
    d. Females usually remain with their natal group.
    e. Both males and females may develop dominance hierarchies.

37. Which is *not* true of social systems?
    a. Social systems can be studied by asking how individuals in a system benefit by it.
    b. Social systems are dynamic; individuals' relationships with one another constantly change.
    c. Relationships within a social system are partly determined by genetic relatedness.
    d. Social systems have evolved primarily because societies have an increased success at obtaining food.
    e. Social systems continue to evolve in relation to animals' sizes, diets, and habitats.

38. The benefits lost by an animal that is *not* able to feed while defending its breeding territory is an example of the
    a. opportunity cost.
    b. energetic cost.
    c. risk cost.
    d. feeding cost.
    e. breeding cost.

39–42. Match the social acts in the list below with the correct descriptions.
    a. Altruistic act
    b. Selfish act
    c. Cooperative act
    d. Political act
    e. Spiteful act

39. Willow leaf beetles from the same clutch feed together. Beetles in groups can initiate feeding sites on tough leaves more easily than lone beetles can, and there are group defenses against predators. *(c)*

40. A tadpole eats a conspecific egg and becomes diseased as a result of doing so. *(e)*

41. A bird sees a predator and gives out a warning call to other birds in the area. The caller may be at increased risk of being eaten by the predator. *(a)*

42. A willow leaf beetle larva eats a conspecific egg and gains a nutritional benefit. *(b)*

43. Which of the following is *not* a consequence of parental investment?
    a. Increased chance of survival of the offspring
    b. Increased care of each offspring

    c. Increased chance of survival of the parent
    d. Increased fitness of the parent
    e. Decreased chance of production of additional offspring

44. Paternal investment by mammals is often less than that of birds. This may be because
    a. young mammals are less likely to be eaten by predators and thus need less care.
    b. paternity is more certain in birds than in mammals.
    c. there is less cost to birds for paternal care.
    d. male mammals do not have functional mammary glands and thus cannot feed the young until they are weaned.
    e. Newborn mammals are more self-sufficient than are just-hatched birds.

45. Fitness for each sex is usually maximized in a mating cycle when males copulate _____ and females copulate _____.
    a. many times; once
    b. once; once
    c. many times; many times
    d. a few times; many times
    e. a few times; once

46. Group living
    a. can result in foraging benefits.
    b. reduces exposure to disease.
    c. is usually advantageous when food is scarce.
    d. has no effect on reproductive success.
    e. usually decreases competition between individuals.

47. A cooperative act benefits
    a. the performer.
    b. spiteful individuals.
    c. the performer and another individual.
    d. selfish individuals and their mates.
    e. nearest kin only.

48. Coefficients of relatedness express
    a. genetic similarities among relatives.
    b. the amount of interaction between mates.
    c. the size difference between males and females.
    d. the relative fitness of conspecifics.
    e. the amount of social activity in a population.

49. Juncos in groups benefit from the vigilance of other group members that scan the surroundings for predators, but also experience increased competition for their seed food when in groups. They should join groups when
    a. predators and food are rare.
    b. predators and food are abundant.
    c. predation risk is low and food is abundant.
    d. the costs of group living are high overall.
    e. other seed-eating bird species are present.

50. Parental investment is an example of _____.
    a. selfish behavior
    b. spiteful behavior
    c. territorial behavior
    d. altruistic behavior
    e. reciprocal behavior

51. Females are usually choosier than males with respect to mates because
    a. **they typically invest more energy per offspring than do males.**
    b. they typically have more total offspring than do males.
    c. males are more variable so females have more to choose from.
    d. only females can nurse young in most animal species.
    e. they pass on more genes to their offspring than do males.

52. Males are most likely to provide care for young when
    a. fertilization is internal.
    b. the males are mammals.
    c. offspring survival is unaffected by the care.
    d. the males mate with many females.
    e. **certainty of paternity is high.**

53. In some birds, individuals other than the parents help in rearing offspring. These helpers are usually
    a. unrelated to the parents.
    b. **close relatives of the parents.**
    c. unrelated females that have lost their own offspring.
    d. members of a different altruistic species.
    e. cuckolding males.

54. When helpers participate in rearing young, the fitness of parental Florida scrub jays
    a. decreases due to competition with the helpers.
    b. decreases due to reduced interaction with their offspring.
    c. increases because the helpers often feed the parents.
    d. **increases due to increased offspring survival.**
    e. is usually unchanged.

55. Groups of grooved-bill anis (ground-dwelling birds) defend communal nests. When one female begins to lay eggs she throws all other eggs in the nest out before laying her own. This is an example of
    a. displacement behavior.
    b. altruistic behavior.
    c. intention behavior.
    d. **selfish behavior.**
    e. helping behavior.

56. In many mammals, including primates,
    a. **females are more likely than males to remain in natal groups.**
    b. males are more likely than females to remain in natal groups.
    c. both males and females usually leave natal groups.
    d. both males and females typically remain in natal groups.
    e. natal groups rarely persist once the young can be independent.

57. Baboons travel in groups that are defended from predators by males. This behavior of the males probably
    a. increases their fitness because it increases their survival.
    b. **increases their fitness because it increases survival of their offspring.**
    c. decreases their fitness because it decreases their survival.
    d. decreases their fitness because it decreases survival of their offspring.
    e. is unrelated to fitness.

58. An example of reciprocal altruism is
    a. female anis throwing each other's eggs out of the communal nest.
    b. **primates grooming one another.**
    c. male bees stealing one another's mates.
    d. male elephant seals defending harems against one another.
    e. parent birds feeding their offspring.

59. It was shown that the female African long-tailed widowbird was more attracted to males with elongated tails. Why don't the male widowbirds have even longer tails?
    a. It was genetically impossible to create longer tails in males.
    b. **The cost of producing the longer tails was more than the benefit.**
    c. The shorter-tailed males were jealous and fought off the longer-tailed males.
    d. The longer-tailed males were sterile.
    e. None of the above

60. Behavioral ecology deals with
    a. how animals make decisions about their survival.
    b. the choice of mates.
    c. how animals select food and shelter.
    d. group interactions.
    e. **All of the above**

61. Which of the following is *not* a criterion a predator uses to choose its prey?
    a. How long it takes to capture the prey.
    b. The energy it receives from eating the prey.
    c. The abundance of the prey.
    d. **Whether another predator is after the same prey.**
    e. Whether the prey contains the proper nutrients for the predator.

51 *Population Ecology*

Fill in the Blank

1. The proportions of individuals in each age group in a population make up its **age distribution.**

2. The stages an individual goes through during its life constitute its **life history**.

3. A group of individuals born at the same time is known as a **cohort**.

4. The number of individuals of any particular species that a habitat or environment can support is that habitat's **carrying capacity**.

5. The measure of the contribution an individual of a particular age is likely to make to its population's growth rate is called **reproductive value**.

6. When comparing two populations whose members differ in size, the most useful measure of their densities is to compare **biomass**.

7. The equation $\Delta N/\Delta t = (b - d)N$ describes **exponential** growth.

8. The total number of individuals of a species per unit of area (or volume) is called the **population density**.

9. Plants, corals, and other organisms that are composed of repeated sets of similar units are called **modular organisms**.

10. When a single bacterium is placed in a culture vessel, the resulting population at first grows very quickly; however, as the population expands and uses up its resources, the growth rate eventually slows down and stops. This S-shaped growth pattern is best modeled mathematically as **logistic** growth.

11. Population regulation that works to decrease numbers when its members are common and increase numbers when they are rare is **density dependent**.

12. If the death rate in a population is unrelated to the number of individuals in an area, the resulting change in population size is said to be **density independent**.

13. A short-term event that disrupts populations, communities, or ecosystems by changing their environment is called a **disturbance**.

14. Killing part of a density-dependent population helps to bring it back to a level where it can achieve the most rapid rate of growth, but a better method than killing is to remove its **resources** to lower carrying capacity.

15. A **population** consists of all the individuals of a species within a given area.

Multiple Choice

1. Which one of the following characteristics of whales is *not* a factor in the whale population's slow response to management practices?
   a. Whales have long prereproductive periods.
   b. Whales produce one offspring at a time.
   c. Whales have long life spans.
   d. Whales have long intervals between births.
   e. Whales are hunted by several nations.

2. Of the following factors that regulate population size, the most density-independent factor is
   a. food supply.
   b. predators.
   c. disease.
   d. available nesting sites.
   e. sudden temperature changes.

3. If a population's birth rate is density-_____ , it will _____ as the population increases.
   a. dependent; increase
   b. dependent; decrease
   c. independent; increase
   d. independent; decrease
   e. dependent or independent; increase

4. Which of the following undergoes the most radical metamorphosis in moving from a larval to an adult stage?
   a. Bee
   b. Caribou
   c. Locust
   d. Turtle
   e. Salmon

5. Agave and yucca plants appear similar and grow in the same environments, but agaves reproduce once and die, while yuccas reproduce several times before dying. Compared to agave, each yucca seed crop should use _____ energy and produce _____ seeds.
   a. more; more
   b. more; fewer
   c. equal amounts of; equal amounts of
   d. less; more
   e. less; fewer

6. Undesirable alleles are least affected by natural selection if they are expressed in
   a. embryos.
   b. the young.
   c. adolescents.
   d. young adults.
   e. the old.

7. For the last thirty years, the average human age of death in the United States has _____ due to _____ .
   a. fallen; new strains of infectious disease
   b. fallen; environmental pollution
   c. **remained stable; deaths resulting from genetic diseases of old age**
   d. risen; control of infectious diseases
   e. risen; elimination of most factors that cause death

8. In which of these populations would the individuals be evenly spaced?
   a. **Penguins incubating their eggs just out of pecking range of neighboring penguins**
   b. Maple seedlings sprouting around the base of the parent tree
   c. Coconut seeds germinating on beaches after being carried by tides and water currents
   d. Beaver offspring living with their parents for several years
   e. Giraffes coming to a water hole to drink

9. If a certain age interval in a population contains ten individuals and has a death rate of 1.000, how many individuals will be in the next older age interval?
   a. **None**
   b. One
   c. Five
   d. Nine
   e. Ten

10. The number of individuals in a population is least affected by the rate of
    a. births.
    b. deaths.
    c. **mating.**
    d. immigration.
    e. emigration.

11. A large species range is usually associated with a species that
    a. has higher population densities at the center of the range.
    b. **has high local population densities.**
    c. is relatively rare.
    d. migrates.
    e. has stringent environmental requirements.

12. Which of the following is *not* true of ecological disturbance?
    a. Adaptations to disturbance occur among most organisms.
    b. Adaptations are most common if the disturbance is frequent.
    c. **Large organisms usually are affected more than small organisms.**
    d. The frequency of some disturbances can be influenced by organisms.
    e. Examples of biological disturbance are tree falls and diseases.

13. In northern Scandinavia, rodents called lemmings have irruptions, or periodic buildups of large populations. The most likely cause of these irruptions is
    a. **an increase in food supply.**
    b. a decrease in predator population.
    c. an increase in favorable nesting sites.

    d. an increase in birth rate.
    e. a decrease in disease rate.

14. A population, such as Sweden's, with a low birth rate and a low death rate
    a. **would have a relatively even distribution of individuals of different ages.**
    b. would have a population dominated by young individuals.
    c. would have a population dominated by old individuals.
    d. would have a population dominated by individuals of intermediate age, with relatively few young or old individuals.
    e. could have almost any age distribution. Birth and death rates do not affect the age distribution.

15. Which of the following is *not* an important part of the life history of all organisms?
    a. Birth: When an organism is born.
    b. Energy-gathering stage: When individuals gather energy and nutrients for growth and reproduction. This stage may encompass the entire life history.
    c. **Parental care stage: When parents provide additional care and protection during early growth.**
    d. Dispersal stage: When individuals move or are moved from their birthplaces to the places where they will live as adults.
    e. Reproductive stage: When offspring are produced.

16. Which of the following is *not* a demographic process that determines the number of individuals in a population?
    a. Death
    b. Birth
    c. **Migration**
    d. Immigration
    e. Emigration

17. For modular organisms it is often preferable to measure the biomass of a population rather than the population density because
    a. modular organisms are easy to distinguish, and most adult members of a population are similar to one another in size and shape.
    b. modular organisms often reproduce quickly and die, making it very difficult to obtain an accurate count of individuals.
    c. modular organisms (with very few exceptions) have very high densities, making it impractical to count individuals. Biomass is measured as a means to avoid counting all of the individuals.
    d. **modular organisms often vary greatly in size and shape (within a species) and individuals are not easy to distinguish.**
    e. The statement is false. For modular organisms it is almost never preferable to measure biomass.

18. In a life table, the number of individuals alive at the beginning of the 1-year to 2-year age interval is 800. During this interval 200 individuals die. The death rate for this interval is
    a. **0.25.**
    b. 200.
    c. 800.
    d. 0.2.
    e. 0.8.

19. Highway 2 runs through the middle of a chicken farm. The chickens' roost is on one side of Highway 2 and their feeding trough is on the other side. Once per day at a random time each chicken crosses the road to get from the roost to the feeding trough, eats its fill, and returns to the roost. Needless to say, the chickens are occasionally struck by cars on the highway and killed. If the random event of being hit by a car is the main form of mortality of these chickens, which survivorship curve would be expected?
    a. A type I curve: nearly all individuals survive for their entire potential life span and die almost simultaneously.
    b. **A type II curve: survivorship is the same throughout the life span.**
    c. A type III curve: survivorship of young individuals is very low but then survivorship is high for the remainder of the life span.
    d. A combination of type I and type III curves: high juvenile mortality but adults survive for their entire potential life span, then die almost simultaneously.

20. A sheep rancher likes to raise her sheep for about 2 years, then sell them off to a local butcher to make mutton chops. Although this rancher takes good care of her sheep, the local varmints (probably the neighbors' dogs) often break into her pasture and kill many of the young lambs. The predators must be fairly small because they do not seem to bother the adults, who usually survive until they are shipped out. What kind of survivorship curve describes this population?
    a. A type I curve: nearly all individuals survive for their entire potential life span and die almost simultaneously.
    b. A type II curve: survivorship is the same throughout the life span.
    c. A type III curve: survivorship of young individuals is very low but then survivorship is high for the remainder of the life span.
    d. **A combination of type I and type III curves: high juvenile mortality but adults survive for their entire potential life span, then die almost simultaneously.**

21. Which of the following is true concerning exponential growth?
    a. **No population can grow exponentially for long.**
    b. Exponential growth slows down as the population nears its maximal size.
    c. Bacterial colonies have been observed to maintain exponential growth for over a month.
    d. Exponential growth is commonly observed in large, slow-growing species such as humans and elephants.
    e. Exponential growth includes a component of environmental resistance.

22. The intrinsic rate of increase, r_{max}, is
    a. the number of individuals added to each generation in a growing population under optimal conditions.
    b. **the difference between the average per capita birth rate, b, and the average per capita death rate, d, under optimal conditions.**

c. the number of individuals added to each generation in a growing population under the conditions that are actually occurring.
    d. the average per capita birth rate, b, under optimal conditions.
    e. the difference between the average per capita birth rate, b, and the average per capita death rate, d, under the conditions that are actually occurring.

23. The population growth equation $\Delta n/\Delta t = r(K - N/K)N$ describes
    a. a population that grows without limits.
    b. **a population that grows rapidly at small population sizes, but whose growth rate slows and eventually stops as the population reaches the number the environment can support.**
    c. a population that rapidly overshoots the number the environment can support and then fluctuates around this number.
    d. a population that grows very rapidly and then crashes when the environmental resources are used up.
    e. a population that is declining in size.

24. Which of the following is *not* an assumption of the logistic growth equation?
    a. An individual exerts its effects immediately at birth.
    b. All individuals produce equal effects.
    c. Resources in the environment are limited.
    d. Each individual depresses population growth equally.
    e. **There is a time lag between gathering resources and reproduction.**

25. Under which circumstances might natural selection favor delaying reproduction until later in life?
    a. Juvenile survival is very poor, and parents can produce large numbers of very small offspring with little cost to themselves.
    b. Juvenile survival is high, and parents always die following reproduction.
    c. **Juvenile survival increases with the experience of the parent, and reproduction greatly reduces the chances for parental survival.**
    d. Juvenile survival is high, and reproduction has little effect on parental survival.
    e. Juvenile survival is high, and adult survival independent of reproduction is very poor.

26. Large numbers of small offspring are favored by natural selection
    a. if large offspring survive so much better than small offspring that more of them survive to reproductive maturity.
    b. in environments where competition among members of the same species is intense.
    c. **in fluctuating environments where disturbances create unoccupied sites.**
    d. in populations that are always near the maximum number the habitat can support.
    e. in mature communities where all of the available sites are occupied.

27. The reproductive value of an individual is greatest
    a. at birth.

b. **just before first reproduction.**
c. halfway through the reproductive stage.
d. just after it is finished reproducing.
e. just before death.

28. Genetic traits that are highly detrimental often do not appear until late in life. A possible explanation for this is that
    a. young individuals are physiologically strong and able to mask the effects of detrimental genetic traits. As an individual ages and becomes physiologically weaker, these traits start to be expressed.
    b. mutations are occurring throughout an individual's lifetime. The probability that an individual will have a highly detrimental mutation increases with age.
    c. traits that enhance the fitness of young individuals become detrimental to older individuals that have different ecological and physiological needs.
    d. reproduction decreases the energy available for the maintenance and growth of the parent. Because older individuals are putting so much energy into reproduction, they are unable to repair the damage caused by genetically deleterious traits.
    e. **natural selection tends to favor the delay of the phenotypic expression of detrimental traits until individuals have completed most or all of their reproduction.**

29. A favorable habitat can change temporarily into an unfavorable one because of environmental changes, such as a drought or the coming of winter. Which of the following is *not* a response or adaptation that organisms use to cope with these temporary, unfavorable conditions?
    a. Migration
    b. A resistant resting stage
    c. **Irruptions**
    d. Hibernation
    e. Storing food

30. Which of the following is an example of density-independent population regulation?
    a. A contagious disease sweeps through a dense population of lemmings.
    b. Jaegers (predators of lemmings) search widely for places where lemmings are abundant and concentrate their hunting in those areas.
    c. An outbreak of lemmings leads to a depletion of the food supply. As a result lemmings are underfed and produce few offspring.
    d. Arctic foxes switch from primarily eating mice to primarily eating lemmings when lemmings are abundant.
    e. **An early cold spell kills 80 percent of the lemmings in a particular area.**

31. A major difference between populations with density-dependent birth or death rates, or both, and populations with density-independent birth and death rates is that
    a. **density-dependent populations have an equilibrium population size; density-independent populations do not have an equilibrium population size.**
    b. density-dependent populations grow more rapidly than density-independent populations.

    c. density-dependent populations never actually occur in nature, whereas density-independent populations are frequently found in nature.
    d. density-dependent populations are more likely to go extinct than density-independent populations.
    e. density-dependent populations are more likely to live in disturbed habitats than density-independent populations.

32. Organisms are likely to have adaptations for tolerating a particular kind of disturbance
    a. if the disturbance is unusually severe.
    b. if the disturbance is rare and occurs at highly unpredictable intervals.
    c. **if the disturbance is frequent relative to the organisms' life span.**
    d. if the disturbance is of very recent human origin.
    e. if the disturbance is unusual in its timing or location.

33. Which of the following statements is true concerning fires as a disturbance?
    a. No plants have successfully adapted to fire as a form of disturbance.
    b. Fires normally destroy all of the seeds of the plants growing in the area.
    c. Fires are extremely unpredictable.
    d. Fires are a recent form of disturbance because they did not occur before humans evolved.
    e. **The frequency of fires may be proportional to the rate at which vegetation accumulates as fuel.**

34. If we wish to maximize the number of individuals in a deer population so that many deer can be harvested, we should
    a. manage the populations so that they are at the number the environment can support to maximize the number of individuals available.
    b. **manage the populations so that they are far enough below the number the environment can support to have high birth and growth rates.**
    c. manage the populations so that the animals are very rare and don't come into contact with each other.
    d. manage the populations so that they slightly exceed the number the environment can support so that the excess can be harvested.
    e. manage the populations so that they greatly exceed the number the environment can support to maximize the number of excess individuals.

35. The income of fishermen that fish the Georges Bank has declined because
    a. in recent years, through careful management, the fish populations have exploded. This has resulted in a glut on the market and a lowering in the price of fish.
    b. fish populations have remained steady, as has the overall yield. However, there are more fishermen sharing the same catch, resulting in a lower income per fisherman.
    c. **through overfishing, fish populations have been reduced. More effective (and expensive) equipment and more effort is used to get the same yield.**

d. Americans are increasingly turning to beef as a source of protein. Even though there are ample fish on the Georges Bank and elsewhere, there is little demand for fish.

e. major new fishing grounds have been found throughout the world. This has resulted in a glut on the market and a lowering in the price of fish.

36. Which would be the most effective way to minimize a rat population in an alley?

    a. Kill as many rats as possible by poisoning and trapping.

    b. **Clean up the alley so that the rats have no garbage to feed on.**

    c. Lure the rats away to another site where they will be less harmful.

    d. Search out and kill very young prereproductive rats.

    e. Release cats into the alley.

37. Preserving a rare, endangered species will usually be very expensive unless

    a. we understand all aspects of its biology and reproduction.

    b. we bring the species into zoos and manage it in a controlled setting.

    c. we continually provide additional resources such as food.

    d. we can make it economically profitable by harvesting the species as a food source.

    e. **we provide it with sufficient suitable habitat.**

38. The efts, or young, of the red-spotted newt (a salamander that breeds in small ponds in the Appalachian Mountains) sometimes emigrate from their pond of birth to find a new pond. Efts apparently travel long distances, because they have colonized most Appalachian ponds. When studying the population dynamics of the red-spotted newt, the appropriate scale of study is

    a. a portion of a single pond where a few of the adults habitually lay eggs.

    b. a single pond.

    c. a few widely spaced ponds.

    d. **all (or most) of the ponds within a large area.**

    e. all (or most) of the ponds in the United States.

39. Desert plants often compete for water. Young plants germinating near a mature plant normally die because they cannot compete successfully for water. Plants that germinate far away from a mature plant are much more successful and often grow to maturity. Based on this, you would expect the mature plants to be

    a. randomly scattered.

    b. aggregated into clumps.

    c. **evenly spaced.**

    d. aggregated into clumps that are evenly spaced.

    e. aggregated into clumps that are randomly scattered.

40. When a time delay is added to the logistic population growth equation

    a. the population growth rate will gradually slow down until it reaches the number the environment can support, at which time the population will stop growing.

    b. the population will grow rapidly until it reaches the number the environment can support, then abruptly stop growing.

    c. the population will slowly decline and eventually go extinct.

    d. the population will grow without limits.

    e. **the population may shoot past the number the environment can support, and eventually fluctuate around that number.**

41. Which of the following would *not* contribute to a clumped pattern of distribution?

    a. Offspring settling close to their place of birth

    b. **Competition among individuals for resources**

    c. Chance

    d. The distribution of suitable food resources

    e. The location of favorable habitat

42. The analysis of population dynamics is simpler for unitary organisms than for modular organisms because

    a. unitary organisms are constructed of repeated units.

    b. unitary organisms are more likely to be sessile.

    c. birth rates are higher for modular organisms.

    d. **individuals of unitary organisms are more easily distinguished.**

    e. the genetic structure of populations of modular organisms is more complex.

43. A straight line relationship with negative slope between age and percent survival describes a survivorship pattern in which

    a. most individuals die at about the same age.

    b. young individuals have the highest death rates.

    c. **the probability of death is about the same for all ages.**

    d. the majority of individuals survive to old age.

    e. death rates increase following reproduction.

44. What pattern of survivorship is most likely for an organism such as the sea urchin, which produces large numbers of very small offspring and provides no parental care?

    a. Most individuals survive for most of their potential life span and then die simultaneously.

    b. The probability of death is low throughout the life span.

    c. Survivorship is high early in life and declines slowly.

    d. **Young individuals have high probability of dying but older individuals have relatively low probability of dying.**

    e. Probability of death is equally high throughout the potential life span.

45. The growth rate of a population in an unlimited environment

    a. is described by an S-shaped curve.

    b. is greater than the intrinsic rate of increase.

    c. **is equal to $(b - d)N$.**

    d. depends on competition for resources.

    e. is infinite.

46. Which of the following is *not* an assumption of the logistic model for population growth?

a. Members of the population compete for resources at high density.
b. All individuals have the same effect on population growth.
c. **Population growth rate increases as the size of the population approaches the carrying capacity.**
d. There is no time lag between birth and the time an organism begins to affect population growth.
e. The environment imposes a carrying capacity on the population.

47. Which of the following is *not* a likely result of devoting more resources to reproduction?
a. Increased number of offspring produced
b. Increased probability of death for the parent
c. **Increased growth rate for the parent**
d. Increased probability of survival for the offspring
e. Decreased probability of survival for the parent

48–51. Use the graph below to answer Questions 48–51.

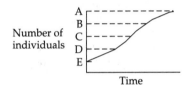

48. At what population size is the growth rate of this population greatest?
a. A
b. B
c. **C**
d. D
e. E

49. At what population size is the term $(K - N)/K$ largest?
a. A
b. B
c. C
d. D
e. **E**

50. At what population size is the rate of population growth beginning to slow down?
a. A
b. **B**
c. C
d. D
e. E

51. What would be the most effective practical means of managing this population to maximize the number of individuals harvested over a long period of time?
a. Increase the carrying capacity
b. Decrease the carrying capacity
c. Harvest until the population size is equal to B
d. **Harvest until the population size is equal to C**
e. Increase r

52. Which of the following is true of the carrying capacity (K)?
a. When $N = K$, the birth rate in a population is zero.
b. The rate of population growth in an unlimited environment is proportional to K.
c. K is always determined by the amount of food in an environment.

d. **In a population at its K, the birth rate equals the death rate.**
e. K changes over time for each population.

53. Which of the following would be unlikely to contribute to large fluctuations in population size?
a. Frequent disturbance
b. A high death rate
c. **A density-dependent birth rate**
d. A high birth rate
e. Seasonal variation in temperature

54. A population of trees consists of large numbers of seedlings and saplings and a few reproducing adults. Which of the following is likely to be true for this population?
a. The birth rate is density-dependent.
b. The population is close to K.
c. The death rate is low.
d. **Birth and death rates of young are high.**
e. Its members are evenly spaced.

55. Which of the following is *not* true of dispersal?
a. Individuals may disperse as either juveniles or adults.
b. All organisms have a dispersal stage in their life history.
c. Dispersal can be an adaptation to disturbance.
d. **Organisms disperse only once in their life cycle.**
e. Organisms can disperse in time as well as in space.

56. A graph showing the proportion of organisms of an initial group alive at different times throughout the life span is
a. a cohort.
b. a life table.
c. **a survivorship curve.**
d. an age structure diagram.
e. a population growth curve.

57–59. Use the following life table for a hypothetical cohort of organisms to answer Questions 57–59.

Age interval (yr)	Number alive at beginning	Number dying	Death rate	Number of offspring
0–1	200	150	0.75	0
1–2	50	10	0.20	0
2–3	40	20	0.50	3
3–4	20	(?)	0.50	9
4–5	10	10	1.00	4
5–6	0	—	—	—

57. During which of the following age intervals does this organism have the highest risk of mortality?
a. **0–1**
b. 1–2
c. 2–3
d. 3–4
e. Cannot tell from the information in the table

58. How many individuals died during the interval 3–4?
a. 5
b. **10**
c. 20
d. 40
e. 50

59. From the life table, you can conclude that
    a. individuals begin reproducing in their first year of life.
    b. the oldest individuals have the largest clutch size.
    c. **on average, clutch size is larger for age interval 3–4 than for the interval 4–5.**
    d. clutch size increases continuously with age.
    e. individuals of all ages produce about the same number of offspring.

60. The intrinsic rate of increase is
    a. the rate of population growth in a limited environment.
    b. the maximum possible birth rate for a population.
    c. **average birth rate minus the average death rate under optimal conditions.**
    d. the average clutch size for individuals in a population in an unlimited environment.
    e. the rate at which a population colonizes a favorable new habitat.

61. A delay between the time an organism is born and the time it begins to affect other members of the population causes
    a. **population size to oscillate around the carrying capacity.**
    b. correlation between birth rate and death rate in the population.
    c. a decrease in the carrying capacity of the population.
    d. fluctuation in the intrinsic rate of increase.
    e. low mortality rates of the young.

62. A population whose dynamics are dominated by periodic recovery from crashes will
    a. have a lower carrying capacity than a population *not* subject to crashes.
    b. **be well below carrying capacity most of the time.**
    c. live in a very stable environment.
    d. never reach exponential growth rates.

63. Many fish populations can be heavily harvested on a sustained basis because
    a. there are many "excess" individuals since populations frequently overshoot their carrying capacity.
    b. they are not limited by the availability of suitable habitat.
    c. **only a small number of females are needed to produce enough eggs to maintain population size.**

    d. natural sources of mortality are low because their natural predators have gone extinct.
    e. they each reproduce only once and would have died soon if not harvested.

64. A population of seed-eating rodents infests your barn and consumes the corn stored there. Which of the following would likely be the most successful way to control the population of these pests?
    a. Leave corn out in the surrounding fields to lure the rodents away from the barn.
    b. Set large numbers of traps to kill as many rodents as possible.
    c. Introduce a voracious predator.
    d. Introduce a corn-eating competitor.
    e. **Store grain in rodent-proof bins to reduce their access to corn and reduce their carrying capacity.**

65. Growth of the human population from prehistoric times to the present
    a. **roughly fits an exponential curve.**
    b. exhibits large oscillations around K.
    c. follows and S-shaped curve.
    d. has slowed considerably over the last 50 years.
    e. has been strongly influenced by density-dependent regulation.

66. An example of a trade-off would be
    a. Pacific salmon reproduce only once and then die.
    b. The larvae of red-spotted newts transform into efts and live on land before returning to the water to reproduce.
    c. Dall sheep have low survivorship of young individuals, then survivorship increases in the middle of the life span and decreases again at the end of the life span.
    d. **Humans and whales take many years to come to reproductive age but then survival rates for the adults are higher than animals that reproduce at a younger age.**
    e. The introduction of predators to reduce a population to increase its carrying capacity.

67. Currently the most important carrying capacity of Earth for humans is set by
    a. limitations on food resources.
    b. **the ability to absorb waste products of fuel consumption.**
    c. disease.
    d. predation.
    e. space limitations.

52 Community Ecology

Fill in the Blank

1. If one participant in an interaction is harmed but the other is unaffected, the interaction is **amensalism**.

2. If both partners in an interaction benefit, the interaction is **mutualism**.

3. If two organisms use the same resource, and the resource is insufficient to supply their combined needs, then the interaction is **competition**.

4. A wasp lays an egg in a caterpillar. The resulting larva totally consumes and kills the caterpillar, then pupates and becomes an adult. This wasp is a **parasitoid**.

5. An animal that is much smaller than its host, and may or may not kill it, is a **parasite**.

6. A palatable species takes on the appearance of an inedible or noxious species. This is known as **Batesian mimicry**.

7. Two or more unpalatable species take on a similar (often bright) appearance. This is known as **Müllerian mimicry**.

8. When two species of *Paramecium* are placed together in a test tube, and one species drives the other to extinction, this is an example of **competitive exclusion**.

9. Growth on bare rock that at the beginning supports no organisms, and ends up supporting a mature forest, is **succession**.

10. When the dead body of a freshly killed mouse decomposes, the change in the community of decomposers in the body is called **degradative succession**.

11. Animals such as cows that eat only plant tissues are called **herbivores**.

12. All of the organisms that live in a particular area constitute an ecological **community**.

13. An **ecological niche** is defined as the set of environmental conditions under which a species can live.

14. When a species has a larger influence on the community than is expected by its abundance, it is called a **keystone** species.

15. Animals that eat other animals are called **carnivores**.

16. When two species have mutually influenced each other's evolution, they are said to have **coevolved**.

Multiple Choice

1. The relationship between a human and the fungus that causes athlete's foot is _____ when the fungus feeds only on dead skin cells, but becomes _____ if the fungus penetrates the skin and feeds on living cells.
   a. amensalistic; commensalistic
   b. commensalistic; amensalistic
   c. a host–parasite interaction; commensalistic
   d. amensalistic; a host–parasite interaction
   e. **commensalistic; a host–parasite interaction**

2. Intraspecific competition may affect individuals in all of the following ways, except via
   a. a reduced growth rate.
   b. a lower reproductive rate.
   c. **reduced requirement for resources.**
   d. exclusion from better habitats.
   e. death.

3. Two mouse species are observed to inhabit the same forested area. Which of the following is *not* a possible reason why the two species are able to coexist?
   a. One species eats mostly berries and the other eats mostly nuts.
   b. Predation by hawks prevents either mouse population from becoming large.
   c. The forest environment is very heterogeneous.
   d. **The preferred habitat for both species is fields, but a third mouse species has excluded them from the fields.**
   e. The two species nest at different times of the year.

4. Which of the following is *not* true of the compound organisms called lichens?
   a. One member provides nutrients and support.
   b. One member is photosynthetic.
   c. They are partly composed of cyanobacteria or green algae.
   d. The relationship of the two members is mutualistic.
   e. **They are restricted to favorable habitats like tropical rainforests.**

5. Which characteristic of moose is most significant in determining that moose are a "keystone species" on Isle Royale in Lake Superior?
   a. **A few moose have a large effect on forest succession.**
   b. Moose are the largest animals in the ecosystem.
   c. Moose browse on a large variety of tree species.
   d. Moose use both terrestrial and aquatic habitats.
   e. Moose care for their young longer than any other species in the ecosystem.

6. If the removal of a species from a community has a greater effect on the structure and functioning of the community than one would have predicted from the species' abundance, then that species is most likely
   a. **a keystone species.**
   b. a carnivore.
   c. a herbivore.
   d. an early successional species.
   e. a mimic of another species.

7. The fact that many tree species and mycorrhizal fungi cannot survive unless they are associated with one another indicates that their relationship is
   a. amensalistic.
   b. commensalistic.
   c. competitive.
   d. **mutualistic.**
   e. a parasite–host interaction.

8. A female fig wasp enters the syconium of a fig, pollinates the flowers, and lays eggs in the ovaries of some of the flowers. The young larvae grow up, eat (and kill) some, but not all, of the seeds, and complete their life cycle. The fig is completely dependent on fig wasps to pollinate its flowers, and the fig wasp requires figs to complete its life cycle. The interaction between figs and fig wasps has aspects of
   a. mutualism and host–parasite interaction.
   b. commensalism and mutualism.
   c. amensalism and host–parasite interaction.
   d. predator–prey or host–parasite interaction.
   e. **a and d**

9. A species of wasp (not the fig wasp) lays its eggs in an already fertilized syconium of a fig. The wasp does this by inserting her long ovipositor through the wall of the syconium and laying her eggs in the developing seeds. The larvae of this wasp eat (and kill) the seeds of the fig. This wasp does not pollinate the fig or in any other manner benefit the fig. The interaction between figs and this wasp has aspects of
   a. mutualism.
   b. commensalism.
   c. amensalism.
   d. **predator–prey or host–parasite interaction.**
   e. competition.

10. Which of the following would *not* be considered a resource for a terrestrial animal?
    a. Food
    b. **Humidity**
    c. Hiding places
    d. Nest sites
    e. Water

11. The barnacle *Chthamalus stellatus* is capable of growing at much lower depths than it is actually found in nature. Experimental studies have determined that the lower limit of *Chthamalus* is determined by another barnacle, *Balanus balanoides*. The rapidly growing *Balanus* is able to smother, crush, or undercut the slower-growing *Chthamalus*. Because of this, *Chthamalus* cannot survive in the lower depths where *Balanus* is found. The interaction between these two barnacles is an example of
    a. predator–prey interaction.
    b. mutualism.
    c. commensalism.
    d. amensalism.
    e. **competition.**

12. A resource required by all animals is oxygen. Which of the following statements is true concerning the effects of this factor on the distribution and abundance of animals?
    a. Because oxygen is such a fundamental requirement of all animals, it is critical in determining the distribution and abundance of terrestrial animals.
    b. **Oxygen is unlikely to be an important factor in the structure of terrestrial communities because it is nearly always available at levels greater than the minimum required for survival.**
    c. Oxygen is unlikely to be an important factor in the structure of aquatic communities because it is nearly always available at levels greater than the minimum required for survival.
    d. In terrestrial communities, many factors, including oxygen levels, affect the distribution and abundance of animals. The structure of terrestrial communities is in part, but not completely, determined by oxygen levels.
    e. In terrestrial communities, oxygen levels only become an important factor in determining community structure when animals are near the edges of their range. When animals are stressed by other factors, variation in oxygen levels becomes a critical factor.

13. In a homogeneous environment, when prey are above a certain threshold, predators can find enough to eat. As a result of this, they will produce more offspring, which will consume more prey. Eventually this will drive the prey numbers below a threshold value, and the predators will begin to starve. Nevertheless, the predators will continue to eat prey even when they are scarce, reducing the prey to an even lower level. Eventually some predators will starve, and their numbers will be reduced, allowing the numbers of prey to increase above the threshold where the predators can again reproduce. This process results in
    a. extinction of the prey, but not the predator.
    b. oscillations in the density of the prey only.
    c. oscillations in the density of the predator only.
    d. **oscillations in the densities of both the predator and the prey.**
    e. stabilization of the predator and prey densities at an intermediate value.

14. Which of the following is *not* an example of an adaptation of prey to avoid predation?
    a. Many tropical katydids look like dead leaves.
    b. **Male stickleback fish develop very bright colors during the breeding season.**
    c. Several moth caterpillars have sharp bristles that contain toxic chemicals.
    d. Many moths have large eye spots that they flash when they are disturbed.
    e. Gazelles (African antelopes) are very fast runners.

15. The edible viceroy butterfly looks like the inedible monarch butterfly. This similarity most likely evolved because
    a. the coloration used by the viceroy butterfly and the monarch butterfly is ideal for absorbing heat on cold mornings.

b. these two butterflies are closely related and their similar coloration reflects a common ancestry.

c. there are only a limited number of color patterns a butterfly can have. It is highly likely that there will be two butterflies with similar color patterns simply by random chance.

d. the larvae of these two butterflies live on the same plant. The color pattern is a consequence of the food they eat.

e. predators frequently mistake viceroy butterflies for monarch butterflies, which they have learned to avoid.

16. Which of the following is *not* true of parasites?

a. The more slowly they kill their host, the better their chances to be transferred to another host.

b. They can only be transferred from one living host to another living host.

c. If a parasite grows too slowly in its host, it could be outcompeted by another parasite.

d. Parasites can have more than one species of host.

e. Parasites do not always kill their host.

17. Which of the following is *not* a chemical defense used by plants as a defense against herbivory?

a. Chemicals that interfere with the transmission of nerve impulses to muscle

b. Chemicals that imitate insect hormones and thereby block insect metamorphosis

c. Unusual amino acids that become incorporated into proteins and interfere with their functioning

d. Chemicals that are extremely difficult to digest

e. Odorous chemicals that attract pollinating insects

18. In a famous series of experiments, G. F. Gause mixed two species of *Paramecium* together in test tubes containing various different culture media. In the experiments in which the environment within the test tubes was uniform, Gause observed that

a. the two species frequently coexisted, with intermediate numbers of both species.

b. one species was always driven to extinction. Which species won the competition depended on the culture conditions.

c. under certain conditions, the two species would coexist. If the conditions deviated from these optimal conditions, one or the other species would be driven to extinction.

d. *Paramecium caudatum* was able to competitively exclude all other species, regardless of conditions.

e. each species was unaffected by the presence of the other species.

19. An experiment was performed to investigate the interactions between the common meadow vole, *Microtus pennsylvanicus* (primarily a grassland species), the red-backed vole, *Clethrionomys gapperi* (a woodland species), and the deer mouse, *Peromyscus maniculatus* (a woodland species). In this experiment, plots 0.2 hectare in size that were half-wooded and half-grassland were fenced off. In half of the plots, all three species were added. In the other half, only the red-backed vole and the deer mouse were added (the common meadow vole was excluded). At the end of the experiment, it was found that in plots where all three species were present, the common meadow vole was found exclusively in the grassland, and the red-backed vole and the deer mouse were found exclusively in the woodland. In the plots with only two species present, the red-backed vole and the deer mouse were found in both the woodland and the grassland portions of the plots. These results indicate that

a. the common meadow vole competitively excluded the red-backed vole and the deer mouse from the grassland.

b. the red-backed vole and the deer mouse avoided the grassland because of the higher predation in that habitat.

c. the red-backed vole and the deer mouse competitively excluded the common meadow vole from the woodland.

d. the common meadow vole does not eat woodland species of seeds.

e. the common meadow vole is a more efficient forager than the red-backed vole and the deer mouse.

20. The sea star *Pisaster ochraceous* is an abundant predator on the rocky intertidal communities on the Pacific coast of North America. The sea star feeds preferentially on the mussel *Mytilus californianus*. In the absence of the sea star, the mussel is a dominant competitor. Predation by the sea star, however, creates bare areas. For this reason

a. the number of species present will be changed by the presence of the sea star, but the direction of change cannot be predicted.

b. The variety of species should be unaffected by the presence or absence of the sea star.

c. The variety of species present should be greater when the sea star is present than when it is absent.

d. The variety of species present should be less when the sea star is present than when it is absent.

e. *Mytilus* should go extinct when the sea star is present.

21. Cattle egrets follow cattle around. By doing so, they are able to capture insects more efficiently, because the cattle disturb the insects as they walk, making them easier to catch. There is no cost or benefit to the cattle from this interaction. This interaction is an example of

a. commensalism.

b. amensalism.

c. mutualism.

d. parasitism.

e. competition.

22. Legumes, such as soybeans, form root nodules that become infected by *Rhizobium* bacteria. These bacteria convert nitrogen into a form that is usable by plants. The plants benefit because they have a source of nitrates, which are frequently limited in terrestrial environments. The bacteria benefit because they receive nutrients and energy from the plants. This interaction is an example of

a. commensalism.

b. amensalism.

c. mutualism.

d. parasitism.

e. predation.

23. Which of the following is *not* an example of mutualism?
    a. Ants protect aphids from predators and "milk" the plant-sucking insects for honeydew (partly digested plant sap).
    b. Cleaner wrasse (a fish) enter the mouths of barracuda (a predatory fish) to remove parasites, which the wrasse eat.
    c. Staphilinid beetles live inside of ant colonies and eat ant eggs, larvae, and pupae. They avoid being killed by the ants by feeling and smelling like ants.
    d. Honeyguides (birds that eat beeswax) guide humans to bees' nests. The humans chop open the nests, take the honey, and leave the wax for the honeyguide.
    e. Termites are able to eat wood because they have protists living in the protected environment of their guts that help them digest cellulose.

24. Which of the following is an example of diffuse coevolution?
    a. Figs are pollinated only by fig wasps, which lay their eggs only in fig syconia.
    b. *Pseudomyrmex* ants live in the thorns of *Acacia cornigera* and protect the acacias from herbivorous insects, mammals, and vines.
    c. The edible viceroy butterfly looks like the inedible monarch butterfly.
    d. Most bird-pollinated flowers are red, a color that attracts many birds.
    e. *Paramecium bursaria* and *P. aurelia* are able to coexist in the same test tube if there is deoxygenated water at the bottom of the container.

25. Which of the following could *not* be used as an example of succession?
    a. A landslide exposes bare rock. Eventually a forest is reestablished on the site.
    b. An alien weed is introduced into the Hawaiian Islands. It eventually displaces a native plant, driving the native plant to extinction.
    c. A freshly killed mouse decomposes.
    d. A tree falls in the forest. Eventually a new tree replaces it.
    e. The leaf litter under a pine tree is in layers, with freshly fallen litter at the top and more decomposed layers deeper down.

26. Two hypothetical islands in one of the Great Lakes were clear-cut and allowed to regenerate. Moose were present on one of the islands, but absent on the other island. Other than this, the two islands were identical. Moose preferentially feed on early successional deciduous plants, such as mountain ash, mountain maple, aspen, and birch. They rarely eat conifers, such as white spruce and balsam firs, species that replace deciduous trees during succession. In comparing succession on these two islands, you would expect that
    a. the successional change from deciduous trees to conifers would be slower on the island with moose than on the island without moose.
    b. the successional change from deciduous trees to conifers would not occur on the island with moose, but would occur on the island without moose.
    c. the successional change would be from deciduous trees to conifers on the island with moose, and from conifers to deciduous trees on the island without moose.
    d. the successional change from deciduous trees to conifers would be faster on the island with moose than on the island without moose.
    e. the moose should not affect the order or rate of successional change.

27. The presence of only a few species of small mammals at local sites, despite the large number of such species living on the continent, suggests
    a. that local habitats are very isolated from one another.
    b. that small mammals maintain large territories.
    c. that competition within local habitats prevents the coexistence of similar-sized species.
    d. that small mammals have short dispersal distances.
    e. that small mammal populations frequently go extinct.

28. In the tropics there are several species of inedible butterflies that all have virtually identical bright coloration. The most likely explanation for the formation of these groups of similar-looking species is that
    a. all of these similar-looking species are closely related. They are all inedible, and all look similar because they are closely related.
    b. the coloration pattern is cryptic and allows the butterflies to hide. They all live in similar habitat so they look similar.
    c. inedible species benefit from looking like other inedible species because predators rapidly learn to avoid all similarly colored species when they eat any member of the group.
    d. all of these species have similar means of locating mates, and the coloration is an important aspect of this process.
    e. all of these species live on the same host plant, and the coloration pattern results from the plant that they use as a food source.

29. In the ecological relationship between fig wasps and fig trees described in your textbook, the male wasps
    a. fly from one tree to the next, pollinating the fig flowers.
    b. emerge from fig seeds after the female wasps have emerged.
    c. never leave the fig.
    d. have specialized structures to carry pollen.
    e. defend the fig tree against herbivores.

30. A community ecologist would most likely be concerned with
    a. energy flow through an ecosystem.
    b. population growth of a single species.
    c. interactions between individuals of the same species living together in a small area.
    d. interactions between individuals of different species living together in a small area.
    e. the cycling of matter through biotic and abiotic components of an area.

31–32. Certain woodpecker-like African birds have become specialized for removing and eating ticks and other parasitic insects from the bodies of large herbivores.

31. The relationship between the birds and the ticks is an example of
    a. mutualism.
    b. parasitism.
    c. commensalism.
    ***d.* predator–prey interaction.**
    e. amensalism.

32. The relationship between the birds and the herbivores is an example of
    ***a.* mutualism.**
    b. parasitism.
    c. commensalism.
    d. predator–prey interaction.
    e. amensalism.

33. Elephants and other large herbivores trample many species of plants that are different from the plant species they use as food. The relationship between the elephants and the trampled plant species is an example of
    a. mutualism.
    b. parasitism.
    c. commensalism.
    d. predation.
    ***e.* amensalism.**

34. Which of the following is *not* true of suspension feeders?
    ***a.* They choose prey carefully.**
    b. The prey is usually small.
    c. The prey is usually abundant.
    d. The predator has some sort of specialized filtering device.
    e. Movement of the environmental medium through the filtering device is common.

35. A pine tree with no parasitic bark beetles will exude _____ pitch and be _____ attractive to a bark beetle than an already parasitized tree.
    a. less; less
    b. less; more
    ***c.* more; less**
    d. more; more
    e. an equal amount of; equally

36. Which of the following factors would be least important in disrupting predator–prey oscillations?
    ***a.* Density-dependent interactions between prey and predator**
    b. Spatial patchiness of the habitat
    c. Temporal patchiness of the habitat
    d. A highly motile prey
    e. The introduction of an additional prey species

37. In a typical Batesian mimicry system, the system is stable only if
    a. the mimic and model have a dissimilar appearance.
    b. the predator learns to distinguish mimic from model.

c. **the mimic converges on the model faster than the model diverges due to directional selection.**
d. the mimic converges on the model more slowly than the model diverges due to directional selection.
e. both mimic and model become distasteful.

38. Müllerian mimicry systems make it _____ difficult for the predator to learn which prey species are distasteful or dangerous, and the relative abundance of the various prey species _____.
    a. more; does not matter
    ***b.* less; does not matter**
    c. more; does matter
    d. less; does matter
    e. no more or less; does not matter

39. Which of the following is *not* an example of a defense used by plants against grazers or browsers?
    a. Production of tannins
    ***b.* Increased production of nonwoody tissue**
    c. Production of nicotine
    d. Production of hairs and spines
    e. Production of hormone-like chemicals that interfere with insect metamorphosis

40. What is the type of ecological relationship that can involve either members of the same or different species and in which both participants are harmed?
    a. Mutualism
    b. Parasitism
    ***c.* Competition**
    d. Predation
    e. Amensalism

41. Which of the following was *not* usually observed by G. F. Gause in his studies on interspecific competition?
    a. Competitive exclusion occurred unless environmental heterogeneity was available.
    b. Species usually had greater growth when alone.
    c. Some species that grew very rapidly in a particular environment would be completely eliminated when grown with another species.
    ***d.* The outcome of competition was independent of the environmental conditions in the culture.**
    e. Populations grew exponentially when alone.

42. Many tropical orchid species grow on the surface of tree branches without harming or benefiting the tree. This orchid–tree relationship is an example of
    a. mutualism.
    b. parasitism.
    ***c.* commensalism.**
    d. predation.
    e. amensalism.

43. The endosymbiosis hypothesis for the evolution of eukaryotic cells, which states that modern mitochondria and chloroplasts descended from once free-living bacterial ancestors that infected larger cells, may describe a very early example of
    ***a.* mutualism.**
    b. parasitism.
    c. interspecific competition.
    d. predation.
    e. amensalism.

44. Which of the following relationships is *not* an example of mutualism?
    a. *Rhizobium* bacteria live in the protected environment of legume roots, and provide the legumes with nitrogen.
    b. A prawn cleans a coral reef fish.
    c. **A plant produces "hitchhiker" fruit that are dispersed by a passing animal.**
    d. *Pseudomyrmex* ants live in the hollow spines of an *Acacia* tree.
    e. A plant produces fleshy fruits with digestion-resistant seeds that are eaten by an animal.

45. The white flowers seen on many night-blooming plants probably resulted from _____ and the pollinators are likely to be _____.
    a. **diffuse coevolution; many species of bats**
    b. species-specific coevolution; just one species of bat
    c. diffuse coevolution; many species of bees
    d. species-specific coevolution; just one species of bee
    e. diffuse coevolution; many species of bats and bees

46. Select the following characteristic that is *not* true of ecological succession.
    a. The species composition of the community changes through time.
    b. The physical conditions of the community change through time.
    c. **The rate of change is constant through time.**
    d. Some species disappear quickly; others are more long-lived.
    e. The community becomes more complex through time.

47. Changes in the biological community associated with the transformation of a "cow pie" after its deposition in a pasture is an example of
    a. primary succession.
    b. secondary succession.
    c. **degradative succession.**
    d. diffuse evolution.
    e. coevolution.

48. Considering the effects of beaver and moose on plant communities in which they live, what do these two mammals have in common?
    a. Both create more open land than would exist without them.
    b. **Both are important influences on ecological succession in the area.**
    c. Both increase species richness.
    d. Both overfertilize the ponds, which leads to eutrophication.
    e. Both browse on all woody plants in the community.

49. Which of the following is *not* true of an ecological niche?
    a. Predators nearby can narrow a particular species' niche.
    b. A species' niche can grow or shrink according to conditions.
    c. **The area if a species' niche is not determined by the presence of another species.**
    d. Physical and biotic factors can determine the area of a species' niche.
    e. None of the above

50. Laboratory competition experiments often cannot fully imitate what happens in nature because
    a. **interactions with other species not present can affect the results.**
    b. resources found in nature cannot be controlled in the lab.
    c. only interspecific, not intraspecific, competition happens in the lab.
    d. *a* and *b*
    e. All of the above

51. Which of the following is an example of an indirect effect of species interactions?
    a. Lynx drive arctic hares to extinction and then die from starvation themselves.
    b. Some birds cannot survive because all available nest sites are taken.
    c. The barnacle *Chthamalus* has a larger niche when the barnacle *Balanus* is absent.
    d. Fig wasps can only reproduce in one species of fig.
    e. **The number of cases of hantavirus increase after a heavy winter snow, making piñon pinecones, which are food for the virus-carrying deer mice, more abundant.**

52. In terrestrial communities, the major modifiers of the physical environment are
    a. **plants.**
    b. vertebrate animals.
    c. invertebrate animals.
    d. protists.
    e. fungi.

53. Populations of a predator and its prey are found on two islands. On the first island, suitable habitat for the prey is uniformly distributed throughout the island, and the prey is found as a single uniform population. On the second island, suitable habitat for the prey is patchily distributed, and the prey is found in isolated patches. All else being equal, you would expect
    a. the densities of the prey and predator populations to be more stable through time on the first (uniform) island than on the second (patchy) island.
    b. **the densities of the prey and predator populations to be more stable on the second (patchy) island than on the first (uniform) island.**
    c. the densities of the prey populations to be more stable on the second (patchy) island, and the densities of the predator populations to be more stable on the first (uniform) island.
    d. both islands to exhibit approximately equally strong oscillations in the predator and prey populations.
    e. the first (uniform) island to have very stable prey and predator populations, and the second (patchy) island to have strong fluctuations.

54. Resources whose supply is less than the demand made upon them are called
    a. niche resources.
    b. **limiting resources.**
    c. primary resources.
    d. competitive resources.
    e. intraspecific resources.

COMMUNITY ECOLOGY **267**

55. In experiments with two species of the plant *Desmodium*, when both were grown together, *D. glutinosum* grew better than *D. nudiforum*. When they were grown separately, however, *D. nudiforum* grew better than *D. glutinosum*. This shows that
    a. *D. glutinosum* was more depressed by interspecific competition than was *D. nudiforum*.
    b. *D. glutinosum* was unaffected by intraspecific competition.
    c. *D. nudiforum* was more depressed by interspecific competition than was *D. glutinosum*.
    d. *D. nudiforum* was more depressed by intraspecific competition than was *D. glutinosum*.
    e. Both species were equally depressed by intraspecific competition.

56. The textbook describes an experiment on dragonfly nymphs and tadpoles in Lake Superior. Which of the following was a conclusion of this experiment?
    a. Tadpoles are the only prey of dragonfly nymphs.
    b. Dragonfly nymphs eat both tadpoles and dystiscid beetles.
    c. Beetles and dragonfly nymphs eat different sizes of tadpoles.
    d. Dragonfly nymphs can completely eliminate tadpoles in certain areas.
    e. Researchers were able to determine which ponds dragonfly nymphs select to prey upon tadpoles.

57. The Rio Negro river in Brazil is darkly colored due to the release of a large amount of a substance from a plant. This substance, part of the plant's defense system, is
    a. nicotine.
    b. tannin.
    c. acute toxin.
    d. nectar.
    e. elaiosome.

53 *Ecosystems*

Fill in the Blank

1. In temperate lakes, the depth where the temperature abruptly changes from the warm surface temperatures to the cold temperatures deeper down is the **thermocline**.

2. The cycling of water between the oceans, atmosphere, and land is known as the **hydrological cycle**.

3. The layer of the atmosphere that we live in (the one closest to the ground) is the **troposphere**.

4. The movement of an element through living organisms and the physical environment is called its **biogeochemical cycle**.

5. The addition of nutrients, especially phosphorus, to a lake will result in blooms of algae that die and decompose, using up all of the oxygen in the lake. This process is called **eutrophication**.

6. With a very few exceptions, all of the energy available to living organisms ultimately comes from **the sun**.

7. The total amount of energy assimilated by photosynthesis is called **gross primary production**.

8. The amount of energy assimilated by photosynthesis after the energy used by the plants for maintenance and biosynthesis is subtracted is called **net primary production**.

9. All organisms that get their energy from a common source (e.g., all herbivores) constitute a **trophic level**.

10. A set of linkages in which a plant is eaten by an herbivore, which in turn is eaten by a carnivore, and so on, is called a **food chain**.

11. The leeward side of a mountain where winds descend and warm and pick up, rather than release, moisture is the mountain's **rain shadow**.

12. The area on Earth where the solar energy flux is greatest, and shifts with the season, is the **intertropical convergence zone**.

13. Organisms that reduce the remains of other organisms to mineral nutrients that can be taken up by plants are **detritivores**.

14. The agricultural method that combines cultivation methods such as crop rotation, mixed plantings, and mechanical tillage of soil with biological controls such as pest-resistant strains of crops, with the use of predators and parasites of pests to reduce the application of chemical pesticides is called **integrated pest management**.

15. A network of linkages connecting a set of plants with a set of herbivores and carnivores is a **food web**.

16. Human activity has most seriously disturbed the **carbon** biogeochemical cycle.

17. The troposphere is transparent to visible light and traps most outgoing infrared light. This results in **warming** of the earth.

18. The amount of heat that an area of Earth receives depends on the **angle** of the sun.

Multiple Choice

1. The poles are cooler than the equator because
   a. the equator receives more hours of sunlight than the poles do.
   b. the equator is closer to the sun than the poles are.
   c. the angle at which the sun strikes Earth is sharper at the poles (the sun is lower on the horizon).
   d. the long polar nights cool Earth much more than the short equatorial nights.
   e. winds tend to move heat from the poles toward the equator.

2. Rain tends to be much more common in mountains than in the adjacent lowlands because
   a. as air travels up the mountains, it cools. Cool air holds less moisture than warm air, and the excess moisture is dropped as rain.
   b. as air travels up the mountains, it is compressed. The mountains squeeze the water out of the air, which is dropped as rain.
   c. because of their great height, mountains are often in the clouds. Precipitation is much greater in these clouds.
   d. as air moves down the mountains, it expands and is unable to form clouds, so no rainfall occurs.
   e. the rainfall is equal in both mountains and lowlands; however, because the rain has farther to travel from the clouds to the lowlands, more of it evaporates before it reaches the ground in the lowlands.

3. A common ecosystem on the leeward (away from the wind) side of mountains and also at 30° north or south latitude is
   a. coniferous forest.
   b. desert.
   c. rainforest.
   d. peatlands.
   e. tundra.

4. The prevailing winds in the continental United States are from the west. This is a consequence of
   a. the position of the sun.
   b. the location of the major mountain ranges in the United States.
   c. the Pacific Ocean being warmer than the Atlantic Ocean.
   d. the currents in the Atlantic and Pacific Oceans.
   e. **the rotation of Earth.**

5. When the intertropical convergence zone is over an area,
   a. the trade winds prevail.
   b. the trade winds reverse direction.
   c. air sinks and the area becomes cool and dry.
   d. **air rises and heavy rains fall.**
   e. air stagnates and hot, dry conditions prevail.

6. In general, ocean currents
   a. circulate clockwise in both the Northern and Southern Hemispheres.
   b. circulate counterclockwise in both the Northern and Southern Hemispheres.
   c. circulate clockwise in the Atlantic Ocean and counterclockwise in the Pacific Ocean.
   d. circulate counterclockwise in the Northern Hemisphere and clockwise in the Southern Hemisphere.
   e. **circulate clockwise in the Northern Hemisphere and counterclockwise in the Southern Hemisphere.**

7. London is farther north than the northern coast of Maine; nevertheless, the winters are much more severe in Maine. Why?
   a. **Warm ocean currents from the south flow off the coast of England, whereas cold currents from the north flow off the coast of Maine.**
   b. England is an island that has relatively little contact with land masses farther north and avoids the cold continental northern winds, whereas Maine is part of a large land mass that extends deep into the north and is subject to strong continental northern winds.
   c. Winds in the Northern Hemisphere blow from the southwest in general. In England these blow over the ocean and are *not* obstructed. In North America they are blocked by the Appalachian Mountains and aren't able to warm Maine.
   d. The rotation of Earth pulls cold air from the poles down to Maine. This effect does *not* occur in England because there is a large area of open ocean north of that area.
   e. England is heavily industrialized, and this has led to local warming. Rural Maine is colder because it lacks this industrial warming.

8. North Dakota is known for its hot summers and cold winters. At the same latitude, western Washington has cooler summers and warmer winters. What is the main cause of this difference?
   a. Washington receives a steady westerly wind that leads to a more nearly constant temperature. Because the westerly winds are blocked by mountains, North Dakota receives hot southern winds in

the summer and cold northern winds in the winter.
   b. Western Washington is nearly always overcast. This cloud cover prevents the sun from heating the land in the summer and prevents excess heat loss during the winter. North Dakota is very dry, and is heated and cooled more drastically.
   c. **Western Washington is near the Pacific Ocean, which moderates its climate. North Dakota is in the middle of a large continent and does not receive this buffering.**
   d. Western Washington is at lower elevation than North Dakota. This difference in altitude accounts for the more extreme weather in North Dakota.

9. Which of the following is *not* a major element that makes up living systems?
   a. Carbon
   b. Nitrogen
   c. Oxygen
   d. Hydrogen
   e. **Potassium**

10. Erosion from poor farming practices leads to large amounts of soil being lost into the sea. What is the likely fate of the elements in the soil?
   a. Most of the elements are rapidly recirculated by ocean currents and are soon reintroduced into terrestrial ecosystems.
   b. Most elements dissolve and eventually cycle into the atmosphere.
   c. Most elements dissolve from the soil and migrate into freshwater ecosystems, where they are eventually reintroduced into terrestrial ecosystems.
   d. Tidal forces quickly return most of the soil to the shore, where they build up and gradually increase the size of the land mass.
   e. **Most elements settle to the bottom of the oceans, and remain there for millions of years until bottom sediments are elevated by movements of Earth's crust.**

11. Excluding human-induced effects, most elements enter freshwater systems from
   a. rainfall.
   b. the oceans (through tidal movements).
   c. the actions of animals.
   d. **the weathering of rocks (via groundwater).**
   e. plants and plant parts decaying in streams, lakes, and rivers.

12. In a temperate zone lake in the middle of summer,
   a. most nutrients are near the surface of the lake, and oxygen is evenly distributed throughout all depths of the lake.
   b. **most nutrients are near the bottom of the lake, and most oxygen is near the surface of the lake.**
   c. nutrients and oxygen are evenly distributed throughout all depths of the lake.
   d. nutrients are evenly distributed throughout all depths of the lake, and oxygen is near the surface of the lake.
   e. most nutrients and most oxygen are found near the surface of the lake.

13. At which of the following temperatures is water most dense?
    a. 16°C
    b. 8°C
    c. 4°C
    d. 2°C
    e. 0°C

14. Most of Earth's nitrogen is in
    a. the atmosphere.
    b. the oceans.
    c. fresh water.
    d. soil.
    e. organisms.

15. The gas that is removed from the atmosphere by plants and algae is
    a. nitrogen.
    b. oxygen.
    c. carbon dioxide.
    d. methane.
    e. water vapor.

16. One difference between gases such as oxygen and carbon dioxide and minerals such as phosphorus and calcium that do not have a gas phase is that
    a. there is relatively little variation from site to site in the concentration of gases, whereas solids tend to remain where they are, creating local abundances and shortages of these minerals.
    b. gases such as oxygen and carbon dioxide are never found in a rock (solid) phase or dissolved in water, whereas minerals without a gaseous phase are frequently found as rocks or dissolved in water.
    c. gaseous elements (and compounds) are essential to living organisms, whereas minerals without a gaseous phase are not.
    d. gases are modified by living organisms, whereas minerals without a gaseous phase are not.
    e. the relative proportions of different gases in the atmosphere change radically over short periods of time, whereas the relative proportions of different solid minerals remain fairly constant.

17. Comparing very old soils and very young soils,
    a. very old soils are much more fertile than young soils.
    b. soil age does not affect the soil's fertility.
    c. soil fertility changes with age, but the direction of change depends on many factors.
    d. very old soils are much less fertile than young soils.
    e. very old temperate soils are more fertile than young temperate soils. In the tropics, the reverse is the case.

18. Most of the world's carbon is found
    a. as carbon dioxide in the atmosphere.
    b. in living organisms.
    c. as bicarbonate and carbonate ions dissolved in the oceans.
    d. as carbonate minerals in sedimentary rock.
    e. in the decaying remains of dead organisms.

19. Photosynthesis and respiration are central to which cycle?
    a. The nitrogen cycle
    b. The carbon cycle

c. The phosphorus cycle
d. The sulfur cycle
e. The hydrological cycle

20. In the past 150 years there has been a major new input into the carbon cycle. What is it?
    a. There are more humans that release large quantities of carbon dioxide as respiration.
    b. Increased animal farming has resulted in greater carbon releases.
    c. Industrialization has resulted in the use of fossil fuels such as oil and coal.
    d. Changes in ocean currents have lead to the release of large quantities of carbon dioxide.
    e. Global warming has increased the release of carbon from carbonate minerals.

21. Nitrogen is often in short supply in terrestrial ecosystems. Why?
    a. There is very little free nitrogen in the air.
    b. Atmospheric nitrogen is primarily in the stratosphere and does not come into contact with terrestrial ecosystems.
    c. Atmospheric nitrogen cannot be used by most organisms. It needs to be converted to useful forms by bacteria and cyanobacteria.
    d. Nitrogen solubility in water is very low and therefore atmospheric nitrogen enters cells very slowly.
    e. Atmospheric nitrogen varies widely from location to location. As a result local shortages of nitrogen frequently occur.

22. The phosphorus cycle differs from the carbon cycle in that
    a. phosphorus does not enter living organisms, whereas carbon does.
    b. the phosphorus cycle does not include a gaseous phase, whereas the carbon cycle does.
    c. the phosphorus cycle includes a solid phase, whereas the carbon cycle does not.
    d. the primary reservoir of the phosphorus cycle is the atmosphere, whereas the primary reservoir for the carbon cycle is in rocks.
    e. phosphorus passes through living organisms many times, whereas carbon only flows through living organisms once.

23. In the 1960s, over 250 billion liters of domestic and industrial waste entered Lake Erie annually. In addition, extensive farming resulted in fertilizer runoff into the lake. The result of this, relative to preindustrial conditions, was
    a. an increase in the algae, bacteria, and oxygen levels in the lake.
    b. a decrease in the algae and oxygen levels and an increase in the bacteria levels in the lake.
    c. a decrease in the algae, bacteria, and oxygen levels in the lake.
    d. a decrease in the algae and bacteria levels and an increase in the oxygen levels in the lake.
    e. an increase in the algae and bacteria levels and a decrease in the oxygen levels in the lake.

24. Acid precipitation
    a. is a natural occurrence when clouds are formed by evaporation over land.
    b. is a natural occurrence when clouds are formed by evaporation over oceans.

c. **occurs as a result of the burning of fossil fuels that contain sulfur and nitrogen compounds.**

d. occurs because of carbon monoxide emissions by automobiles.

e. occurs when large tracts are logged or otherwise deforested.

25. The two main human-induced causes of increased carbon dioxide levels are

a. **the burning of fossil fuels and the burning of forests.**

b. the burning of fossil fuels and the raising of cattle.

c. modern agricultural fertilizers and the raising of cattle.

d. pollution-induced mortality of oceanic algae and the burning of forests.

e. modern agricultural fertilizers and pollution-induced mortality of oceanic algae.

26. If carbon dioxide levels were to double, as they may by the middle of the next century, all of the following are expected to occur except which one below?

a. World mean temperature will increase 3–5°C.

b. Droughts in the central regions of continents will become more common and more severe.

c. Precipitation in coastal areas will increase.

d. Polar ice caps will melt, raising sea levels and flooding coastal cities and agricultural areas.

e. **The destruction of the ozone layer by carbon dioxide will allow more ultraviolet light to reach Earth.**

27. Which of the following statements is true regarding the movement of energy and nutrients through ecosystems?

a. Energy flows and nutrients flow.

b. **Energy flows and nutrients cycle.**

c. Energy cycles and nutrients cycle.

d. Energy cycles and nutrients flow.

e. Energy flows and nutrients either cycle or flow.

28. Which of the following changes would *not* result in an increase in net primary production?

a. Increasing precipitation in an arid area

b. Increasing soil fertility

c. **Increasing latitude (moving from the equator toward the poles)**

d. Moving down a mountain to warmer temperatures

e. Increasing the light in aquatic ecosystems

29. On average, how much of the energy assimilated at one trophic level is converted to production at the next trophic level (excluding the conversion of sunlight into chemical energy by plants)?

a. **5–20%**

b. less than 1%

c. 20–30%

d. 30–50%

e. Greater than 50%

30. Which of the following always has a "pyramidal" shape, that is, decreasing values at higher trophic levels?

a. Pyramids of numbers only

b. Pyramids of biomass only

c. **Pyramids of energy only**

d. Both pyramids of biomass and pyramids of energy

e. All three pyramids

31. Which of the following is true concerning nutrients in modern agricultural systems?

a. Nutrient cycles are very tight, and there is little loss of nutrients through runoff.

b. **Nutrient cycles are frequently broken because food is consumed and wastes are disposed of far from where the food was produced.**

c. Modern agricultural systems are normally carefully designed to promote efficient nutrient cycling.

d. The addition of fertilizers, together with the natural cycling of nutrients, has resulted in a buildup in agricultural soils so that they are now far deeper and richer than they were prior to the development of modern agriculture.

e. There normally is less loss of nutrients from agricultural fields than from natural terrestrial ecosystems.

32. A terrestrial ecosystem is directly influenced by all but which of the following?

a. Solar energy fluxes

b. Earlier successional stages

c. Local species richness

d. Global atmospheric circulation

e. **Eutrophication**

33. In comparing summer with winter in an area such as New York City, which of the following would be greater in the winter?

a. Solar energy flux

b. **Amount of reflected solar energy**

c. Angle of the sun at noon

d. Length of the day

e. Energy balance (solar input – heat loss)

34. Rain shadows form on the _____ of mountains and result in _____ rainfall there.

a. windward side; more

b. windward side; less

c. leeward side; more

d. **leeward side; less**

e. peaks; more

35. As a sea breeze is forced to rise by coastal mountains, a process occurs that is the same as that occurring in the area of the _____, except that in the latter the energy to rise is provided by the _____.

a. trade winds; Earth's spin

b. westerlies; sun

c. westerlies; Earth's spin

d. **intertropical convergence zone; sun**

e. intertropical convergence zone; Earth's spin

36. The direction of the westerlies and the trade winds is directly caused by

a. location of the intertropical convergence zone.

b. tilt of Earth's axis.

c. **variation in the rotational speed of Earth at different latitudes.**

d. differences in heat balance between the equator and the poles.

e. the global oceanic circulation.

37. Oceans warm and cool more _____ than land because water has a _____ specific heat.
    a. **slowly; greater**
    b. slowly; lesser
    c. rapidly; greater
    d. rapidly; lesser
    e. slowly; equal (but the oceans are larger)

38. On the west coasts of continents, ocean waters are _____ and precipitation falls mostly in the _____.
    a. cool and nutrient-rich; summer
    b. **cool and nutrient-rich; winter**
    c. cool and nutrient-poor; winter
    d. warm and nutrient-rich; summer
    e. warm and nutrient-rich; winter

39. Fruit trees and other frost-sensitive plants are frequently grown on the downwind side of large lakes. This has the most to do with
    a. the fact that air that has blown over water will usually release its moisture when it encounters colder land.
    b. the high specific heat of land.
    c. wind shadows.
    d. **the difference between maritime and continental climates.**
    e. the moderating effect that land has on the lake.

40. The major differences between ocean and freshwater compartments of the global ecosystem result because
    a. **of the much greater depth of the oceans.**
    b. the oceans only turn over once a year.
    c. freshwater systems only receive input from groundwater.
    d. temperature only varies with depth in freshwater systems.
    e. nutrient recycling is less rapid in freshwater systems.

41. Which global ecosystem compartment would be characterized by very slow movement of materials within the compartment and exchange of gases mostly via organisms?
    a. Oceans
    b. Fresh water
    c. Atmospheric
    d. **Terrestrial**
    e. More than one

42. Which global ecosystem compartment would be characterized by the circulation of materials affected by Earth's rotation on its axis and most of the organisms in the uppermost layer?
    a. **Oceans**
    b. Fresh water
    c. Atmospheric
    d. Terrestrial
    e. More than one

43. Select a true statement comparing the troposphere and stratosphere compartments of the global ecosystem.
    a. **Most water vapor resides in the troposphere.**
    b. Vertical circulation of air occurs mostly in the stratosphere.
    c. Circulation of the stratosphere most directly influences the ocean currents.

d. The stratosphere represents most of the mass of the atmosphere.
    e. Most weather events occur in the stratosphere.

44. Which of the following statements about the hydrological cycle is *not* true?
    a. Most input into the oceans occurs via runoff from rivers.
    b. More water evaporates from the surface of the ocean than falls as rain over the ocean.
    c. Less water evaporates from the surface of the land than falls as rain over the land.
    d. **Water found in sedimentary rock is constantly exchanged with the ocean.**
    e. The energy driving the hydrological cycle is heat from the sun.

45. In biogeochemical cycles, elements that cycle fastest
    a. are found in organisms.
    b. are scarce.
    c. **have a gaseous phase.**
    d. do not become fixed into sediment.
    e. do not accumulate in the higher trophic levels.

46. Which of the following biogeochemical cycles is characterized by a major reservoir that is gaseous and a major inorganic form that can only be utilized by a small group of bacteria and cyanobacteria?
    a. Carbon cycle
    b. **Nitrogen cycle**
    c. Phosphorus cycle
    d. Sulfur cycle
    e. Oxygen cycle

47. Which of the following biogeochemical cycles is characterized by a form that is a major greenhouse gas?
    a. **Carbon cycle**
    b. Nitrogen cycle
    c. Phosphorus cycle
    d. Sulfur cycle
    e. More than one cycle

48. Which of the following biogeochemical cycles lacks a gaseous form?
    a. Carbon cycle
    b. Nitrogen cycle
    c. **Phosphorus cycle**
    d. Sulfur cycle
    e. Oxygen cycle

49. Which of the following biogeochemical cycles has a gaseous form that is released by volcanoes and fumaroles?
    a. Carbon cycle
    b. Nitrogen cycle
    c. Phosphorus cycle
    d. **Sulfur cycle**
    e. Oxygen cycle

50. Which of the following biogeochemical cycles has a major reservoir in sedimentary rocks?
    a. Carbon cycle
    b. Nitrogen cycle
    c. Phosphorus cycle
    d. Sulfur cycle
    e. **More than one cycle**

51. In the human-induced process called eutrophication, the main biogeochemical cycle that is altered is the _____ cycle, and the effect is to create _____ conditions and decrease species diversity.
    a. hydrological; aerobic
    b. phosphorus, anaerobic
    c. hydrological; aerobic
    d. phosphorus; aerobic
    e. nitrogen; anaerobic

52. After nutrient input into a lake is reduced, the time required for the lake to return to a pre-eutrophication condition depends on
    a. the rate of turnover of its waters.
    b. the presence of the appropriate algae-eating fish.
    c. whether there is a thermocline in the lake.
    d. the amount of groundwater reaching the lake.
    e. the presence of an inverted energy pyramid.

53. In the human-induced process called acid precipitation, the main biogeochemical cycles that are altered are the _____ cycles, and one effect in lakes is to _____ populations of nitrifying bacteria.
    a. phosphorus and nitrogen; increase
    b. phosphorus and nitrogen; decrease
    c. nitrogen and sulfur; decrease
    d. nitrogen and sulfur; increase
    e. phosphorus and sulfur; decrease

54. Which of the following occurrences is *not* considered to be a likely consequence of global warming?
    a. Melting of the polar ice caps
    b. Flooding and increased precipitation in coastal areas
    c. Increased ozone depletion
    d. Average worldwide temperature increase of 3–5°C
    e. Droughts in central areas of continents

55. Which of the following is *not* included in either gross or net primary production in plants?
    a. Light reflected from the leaf
    b. Energy fixed into glucose
    c. Energy expended in moving material through membranes
    d. Energy used to excite a chlorophyll electron
    e. Dry weight of a plant

56. Which of the following statements about food chains and energy flow through ecosystems is *not* true?
    a. A single organism can feed at several trophic levels.
    b. The lower the trophic level at which an organism feeds, the more energy available.
    c. Detritivores feed at all trophic levels except the producer level.
    d. Food webs include two or more food chains.
    e. All organisms that are not producers are consumers.

57. Assuming that the energy transfer efficiency between trophic levels is 10%, how much grain would be required to produce 70 kg of human biomass if the grain is eaten by cows and the cows are eaten by humans?
    a. 210 kg
    b. 700 kg

    c. 2,100 kg
    d. 7,000 kg
    e. 70,000 kg

58. Grasslands can support greater grazing rates by herbivores than forests can because
    a. grasslands receive more sunlight.
    b. the net production of grasslands is greater.
    c. grasslands produce less woody tissue.
    d. more of the grassland production is above ground.
    e. manure from large herds of herbivores increases the production of grasslands.

59. An inverted pyramid of _____ may occasionally be observed in _____ communities.
    a. energy; grassland
    b. energy; forest
    c. biomass; marine
    d. biomass; grassland
    e. biomass; forest

60. Which of the following is *not* an objective of integrated pest management?
    a. To eliminate the use of chemicals in agriculture
    b. To develop pest-resistant strains
    c. To use natural biological control methods
    d. To reduce agriculture-caused pollution
    e. To reduce the dependence of agriculture on fossil fuel

61. Rain and snow become acidified when the burning of fossil fuels releases the acid forms of the elements
    a. phosphorus and sulfur.
    b. sulfur and nitrogen.
    c. nitrogen and chlorine.
    d. phosphorus and chlorine.
    e. nitrogen and phosphorus.

62. Water, as in the oceans, heats and cools more slowly than air because
    a. water is thicker than air.
    b. water has a higher specific heat than air.
    c. wind circulation makes the oceans move in a westward direction, away from the sun.
    d. water is closer to the center of Earth, which is warmer.
    e. it is easier for sunlight to travel through air than through water.

63. In the light, a plant fixes 0.12 ml of $14CO_2$ per hour, but in the dark the same plant releases 0.04 ml of the gas per hour. What is the estimated net primary production of this plant?
    a. 0.04 ml/hour
    b. 0.08 ml/hour
    c. 0.12 ml/hour
    d. 0.16 ml/hour
    e. 0.0048 ml/hour

64. At noon on the summer solstice (the longest day of the year in the Northern Hemisphere), the sun is directly over
    a. the equator.
    b. 30° latitude north.
    c. 30° latitude south.
    d. the Arctic Circle.
    e. the Antarctic Circle.

54 Biogeography

Fill in the Blank

1. **Biogeography** is the study of the distribution of organisms over space and time.

2. **Historical biogeographers** study the evolutionary histories of groups of organisms, where they originated, and if and how they spread.

3. **Ecological biogeographers** study present-day interactions among organisms to determine how those relationships influence where species are found.

4. A **vicariant** distribution is one determined by the separation of formerly contiguous areas by a barrier.

5. **Australia** and **South America** are the most distinct biogeographic regions because they have been isolated from other continents for the longest time.

6. One theory of biogeography suggests that the number of species in an area is the result of a dynamic balance between **extinctions** and **colonizations**.

7. The major biome types are identified by their characteristic **vegetation**.

8. A particular community type, such as a hot desert, occurs where the **climate** is suitable for that type of vegetation.

9. Although there are **different** species of organisms in a particular biome on different continents, the species are ecologically **similar**.

10. The **tundra** biome is found in the Arctic and high in the mountains at all latitudes.

11. The most striking difference between Arctic tundra and alpine tundra in the tropics is **day length**.

12. Arctic tundra is underlain by permanently frozen ground called **permafrost**. Animals survive the harsh winters by means of either **migration** or **dormancy**.

13. A modern biogeographic technique is the **area cladogram** that shows the distribution patterns of a taxonomic group.

14. Major ecosystems that vary by nature of their dominant vegetation are called **biomes**.

15. The ocean's **temperature** at different levels is the biggest barrier to colonization by a species.

16. Just as all land mass was once united as Pangaea, the oceans were all united and called **Panthalassa**.

Multiple Choice

1. An endemic species is one that
   a. **is found in only one region.**
   b. evolved in the region where it currently lives.
   c. occurs in two distinct areas separated by a barrier.
   d. has a good fossil record showing its evolution.
   e. disperses readily from one area to another.

2. An endemic taxon may occur for all of the following reasons, except that the taxon
   a. is old and nearly extinct.
   b. has evolved only recently.
   c. lives in an isolated island habitat.
   d. has poor dispersal mechanisms.
   e. **has little genetic variability.**

3. Australia has a greater percentage of endemic species than other continents because Australia
   a. lies near several large chains of islands.
   b. has more varied habitats than other continents.
   c. **has been isolated for the longest time due to continental drift.**
   d. lacks a region near the equator.
   e. has received many immigrant species due to prevailing ocean currents.

4. A study of small flies called midges shows that one genus occurs only in New Zealand, two genera are found in Australia, South America, and New Zealand, and two other genera occur in Australia and South America only. This evidence supports the theory that the land masses drifted apart from one another in which order?
   a. **New Zealand first, then Australia and South America**
   b. Australia first, then New Zealand and South America
   c. South America first, then New Zealand and Australia
   d. All three separated at the same time.
   e. South America and Australia separated; New Zealand was never joined to either.

5. A graph showing relationships among taxa that is based on defining their similarities and differences in a rigid, standardized way is specifically called
   a. an evolutionary lineage.
   b. **a cladogram.**
   c. an historical phylogeny.
   d. a taxonomic model.
   e. a parsimonial distribution.

6. Which of the following is common to both ecological and historical biogeographers?
   a. Using experiments to test hypotheses
   b. Concentrating on current interactions among organisms

   c. **Studying the effects of time and space on species' distributions**
   d. Explaining the distribution of organisms in terms of continental drift
   e. Constructing lineages in terms of a "molecular clock"

7. Which of the following is *not* a modern barrier between major biogeographic regions?
   a. Desert
   b. Mountains
   c. Waterways
   d. High plateau–lowland boundary
   e. **Tundra**

8. In an experiment near the southern tip of Florida, all the arthropods were destroyed on six small islands. Which of the following statements is true about the recolonization of these islands by arthropods?
   a. **Each island regained approximately its original number of species.**
   b. Recolonization rates were very slow on the smaller of the six islands compared to the larger.
   c. The recolonizing species were identical to the original species.
   d. Equilibrium numbers of species had not been reached by the close of the experiment.
   e. Rates of immigration were constant for the first year of recolonization.

9. The geographic distribution of a biome may be affected by the seasonal variation of all of the following, except
   a. temperature.
   b. **latitude.**
   c. precipitation.
   d. fire.
   e. grazing.

10. If one started near Washington, D.C. on the East Coast of the United States and traveled northward toward the North Pole, one would encounter which of the following biomes, in which order?
   a. **Temperate deciduous forest, boreal forest, tundra**
   b. Chaparral, temperate deciduous forest, boreal forest
   c. Savanna, boreal forest, chaparral
   d. Temperate grassland, boreal forest, tundra
   e. Boreal forest, tundra, chaparral

11. Forests in both the Northern and Southern hemispheres are termed "boreal" forests when they are dominated by
   a. **evergreen trees with needles or small leaves.**
   b. cone-bearing gymnosperm trees.
   c. deciduous beeches or related angiosperm trees.
   d. a diverse assemblage of insect-pollinated trees.
   e. evergreen trees adapted to frequent fires.

12. Which of the following biomes may lack seasonal temperature variations?
   a. Tundra
   b. **Tropical evergreen forest**
   c. Boreal forest
   d. Cold desert
   e. Chaparral

13. Which of the following biomes lacks a pronounced seasonal variation in amounts of precipitation?
   a. Tropical deciduous forest
   b. Thorn forest
   c. **Boreal forest**
   d. Chaparral
   e. Cold desert

14. Historical biogeographers use several types of data. Which of the following types of information would *not* be used by a historical biogeographer?
   a. Phylogenies
   b. Fossils
   c. Distributions of living species
   d. Places of origin of a group of organisms
   e. **Modern ecological relationships**

15. Biogeographers often apply the rule of parsimony when interpreting distribution patterns. Assume there are four organisms with the following degrees of similarity: Organisms A and B are very similar, and organisms C and D are very similar. Organisms A and B are similar to fossil E; organisms C and D are similar to a fossil F. Which of the following would be a parsimonious explanation for their evolutionary relationship?
   a. B is more closely related to C than to A.
   b. B is more closely related to C than to D.
   c. B and A evolved from F; C and D evolved from E.
   d. **B and A evolved from E; C and D evolved from F.**
   e. A, B, C, and D evolved from F.

16. If a species lives in two distinct areas now separated by a barrier because it was in those areas before the barrier was imposed, it is said to have a
   a. dispersal distribution.
   b. **vicariant distribution.**
   c. natural distribution.
   d. migration distribution.
   e. static distribution.

17. A vicariant distribution is one determined by
   a. **separation of formerly contiguous land masses.**
   b. dispersal.
   c. extinction and recolonization.
   d. adaptive radiation.
   e. interbreeding of two similar species.

18–22. Refer to the following graph.

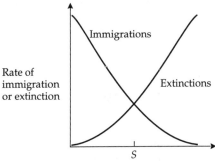

Number of species on island

18. The above figure suggests that the rate of extinction of species
   a. decreases as the number of species increases.
   b. **increases as the number of species increases.**
   c. is independent of the number of species.
   d. varies for different islands.
   e. is large when the number of species is small.

19. The rate of immigration is high when the number of species is low because
    a. there is no competition.
    b. the first colonists to arrive are all "new" species.
    c. there are no predators.
    d. the island is newly formed and attracts species.
    e. there has not been time for extinctions to occur.

20. S represents
    a. the point at which species are numerous enough to compete for resources.
    b. the equilibrium composition of species: no new species can invade and none of the existing species will go extinct.
    c. the minimum number of species that ever existed on the island.
    d. the maximum number of species that ever existed on the island.
    e. the equilibrium number of species: extinctions will equal immigrations.

21. If the number of species on the island is S + 3, we would expect to see
    a. on average, three more species go extinct than invade.
    b. three species go extinct and then no changes in the biota of the island.
    c. on average, three more species invade than go extinct.
    d. three species invade and then no changes in the biota of the island.
    e. Not enough information to tell

22. If the island suddenly doubled in size, we would expect to see
    a. the extinction curve shift to the left and a decrease in S.
    b. the immigration curve shift to the right and an increase in S.
    c. the extinction curve shift to the right and a decrease in S.
    d. the extinction curve shift to the right and an increase in S.
    e. the immigration curve shift to the left and an increase in S.

23–27. Match the biomes from the list below with the following descriptions.
    a. Tundra
    b. Boreal forest
    c. Tropical evergreen forest
    d. Grassland
    e. Chaparral

23. This biome is found in many climates but all of them are relatively dry much of the year. *(d)*

24. This biome is characterized by cold temperatures, permafrost, and little rainfall. It is found in the Arctic and in high mountains at all latitudes. *(a)*

25. This is the richest of all biomes in species of both plants and animals, and has the highest overall energy flow of all ecological communities. The soils in this

biome are usually poor, however, as most of the nutrients are tied up in the vegetation. *(c)*

26. This biome is dominated by coniferous trees. It has Earth's tallest trees, and supports the highest standing biomass of wood of all ecological communities. *(b)*

27. This biome is found on the western sides of continents at moderate latitudes. The dominant vegetation is low-growing shrubs with evergreen leaves. *(e)*

28. Which of the following statements about aquatic and terrestrial communities is *not* true?
    a. Terrestrial photosynthesizers are mostly large and provide a lot of physical structure to communities.
    b. In many aquatic communities, animals rather than plants are the largest and longest-lived organisms.
    c. The most productive aquatic communities, intertidal communities exposed to high wave action, are more productive than tropical rainforests.
    d. There is little competition for space in aquatic shoreline communities.
    e. Most aquatic producers live near the shore or in upper layers of water.

29–32. Match the aquatic habitats from the list below with the following descriptions.
    a. Marine benthic zone
    b. Marine pelagic zone
    c. Marine abyssal zone
    d. Marine littoral zone
    e. Rivers

29. Communities in this habitat may exist on and in mud and sand; others exist on rocky substrates that are sites of intense competition for space. On rocky substrates, algae, barnacles, sponges, and echinoderms are likely to be common. *(d)*

30. Zooplankton and phytoplankton are common in this region; planktonic larvae of organisms from other regions are seasonal. Most fish in this region live in large schools and are streamlined for rapid movement. *(b)*

31. This area is dynamic, and its properties change dramatically with seasons and other factors. Penetration of sunlight may be restricted by suspended solids. Most plants are firmly attached to the bottom to avoid being uprooted by currents. *(e)*

32. This region is below the level of sunlight penetration. *(c)*

33. Different regions on Earth with similar climates have
    a. similar ecological communities with similar species.
    b. similar ecological communities with identical species.
    c. identical ecological communities with identical species.
    d. very different ecological communities with very different species.
    e. similar ecological communities with very different species.

34. Which of the following is *not* a difference between aquatic and terrestrial ecosystems that has been important in the evolution of ecological communities?
   a. **Compared with air, water is a less dense medium that provides little support for organisms living in it.**
   b. Moving water creates greater force than moving air.
   c. Movement through water is more difficult than movement through air.
   d. Sunlight penetrates only short distances through water; thus communities at moderate depths receive little light.
   e. Oxygen is often in short supply in aquatic environments.

35. Australia has been separated from other continents for about 65 million years. South America has been isolated from other continents for about 60 million years. North America and Eurasia have been joined for much of Earth's history. Which continents would you expect to have the most similar biota?
   a. Australia and South America
   b. North and South America
   c. **North America and Eurasia**
   d. Australia and both North and South America
   e. Australia and both North America and Eurasia

36. Which of the following statements is *not* an idea that dominated the early history of biogeography?
   a. The continents are fixed in position.
   b. It is better to focus on a single taxon than to study unrelated groups.
   c. **Information on the ecological relationships of living species can be useful for interpreting biogeographic patterns.**
   d. Dispersal of organisms explains most biogeographic patterns.
   e. Taxa originated in defined local areas.

37. Two adjacent continents are separated by 150 km of ocean. On continent A there are 10 species of organisms closely related to 5 species on continent B. By applying the principle of _____, you would conclude that the species moved from _____.
   a. **parsimony; A to B**
   b. parsimony; B to A
   c. evolution; A to B
   d. evolution; B to A
   e. island biogeography; A to B

38. Based on your knowledge of the history of continental drift on Earth, which of the following pairs of biogeographic regions should have the most similar fauna?
   a. Palearctic, Ethiopian
   b. **Ethiopian, Neotropical**
   c. Nearctic, Neotropical
   d. Australian, Neotropical
   e. Oriental, Ethiopian

39. Which of the following biogeographic regions has the greatest number of endemic taxa?
   a. Nearctic
   b. Neotropical
   c. Ethiopian
   d. **Australian**
   e. Oriental

40. If species arrive on an island at a constant rate, and there is no extinction, the rate of arrival of new species on the island should _____ as the total number of species on the island increases.
   a. increase
   b. **decrease**
   c. remain constant
   d. increase, then decrease
   e. decrease, then increase

41. The number of species on an island at which the arrival rate of new species equals the extinction rate of species already present is called the equilibrium species number. At this number, the species richness _____ and the species composition _____.
   a. is constant; is constant
   b. **is constant; can be variable**
   c. can be variable; is constant
   d. can be variable; can be variable
   e. depends on the size of the island; is constant

42. Using the model of island biogeography developed by Robert MacArthur and E. O. Wilson, which of the following can be predicted?
   a. Species composition on an island
   b. Which species will become extinct
   c. Which species will arrive
   d. **Species richness, given island size**
   e. Rate of evolution of endemic species

43. If ISN and ISF are two small islands of the same size that are near and far from a species pool, and ILN is a large island that is near the same species pool, select the sequence in which these three islands are arranged from the smallest to the largest expected equilibrium species number.
   a. **ISF, ISN, ILN**
   b. ISN, ILN, ISF
   c. ILN, ISF, ISN
   d. ISF, ILN, ISN
   e. ILN, ISN, ISF

44. Which of the following factors would be least important in determining the equilibrium species number of an island?
   a. Island size
   b. Distance to the mainland
   c. Distance to other islands
   d. **Size of the mainland**
   e. Direction of prevailing wind and currents

45. Which of the following statements about habitat islands is *not* true?
   a. A habitat island is surrounded by unsuitable habitat.
   b. **The arrival rates are usually lower for habitat islands than for oceanic islands.**
   c. The model of island biogeography developed by Robert MacArthur and E. O. Wilson also explains the effect of habitat island size and distance from a species pool on the species richness of the habitat island.
   d. Adjacent lakes are habitat islands separated by intervening land.
   e. Intervening areas between habitat islands may be suitable for a brief stopover.

46. Which of the following is *not* true regarding biogeographic regions and biomes?
    a. **Generally, similar biomes are restricted to a sin-
gle biogeographic region.**
    b. Biogeographic regions are based on similarities in
the biota.
    c. Biomes are based on climate differences.
    d. Many biomes are distributed latitudinally.
    e. Both biomes and biogeographic regions exclude
the oceans.

47. Select the biome to which the following characteristics apply: found in dry regions of different temperature; much below-ground biomass.
    a. Tundra
    b. Boreal forest
    c. Temperate deciduous forest
    d. **Grassland**
    e. Cold desert

48. Select the biome to which the following characteristics apply: dominated by coniferous, evergreen trees;
short growing seasons.
    a. Tundra
    b. **Boreal forest**
    c. Temperate deciduous forest
    d. Grassland
    e. Cold desert

49. Select the biome to which the following characteristics apply: presence of permafrost; high latitude or
altitude distribution.
    a. **Tundra**
    b. Boreal forest
    c. Temperate deciduous forest
    d. Grassland
    e. Cold desert

50. Select the biome to which the following characteristics apply: wide range of temperatures and ample
precipitation year-round; animal-dispersed pollen
and fruit.
    a. Tundra
    b. Boreal forest
    c. **Temperate deciduous forest**
    d. Grassland
    e. Hot desert

51. Select the biome to which the following characteristics apply: often located in rain shadows; most precipitation falls in winter.
    a. Tundra
    b. Boreal forest
    c. Temperate deciduous forest
    d. Grassland
    e. **Cold desert**

52. Select the biome to which the following characteristics apply: found in areas with descending warm, dry
air; most precipitation falls in summer.
    a. Chaparral
    b. Thorn forest
    c. Tropical savanna
    d. Tropical deciduous forest
    e. **Hot desert**

53. Select the biome to which the following characteristics apply: greatest species richness; highest energy
flow.
    a. Chaparral
    b. Thorn forest
    c. Tropical savanna
    d. Tropical deciduous forest
    e. **Tropical evergreen forest**

54. Select the biome to which the following characteristics apply: found on the equatorial side of hot deserts;
short, intense summer rainy season.
    a. Chaparral
    b. **Thorn forest**
    c. Tropical montane forest
    d. Tropical deciduous forest
    e. Tropical evergreen forest

55. Select the biome to which the following characteristics apply: cool, wet winters and hot, dry summers;
dominated by low-growing shrubs with tough evergreen leaves.
    a. **Chaparral**
    b. Tropical savanna
    c. Tropical montane forest
    d. Tropical deciduous forest
    e. Tropical evergreen forest

56. Select the biome to which the following characteristics apply: longer rainy season and taller trees than
neighboring thorn forest biome; many trees flower in
the dry season, when they are leafless.
    a. Boreal forest
    b. Temperate deciduous forest
    c. Tropical montane forest
    d. **Tropical deciduous forest**
    e. Tropical evergreen forest

57. Select the biome to which the following characteristics apply: mostly wind-dispersed pollen and fruit;
low species diversity.
    a. **Boreal forest**
    b. Temperate deciduous forest
    c. Tropical montane forest
    d. Tropical deciduous forest
    e. Tropical evergreen forest

58. Which of the following pairs of biomes is maintained
by grazing?
    a. Cold desert and hot desert
    b. Chaparral and thorn forest
    c. **Grassland and tropical savanna**
    d. Temperate and tropical deciduous forests
    e. Cold desert and tundra

59. Which of the following is *not* true of aquatic ecosystems?
    a. **Aquatic producers have extensive support tissue.**
    b. Aquatic producers are very small and provide lit-
tle structure to the ecosystem.
    c. Aquatic producers have adaptations for staying in
the uppermost layers.
    d. Animals are usually larger and live longer than
plants.
    e. Most open-water producers are cyanobacteria and
algae.

60. Which of these physical events is *not* a major influence in the distribution of organisms?
    a. Continental drift
    b. Earthquakes
    c. Sea level changes
    d. Glacial retreat
    e. Formation of mountains

61. _____ and _____ regions have more species richness than other regions.
    a. Tropical; flatland
    b. Tropical; mountainous
    c. Tropical; peninsular
    d. High latitude; peninsular
    e. High latitude; mountainous

62. The Yukon and Mackenzie rivers are not connected, but share the same species of fish. How is this possible?
    a. The eggs of the fish were transported overland by other animals.
    b. The fish species simultaneously arose in each river.
    c. The fish live in salt water for part of their lives and return to either river to spawn.
    d. The rivers used to be connected by a glacier that retreated, leaving the same species of fish in each river.
    e. None of the above

63. Poisonous sea snakes are found in the Pacific Ocean but not in the Caribbean Sea because
    a. the water temperature differs between them too much for the snakes to live in both.
    b. the ocean's currents are counterclockwise so the eggs and larvae are carried the other direction.
    c. the Panamanian Isthmus rose and blocked the dispersal of the snakes.
    d. the Caribbean coastal zones don't get proper nutrients for the snakes.
    e. sea snakes cannot traverse deep ocean waters to get to the Caribbean.

55 *Conservation Biology*

Fill in the Blank

1. When ecosystems, such as agricultural lands, are managed so as to divert most of their primary production to certain species intended for human use, we say that their production is **co-opted**.

2. An important component of a population vulnerability analysis is the amount of variation in birth and death rates. This variation is termed **demographic stochasticity**.

3. An important component of a population vulnerability analysis is the amount of genetic variance in a population. This is termed **genetic stochasticity**.

4. In the 1950s and '60s peregrine falcons were extirpated from much of the eastern United States. The main cause of this was **pesticides (DDT; dieldrin)**.

5. Nearly all of the mammals and birds native to Madagascar are found only on that island. Species with distributions like this are said to be **endemic**.

6. Species that influence the structure and functioning of an ecological community to a much greater extent than would be expected simply from their abundance are **keystone species**.

7. **Edge effects** are the phenomena that are influenced by adjacent habitats, and increase as patch size decreases.

8. A key development in conservation practice is the setting aside of large, managed areas of land that include at their core an undisturbed natural area, and surrounding that core, buffer areas in which economic activities that do not destroy the ecosystem are permitted. These areas are called **megareserves**.

9. The subdiscipline of conservation biology that is concerned with converting disturbed areas back into natural areas (e.g., converting cropland into a native prairie) is called **restoration ecology**.

10. In the past (over a thousand years ago) the primary means by which humans caused the extinction of animals was **overhunting**.

11. The most important cause of extinction in the next century is almost certain to be **habitat destruction (or habitat loss)**.

12. When estimating the risks faced by small populations, the smallest population size that is sufficient to survive random environmental variation is the **minimum viable population**.

13. When immigrants prevent species extinction in a local population, this is known as the **rescue effect**.

14. The passenger pigeon and the Steller's sea cow became extinct due to **overhunting** by humans.

15. Species that require large areas of habitat and help the survival of most other species in an ecosystem are called **umbrella species**.

Multiple Choice

1. Many scientists believe that during the next two decades,
   a. we will have stabilized the world's environments, and no species that are currently living will have gone extinct.
   b. we will have developed technology that will allow us to re-create all of the species that have gone extinct.
   c. all species except *Homo sapiens* will be extinct.
   d. less than 1 percent of the currently living species will have gone extinct.
   e. **from 10 to 50 percent of the currently living species will have gone extinct.**

2. In recent years the number of species driven to extinction has increased dramatically. Which of the following is *not* a reason for this?
   a. Overexploitation
   b. Habitat destruction
   c. Introduction of predators
   d. **Protection of natural predators**
   e. Introduction of diseases

3. When early Polynesian people first settled in Hawaii, they drove at least 39 species of endemic land birds to extinction. The primary reason for this was
   a. introduction of predators.
   b. **overhunting.**
   c. habitat destruction.
   d. introduction of diseases.
   e. competition for food.

4. About 20,000 years ago there was a major extinction of large mammals in North America. The main cause of this extinction was probably
   a. the end of the Ice Age, causing a change in environments.
   b. competition from new species of mammals, due to a recently formed connection with South America.
   c. **overhunting by humans that had recently arrived in North America over a land bridge.**
   d. an asteroid that struck Earth at that time.
   e. a severe period of drought.

5. Human agricultural ecosystems (including forestry) account for
   a. 5 percent of Earth's land surface.

b. 10 percent of Earth's land surface.

c. **30 percent of Earth's land surface.**

d. 60 percent of Earth's land surface.

e. 80 percent of Earth's land surface.

6. Nearly half of the small marsupials and rodents of Australia have been extirpated in the last 100 years. The main cause of this is
   a. overhunting.
   b. the introduction of diseases.
   c. habitat destruction, especially in the remote outback.
   d. **the introduction of rabbits and foxes.**
   e. global warming.

7. In North America, the American chestnut and the American elm have been virtually eliminated from large areas of the East and Midwest. The main cause of this is
   a. harvesting for lumber.
   b. the extinction of birds that dispersed their fruits.
   c. habitat destruction.
   d. **the introduction of diseases.**
   e. global warming.
   e. the role of the population in the greater community.

8. When a species goes extinct in one area, it is often desirable to reintroduce the species from populations in other areas. A major problem with this approach is that
   a. genetic diseases can easily be introduced when the species is reintroduced.
   b. **populations are often adapted to local conditions and may *not* survive when moved to a very different location.**
   c. the community will have adapted to the extinct species' absence. Reintroduction may seriously disrupt the community.
   d. it is very difficult to get an adequate sample of individuals to properly reestablish the population.
   e. animals learn their home ranges and frequently attempt to return to their native habitat.

9. Captive propagation in zoos and botanical gardens has been quite successful for several endangered species. Nevertheless, captive propagation is considered to be only a partial or temporary solution to the biodiversity crisis. Which of the following is *not* a reason for this?
   a. There is not enough space in existing zoos and botanical gardens to maintain populations of more than a small fraction of rare and endangered species
   b. **Captive propagation projects in zoos have been influential in raising public awareness of the biodiversity crisis.**
   c. A species in captivity can no longer evolve together with the other species in its ecological community.
   d. Extensive research on nutrition, reproduction, etc., is required for each species.
   e. Small zoo populations tend to have low genetic diversity.

10. In 1942 there were 350 pairs of peregrine falcons that bred in the United States east of the Mississippi River. By 1960, the breeding population had entirely disappeared. Today peregrines are once again found in this area. Their reestablishment occurred because
   a. peregrines migrated in from areas west of the Mississippi.
   b. nonbreeding birds began to pair when nesting conditions again improved.
   c. the eastern United States went through a very warm period. Recently the climate has begun to cool, and peregrines are migrating down from Canada.
   d. extensive natural forest areas have been restored on the eastern seaboard, allowing peregrines to once again nest successfully.
   e. **captive-reared birds have been systematically released in the eastern United States. The population has become reestablished from these birds.**

11. Concerning species of plants that evolved as rare species for natural reasons, which of the following statements is *not* true?
   a. **Rare species tend to spread pollen widely over the bodies of pollinators to ensure that all pollinations are successful.**
   b. A rare species' success in interspecific competition will determine its survival more than its success in intraspecific competition.
   c. Herbivores and pollinators are less likely to evolve specialized relationships with rare species.
   d. Rare species are more likely to defend themselves against generalized herbivores than specialized herbivores.
   e. Rare species may evolve traits that enable them to reproduce even though they are widely separated from conspecific individuals.

12. Global warming is expected to increase the average temperatures in North America by 2–5°C within the next century. Why is this expected to have serious consequences for deciduous forest trees such as the American beech?
   a. This great increase in temperature is expected to lead to severe droughts that will kill many tree species.
   b. This increase in temperature will increase the reproductive rates of herbivorous insects leading to increased mortality of trees.
   c. **Trees will have to shift their ranges 500 or more kilometers north in the next century to maintain the same average temperature. This is far greater than the movement that can be expected using only natural dispersal of seeds.**
   d. The higher temperatures will mean longer growing seasons and faster growth of trees. This will alter the population dynamics of forests, possibly driving many species extinct due to competition.
   e. The increased temperatures will lead to an increase in the number of fires. This will dramatically change the forest dynamics. Fire-intolerant species such as American beech will be driven to extinction.

13. Which of the following is *not* a difficulty for species conservation associated with the projected increase in global temperatures of 2–5°C in the next century?
    a. The projected rate of increase is much faster than climatic changes in the past.
    b. Because of habitat fragmentation, it may *not* be possible for species to simply extend their ranges northward as the climate warms.
    c. Climatic zones will *not* simply shift northward; new climates will develop, and some existing ones will disappear.
    d. Associations between climates and soil types that maintain particular species will often be separated as the climate zones shift northward.
    e. **Global warming will enlarge the atmospheric ozone hole, causing increased stress on species.**

14. Species native to islands such as Madagascar are particularly susceptible to the effects of habitat destruction because
    a. island populations are frequently more susceptible to disease.
    b. **many species found on islands are found nowhere else.**
    c. habitats are more easily destroyed on islands than on continents.
    d. habitat fragmentation is more serious on islands because islands normally have a single continuous habitat.
    e. island populations tend to have more mutualistic relationships than mainland populations.

15. In Peruvian forests it is essential to preserve the dozen or so species of figs and palms because
    a. **these species support an entire community of frugivorous birds and mammals. Loss of these species would eliminate most of the frugivores.**
    b. these trees are the only trees in the forest of proven economic value. They must be preserved so that they can be harvested in future years.
    c. these trees are particularly important in preventing erosion.
    d. indigenous peoples use these trees in their religious ceremonies. Maintaining these species is essential for preserving these indigenous religions.
    e. these species show special promise in developing new medicines.

16. A lumber company proposes to clear-cut a large area of forest, but they aim to leave small patches of forest to provide habitat for forest animals. Which of the following is *not* a problem with this method of conserving natural habitat?
    a. Small patches cannot support populations of species that require large areas.
    b. Small patches can harbor only small populations of the species that can survive there.
    c. **Many small patches will disperse the populations of the affected species over a larger geographical area.**
    d. Species that live in the clear-cut areas can invade the edges of the patches to compete with or prey on the species living there.
    e. In small patches temperatures will be higher, winds stronger, and the environment will generally be more extreme than it was in the original forest.

17. In the northwestern United States, the spotted owl is endangered and faces a high risk of extinction in the next 50 years. What is the primary cause of the decline in spotted owls?
    a. Increased use of pesticides in the Northwest has drastically reduced the hatching success of spotted owls.
    b. Spotted owls are hunted for their fine feathers.
    c. **Logging of old-growth forests has reduced the suitable habitat of the owls and their primary prey, flying squirrels.**
    d. Increased industrialization in the Northwest has increased pollution, causing the decline in spotted owls, which are very sensitive to pollution.
    e. Spotted owls depend on recently burned forest for their survival. Increased control of forest fires has reduced their population size.

18. A major difficulty faced by tropical countries such as Costa Rica that are trying to maintain a national park system is that
    a. most of the land in these countries has already been destroyed and there is little to preserve.
    b. national park land is so remote that the parks are frequently little more than marks on a map.
    c. **the high rates of population growth put heavy pressures on parks from agricultural settlers and poachers.**
    d. the high rainfall in these areas makes it difficult to maintain an active conservation program.
    e. tropical ecosystems are not in any current danger and there is little incentive to establish parks.

19. La Amistad Reserve in Costa Rica is a 500,000-hectare reserve. A unique feature of this reserve is that
    a. it is a large area that is kept completely natural.
    b. the entire reserve is being managed for sustained agriculture and lumbering.
    c. **it includes both areas managed for economically valuable products and core areas that are kept completely natural.**
    d. only indigenous peoples are allowed to enter the reserve.
    e. this area is being developed intensively for tourism to take pressure off other national parks in the country.

20. In the project to restore the tropical dry forest of Guanacaste National Park in Costa Rica, the single most important threat to the ecosystem being restored is
    a. **fire.**
    b. grazing.
    c. global warming.
    d. development.
    e. introduced animals such as the mongoose.

21. Which of the following is *not* an essential service for humans that is provided by natural ecosystems, such as wetlands?
    a. **Absorption of oxygen and release of carbon dioxide**
    b. Reduction of erosion and water runoff
    c. Treatment and purification of wastewater
    d. Production of fish, waterfowl, and other wildlife
    e. Absorption of pollutants such as sulfates

22. The natural ecosystems in Kenya contain animals such as elephants, rhinoceroses, lions, leopards, and buffalo. This ecosystem provides a major source of revenue for that country because
    a. these animals are harvested for valuable products such as ivory, rhinoceros horns, and furs.
    b. nature tourism based on these ecosystems is a major industry in Kenya.
    c. there is a constant source of virgin land that can be exploited for agriculture by Kenya's growing population.
    d. animals from these ecosystems are hunted for meat.
    e. these animals are sold to zoos throughout the world.

23. Which of the following statements is true concerning the relationship between humans and the rest of the living world?
    a. Modern technology has made it so that we no longer depend on other living organisms.
    b. We are dependent on artificial ecosystems, such as agroecosystems, but gain no benefit from natural ecosystems.
    c. We are dependent on natural ecosystems at present, but the technology exists to completely replace natural ecosystems so that we will no longer depend on them.
    d. Our survival is tightly linked to the survival of natural ecosystems throughout the world.
    e. Nonindustrial human societies are dependent on natural ecosystems, but industrial societies are not.

24. Traditional agroecosystems differ from modern agroecosystems in that
    a. traditional agroecosystems have a greater fossil fuel usage rate, while achieving a lower overall yield.
    b. traditional agroecosystems are far more destructive to the soil than modern agroecosystems.
    c. traditional agroecosystems make greater use of fertilizers and insecticides.
    d. traditional agroecosystems are more diverse, because many more economically valuable species are grown together and because these species support many other species incidentally.
    e. traditional agroecosystems achieve a lower yield because the land is *not* carefully tuned to the crops being grown and because no pesticides are used. Because of these poor agricultural techniques, more land is necessary, leading to more species being driven to extinction.

25. Populations of some Galapagos tortoises are maintained today only by humans who remove eggs and rear the young tortoises in captivity. Why?
    a. These populations are harvested by humans for meat and eggs.
    b. These primitive animals have such a slow reproductive rate that they have not been able to adapt to global warming.
    c. There is such a high demand for these tortoises in zoos that they are raised this way to maximize their reproductive rate.
    d. This is the best way to assure that these endangered animals are properly counted and marked for ongoing scientific studies.

e. In nature rats and pigs excavate all of the nests and devour all of the eggs that are laid.

26. Which of the following characteristics of a population is *not* included in the determination of a population vulnerability analysis (PVA)?
    a. Genetic variability
    b. Evolutionary history
    c. Morphology
    d. Behavior
    e. Physical environment

27. Species that were once common but are currently rare as a result of human activities are often in greater danger of extinction than species that have always been rare, because species that have been rare for a long time
    a. may evolve traits that enable them to reproduce even though they are widely separated from conspecific individuals. Common species that suddenly have their population sizes reduced lack these adaptations.
    b. are much better at hiding than common species are. Common species that suddenly have their population sizes reduced are more subject to predation.
    c. have evolved to be good at intraspecific competition. Common species that suddenly have their population sizes reduced would have to switch from a system where interspecific competition is important to one where intraspecific competition is important, and might be eliminated by competition with other species.
    d. tend to be long-lived. Common species that suddenly have their population sizes reduced are usually not long-lived and would be unlikely to be able to maintain their population sizes.
    e. have a network of mutualistic relationships with other rare species that tends to minimize the effects of disturbance. Common species lack such extensive mutualistic networks.

28. As of the mid-1990s, the number of humans on Earth is closest to
    a. 0.5 billion.
    b. 1.0 billion.
    c. 2.5 billion.
    d. 6.0 billion.
    e. 10.5 billion.

29. Hawaii has many species of long-lived lobelia plants and also many species of honeycreeper birds that pollinate the plants. Assume that lobelia species A is pollinated exclusively by honeycreeper species A. If honeycreeper species A becomes extinct, what is the most likely fate of lobelia species A?
    a. The lobelia species will become extinct for lack of a pollinator.
    b. An empty niche will be created by the bird extinction, and other honeycreeper species will fill it by pollinating the plants.
    c. The plant flower shape will evolve toward a more generalized wind-pollinating form.
    d. The plants will evolve asexual methods of reproducing.
    e. The species will switch to self-crossing (pollen landing on the female structures in the same flower) in order to survive.

30. North American songbirds can be grouped into migrant species that spend the winters in the tropics and subtropics and resident species that do *not* migrate. A study of songbird populations of the eastern United States reveals that many populations of
    a. resident birds are increasing as the birds adapt to suburban settings and the loss of natural habitat removes predators.
    b. resident birds are declining due to increased rates of disease as they are crowded into smaller habitats.
    c. **migrant birds are declining due to increased predation in the north and habitat loss in the south.**
    d. migrant birds are increasing due to protection in megareserves in the south.
    e. resident birds are increasing due to plentiful winter food supplies from numerous birdfeeders.

31. Which of the following is *not* true of cowbirds?
    a. Cowbirds are native to North America.
    b. Female cowbirds lay their eggs in the nests of other bird species.
    c. If the parent birds are small, a nest may contain only a lone cowbird nestling.
    d. The habitat of adult cowbirds is pastureland used by domestic livestock.
    e. **Cowbird populations have generally decreased over the past two decades due to loss of habitat.**

32. Mitigation projects that attempt to create replacement wetlands
    a. are usually carried out by experienced wetland ecologists.
    b. are becoming less common due to lack of earlier success.
    c. must be located at the site of the previous wetland.
    d. currently show high success rates due to recent advances in modeling wetlands.
    e. **provide developers a means to get building permits that otherwise would be denied for sites that contain wetlands.**

33. If global warming causes average North American temperatures to raise 2–5°C in the next century, the most likely effect on the American beech will be that
    a. **the species will not be able to extend its range northward fast enough to remain in a suitable climate.**
    b. the trees will produce more seeds due to increased metabolic rates and become more common in forest communities.
    c. adult trees will die first, while shaded seedlings in the understory will survive longer.
    d. as animal seed predators migrate northward, the beech population will increase due to greater germination rates.
    e. the trees will grow faster, begin reproduction sooner, and die at a younger age.

34. The number of species that become extinct due to habitat destruction is greatest in _____ ecosystems with many _____ species.
    a. temperate; migratory
    b. **tropical; endemic**
    c. temperate; keystone
    d. tropical; migratory
    e. temperate; endemic

35. Major causes of human-related extinctions of species include all of the following, except
    a. climate modification.
    b. overexploitation.
    c. habitat destruction.
    d. **captive propagation.**
    e. introduction of predators and diseases.

36. Which of the following is *not* true of the textbook's examples of extinctions that occurred at least 1000 years ago?
    a. Many exterminated species were large mammals.
    b. Humans had recently arrived in the area.
    c. **Habitat destruction by humans was the cause.**
    d. Many flightless birds were exterminated.
    e. The exterminated species were initially numerous.

37. Select the following category that has contributed least to the co-option of net primary production by humans.
    a. Human-occupied areas
    b. Grazing land
    c. Cultivated land
    d. Managed forests
    e. **Megareserves**

38. Which of the following categories is likely to be the most important cause of species extinctions in the next century?
    a. **Habitat destruction and co-option**
    b. Overexploitation
    c. Introduced predators
    d. Introduced disease
    e. Climate modifications

39. Which of the following is *not* an introduced pest, predator, or competitor that has lead to a notable extermination?
    a. Rabbits and foxes in Australia
    b. **Collared lizards in the Ozarks**
    c. Black rats in the Galapagos Islands
    d. The chestnut blight in North America
    e. Pigs in the Galapagos Islands

40. A population vulnerability analysis is influenced by all of the following population factors except one. Select the exception.
    a. Its minimum viable population
    b. **Its adaptability to captive propagation**
    c. Its genetic variability
    d. Its mating system
    e. Its sex ratio

41. Which of the following statements about captive propagation is *not* true?
    a. Separating the organism from its ecological community can lead to difficulties.
    b. Captive propagation may cause inbreeding depression.
    c. **Successful captive propagation programs may make it possible to reduce the habitat needed for an endangered species.**
    d. Artificial insemination has been successfully used in some captive propagation programs.
    e. Captive propagation can be used until the external threat to the species is corrected.

42. Which of the following statements about the extirpation and subsequent reintroduction of the peregrine falcon into the eastern United States is *not* true?
    a. Extirpation was caused by high adult mortality due to DDT accumulation.
    b. The habitat is now safe for reintroduction of the peregrine falcon.
    c. Released birds usually remain in the same area to breed in subsequent years.
    d. Thinned eggshells caused the eggs to easily break.
    e. Captive propagation efforts are expensive.

43. Which of the following statements about the extirpation and subsequent reintroduction of the collared lizard to the prairie glades of the Ozarks is *not* true?
    a. Collared lizards have existed in the Ozarks area for about 8,000 years.
    b. A combination of climate change and co-option has resulted in the extirpation of the collared lizard from many glades in the Ozarks.
    c. Collared lizards in the southwestern United States are genetically distinct from the Ozark lizards.
    d. Lizards from a single glade are genetically diverse.
    e. Reintroduction efforts use about 10 mature lizards selected from at least five different glades.

44. All of the following characteristics except one are usually associated with rare species. Select the exception.
    a. A recent arrival in a new area
    b. An endemic species on an island
    c. A keystone species
    d. A top predator
    e. A species requiring a large foraging area

45. Which of the following statements about keystone species is *not* true?
    a. Extinction of a keystone species could cause the extinction of many other species.
    b. Keystone species are common among plants and their pollinators, such as fig species and frugivorous bats in Peruvian forests.
    c. Keystone species are usually the most conspicuous and dominant species in their biological communities.
    d. Mutualistic relationships depend on keystone species.
    e. A measure of a keystone species is community importance.

46. In forest ecosystems, edge effects within a patch should _____ and the environmental variation should _____ with a decrease in habitat patch size.
    a. increase; decrease
    b. increase; increase
    c. increase; remain constant
    d. decrease; increase
    e. decrease; decrease

47. Which of the following statements about the size effects of habitat fragmentation is *not* true?
    a. Species with large home ranges and poor dispersal sizes disappear as patch size decreases.
    b. Edge effects increase uniformly as patches become smaller.
    c. Death related to dispersal between suitable habitat patches increases as patches become smaller.
    d. The adverse effects of smaller habitat patches decrease if the patches are connected to larger habitat areas.
    e. Ten one-hectare habitat patches are not the same as one ten-hectare patch in terms of the biological community that can be supported.

48. Which of the following statements about megareserves is *not* true?
    a. The core of a megareserve consists of completely undisturbed habitat.
    b. Habitat fragmentation would probably be greatest in the outer zone of a megareserve.
    c. Agriculture and plantation forestry occurs in the central buffer zone of a megareserve.
    d. Costa Rica has pioneered the development of megareserves.
    e. Nature tourism is an important resource in a megareserve.

49. A forest reserve is most similar to
    a. the core area of a megareserve.
    b. the buffer zone of a megareserve.
    c. the outer zone of a megareserve.
    d. a national park.
    e. an entire megareserve.

50. Which of the following statements about restoration of the tropical dry forest in Guanacaste National Park of Costa Rica is *not* true?
    a. This is the world's largest restoration project.
    b. Destruction of the tropical dry forest in this area has been mainly through burning for human habitation.
    c. Information on seed dispersal of forest trees is an important part of this project.
    d. Cattle grazing reduces the threat of fires to neighboring undisturbed tropical dry forest.
    e. Eventually grazing will be eliminated after succession reaches a certain state.

51. Ecosystems carry out all of the following activities, except
    a. absorption of CO_2.
    b. emission of O_2.
    c. cleansing of water.
    d. renewal of the fossil fuel supply.
    e. regulation of stream flow.

52. Which of the following statements about the spotted owl is *not* true?
    a. Over 70% of the spotted owl's habitat has been co-opted.
    b. The spotted owl preys primarily on flying squirrels.
    c. As the amount of old growth forest is reduced, existing spotted owls will need to reduce their home ranges.
    d. As the amount of old growth forest is reduced, juvenile spotted owl mortality during dispersal will increase.
    e. Clear-cutting every 50–80 years will not permit spotted owl habitat to develop.

53. Which of the following is *not* a reason to protect bio-diversity?
    a. The aesthetic value
    b. Because of mutualistic relationships, whole communities could be endangered by the extinction of one species.
    c. Important medicinal compounds can be found only in certain species.
    d. Extinction is irreversible.
    e. **None of the above**

54. Many species of butterflies in Papua New Guinea were threatened by overexploitation and habitat degradation. What was the solution that was used to increase the butterfly numbers?
    a. A megareserve was created.
    b. **Butterfly farms were started that were run by and benefited the native villagers.**
    c. All the butterfly predators were eliminated.
    d. The butterflies were moved to another location more suitable to their continued existence.
    e. No solution has been found yet that is feasible.

55. A population viability analysis
    a. can predict when a species will become extinct.
    b. **estimates the risk of extinction based on size of a species population in a specific time period.**
    c. predicts if a species will survive in a certain habitat.
    d. is the smallest population size necessary for a species to maintain or increase in number.
    e. predicts if local disturbances will cause the extinction of a species.

56. Fynbos, the endemic shrubs of the hills of Western Cape Province in South Africa, and the primary vegetation of the watershed that provides the region's water supply, have been invaded by alien plants that grow taller and faster. A movement to get rid of these invaders by cutting and digging them out has begun. Which of the following is *not* a reason for removing these alien plants?
    a. It is cheaper than obtaining water by another method.
    b. Tourism money comes into the country from people wanting to see these plants.
    c. **It is the most technologically advanced method of supplying water to the region.**
    d. The fynbos can be sold as cut flowers.
    e. The intensity and severity of fires is lessened when the fynbos are present.

57. An example in the textbook described a study in Belize that demonstrated that both species preservation of forests and economic development can be supported by harvesting
    a. agricultural plants, such as corn and squash.
    b. lumber from logging of the forests.
    c. **medicinal plants.**
    d. cattle.
    e. parrots that are sold as pets.

58. Which of the following organismal groups would be least likely to exhibit extensive endemism?
    a. Plants in a small, isolated continental region
    b. Snails in a series of isolated lakes
    c. **Birds in a small, isolated continental region**
    d. Reptiles in a small, isolated continental region
    e. Invertebrates in tropical montane forests